CONTENTS

6-56

Printed in U. S. A.

RCA
Transmitting Tubes

THIS MANUAL has been prepared to assist those who work or experiment with transmitting tubes and circuits. It will be found valuable by engineers, service technicians, radio amateurs, students, experimenters, and all others technically interested in transmitting tubes.

Power types having plate-input ratings up to four kilowatts and associated rectifier types are included in this Manual. In the TUBE TYPES Section, detailed information is given on all important RCA types in this category. Essential basic data for discontinued RCA types are included for reference purposes.

In addition to the tube types covered in this Manual, the TUBE DIVISION OF RADIO CORPORATION OF AMERICA offers a variety of high-power and super-power tubes for transmitting and industrial applications. Other lines of RCA electron devices include:

RECEIVING TUBES
Rectifiers, Diode Detectors, Voltage and Power Amplifiers, Converters, Oscillators, and Mixers

TELEVISION CAMERA TUBES
Iconoscopes, Monoscopes, Vidicons, and Image Orthicons

PHOTOTUBES
Single-Unit, Twin-Unit, and Multiplier Types

PICTURE TUBES
Black-and-White and Color

THYRATRONS & IGNITRONS

CATHODE-RAY TUBES
Special-Purpose Kinescopes, Storage Tubes, and Oscillograph Types

SPECIAL TYPES
"Special Red" Tubes, Vacuum-Gauge Tubes, Magnetrons, Traveling-Wave Tubes, and Receiving-Type Tubes for Industrial Applications

SEMICONDUCTOR DEVICES
Transistors and Diodes

For Sales Information,
write to
Sales

For Technical Information,
write to
Commercial Engineering

TUBE DIVISION
RADIO CORPORATION OF AMERICA
Harrison, N. J.

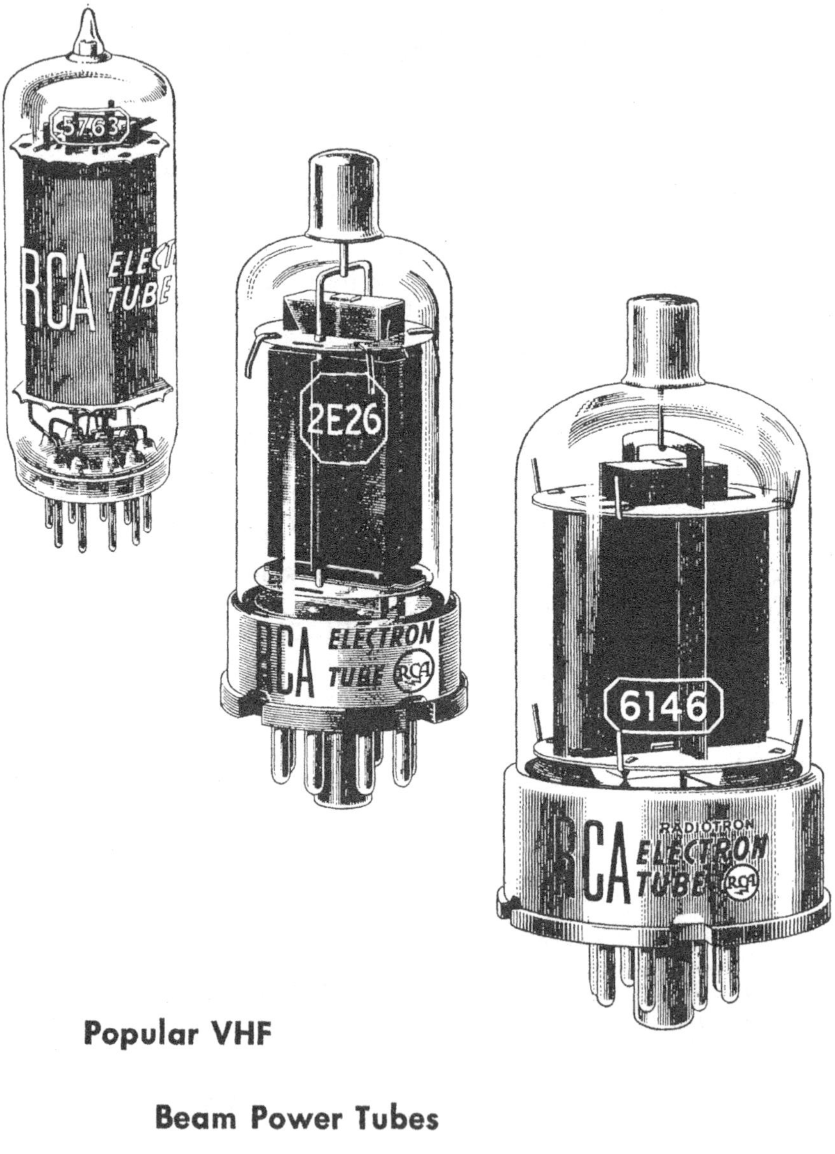

Popular VHF

Beam Power Tubes

for fixed-station and mobile service

RCA Transmitting Tubes

Power-Tube Fundamentals

Power tubes are devices for controlling the transfer of energy in electrical circuits. In this respect they are similar to rheostats, switches, and other circuit-type control devices. Tubes, however, permit much more rapid, precise, and efficient control of electrical energy than mechanically operated devices.

The transfer of electrical energy through a circuit involves control of two factors, rate and direction. The rate of energy transfer is determined by the number of individual electron charges moving unidirectionally through the circuit in a given interval of time and is proportional to the applied voltage. The direction in which the electron charges move is determined by the polarity of the applied voltage.

Electron charges may be transferred through a circuit element by several methods. In one method, kinetic energy is transferred between adjacent electrons within the molecular structure of a conductor. This method is employed in switches, rheostats, and other devices which utilize conductive materials as control electrodes. Because the currents through such devices are controlled by mechanical means, the speed with which the amount or direction of current can be changed is limited by friction and inertia.

In a second method, individual electrons are transferred through a low-density, nonconductive medium, such as a vacuum or a low-pressure gas. This method is used in tubes and has the advantage that both the rate and the direction of current flow may be controlled by electric fields. Because these fields, as well as the electrons, have negligible inertia, tubes can effect changes in the value and direction of electric current at speeds considerably higher than those obtainable with mechanically operated devices.

In electrical circuits, control of the direction of current flow is necessary when the power source produces ac voltages and currents and the load requires a unidirectional current. Tubes which are used primarily to control the direction of current flow are known as **rectifiers**. All such tubes, however, are also rate-control or rate-limiting devices in the sense that they have a finite current-carrying capability.

Rate-control requirements in electrical circuits range from occasional on-off switching to continuous variations occurring several billion times per second. Tubes which provide this form of control are known generically as **amplifiers**. Power-tube amplifiers are capable of controlling relatively large amounts of energy. All triode and multigrid power tubes are inherently rectifiers as well as amplifiers because they deliver unidirectional current regardless of the kind of energy furnished by the power source.

Basic Considerations

In its simplest form, an electron tube consists of a **cathode** (the negative electrode) and an **anode** or **plate** (the positive electrode) in a sealed envelope. More complex types may also contain one or more additional electrodes. The purpose of the cathode is to furnish a continuous supply of free electrons; the plate collects these electrons. The rate at which electrons are collected by the plate (the **plate current**) is determined by the number of free electrons available and by the polarity and the strength of the electric field between the plate and cathode. Power tubes and rectifiers are usually operated so that the number of electrons available is constant. Consequently, the rate of collection or current flow is determined principally by the characteristics of the internal electric field.

The internal electric field is established by connection of a source of potential between the plate and cathode. When the plate is at a negative potential with respect to the cathode, the internal field tends to prevent electrons from leaving the vicinity of the cathode, and there is no transfer of energy through the tube. When the plate is operated at a positive potential with respect to the cathode, the field causes a movement of electrons to the plate. The current through the tube is then determined by the strength of the field, or the **plate voltage.**

Vacuum Tubes

Under normal operating conditions, the velocity of the electrons emitted by the cathode of a vacuum tube is just sufficient to insure their release from the emitting surface. If no accelerating field is applied, these electrons tend to return to the cathode when their escape energy has been expended. However, the intense negative field created by new electrons reaching the emitting surface repels those previously emitted and they accumulate in the space surrounding the cathode. This accumulation of electrons is called the **space charge**.

The approximate distribution of the space-charge electrons in the absence of an accelerating field is shown in Fig. 1. The concentration is greatest in

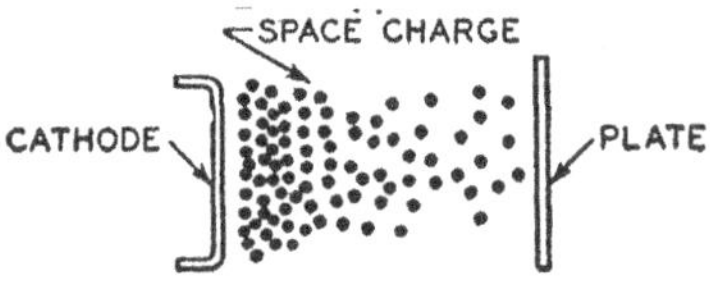

Fig. 1

the region nearest the cathode. The general relationship between plate voltage (E_b) and plate current (I_b) in a two-electrode vacuum tube is shown in Fig. 2. At very low positive plate voltages (region E_0 to E_1), only the loosely bound electrons on the outer surface of the space charge are attracted to the plate, and the plate current does not change uniformly with equal increments in plate voltage. Over a higher range of plate voltages (region E_1 to E_2), the relation between plate voltage and plate current is nearly linear. When operated

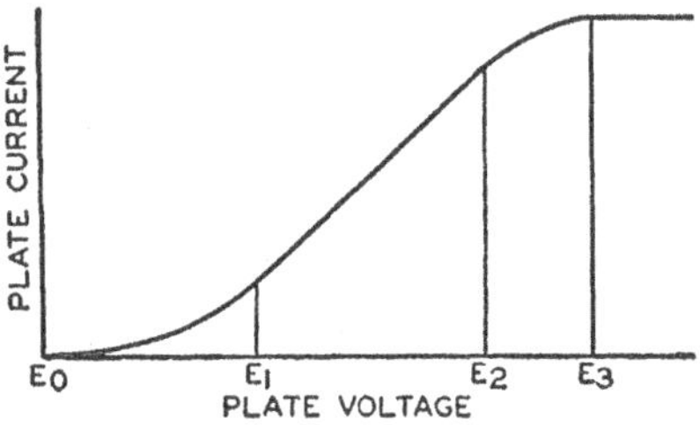

Fig. 2

in this region, a two-electrode vacuum tube has substantially constant internal resistance (called plate resistance, or r_p), and the plate current follows the normal Ohm's-Law relationship.

At plate voltages higher than E_2, an increase in plate voltage does not produce a proportional increase in plate current because practically the full emission capabilities of the cathode are being utilized. The voltage at which essentially all of the electrons emitted by the cathode are collected by the plate is known as the **saturation** voltage and is indicated in Fig. 2 by E_3.

Two-electrode vacuum tubes are extremely useful as power rectifiers. Because they are entirely nonmechanical in operation, they can be used over a wide range of frequencies. They can operate at both very high and very low temperatures, and can be designed to withstand very high inverse voltages. The substantially linear relationship between plate voltage and plate current in such tubes is also useful as a means of obtaining virtually distortionless rectification (detection) of radio signals.

Like all rectifiers, the two-electrode vacuum tube is a special form of switching device and, therefore, does not provide any power gain. However, the control of circuit currents by means of electric fields can be extended to include amplification, oscillation, and other functions involving actual power gains by the addition of a third electrode called a **grid** between cathode and plate. When the grid is placed relatively near the cathode, the application of small voltages to the grid can produce the same change in the internal field, and thus in the plate current, as large changes in plate voltage. Large amounts of plate-circuit power can thus be controlled with relatively little energy. Special control characteristics may be obtained by the use of two or more grids or control electrodes in a tube. The construction and characteristics of the principal types of multi-electrode tubes in general use are described in detail later in this section.

Electrons accelerated by even moderately high plate voltages may acquire enough kinetic energy so that they dislodge equal or greater numbers of electrons when they strike the plate. Emission produced in this manner is known as secondary emission.

Like primary electrons, secondary electrons are attracted to a positive electrode in the tube. In a two-electrode tube, they return to the plate and their only effect is to produce a weak negative field similar to a space charge which tends to repel some of the primary electrons approaching the plate. Although an increase in plate voltage beyond the saturation value does not increase the plate current of a tube, it produces a proportional increase in the velocity with which electrons move to the plate, and thus increases secondary emission.

Although secondary emission is frequently employed in special multi-electrode tubes, it may produce effects which interfere with normal operation of power-tube amplifiers. These effects and the methods used to overcome them are discussed in detail later in this section.

Gas Tubes

In a vacuum tube, space charge inhibits the release of electrons from the cathode, and thus limits the plate current at low and moderate plate voltages. Although the space-charge effect may be reduced by a reduction in the spacing between plate and cathode, it cannot be entirely eliminated by this method. The negative space charge can be neutralized, however, by other methods—for example, by the introduction of a controlled amount of **mercury vapor** or **inert gas** in the tube.

When a gas is present in a two-electrode tube, free electrons in the gas are attracted to the positive anode and add to the anode current. Positive ions created continuously by collisions between gas atoms and the free electrons neutralize the space charge so that large currents may be drawn at low anode voltages. In addition, the space-charge neutralization effectively increases the thermal efficiency of the cathode. These advantages make gas tubes particularly suitable for use as power rectifiers. The use of gas tubes, however, requires precautions in circuit design, physical installation, and operation which are not necessary with vacuum tubes. These additional requirements are discussed in the *Rectifier Considerations* Section.

Generic Tube Types

In tube terminology, generic type names such as "diode," "triode," "tetrode," and "pentode" indicate the number of electrodes directly associated with the emission, control, or collection of electrons. Auxiliary elements such as heaters, internal shields, or metal-envelope shields, even when provided with separate electrical connections and shown in the tube symbol, are not counted in establishing generic-type classifications.

Diodes

The diode types listed in this Manual are used principally as rectifiers in equipment for converting low-frequency alternating current from commercial power lines or local sources to direct current.

Tubes which contain a single diode unit, such as the 836 or 866-A, are known as half-wave rectifiers because they are capable of conducting current during only one half of each ac cycle. Tubes which contain two diode units, such as the 5R4-GY, are called full-wave rectifiers because they can be connected so as to conduct current during both halves of each ac cycle. Fig. 3 shows graphical symbols for a filament-type half-wave rectifier and a heater-cathode-type full-wave rectifier.

Gas rectifiers have a very small internal voltage drop which is practically independent of load current and are, therefore, desirable for applications requiring relatively constant output voltage with varying loads. In mercury-vapor types, and to a smaller degree in inert-gas types, the voltage drop is affected by bulb temperature. Control of bulb temperature and other special considerations involved in the operation of gas rectifier tubes are discussed in the *Rectifier Considerations* Section.

In a vacuum rectifier, the internal voltage drop is approximately proportional to the load current. Consequently, rectifiers of this type, such as the 5R4-GY, 836, and 1616, do not provide as good regulation of output voltage as gas types in applications involving varying load currents. Vacuum rectifiers, however, are not affected by ambient temperature and do not require special installation and circuit considerations. Certain heater-cathode-type vacuum rectifiers, such as the 836, have very low internal resistance and are capable of providing voltage regulation almost as good as that obtainable with gas types.

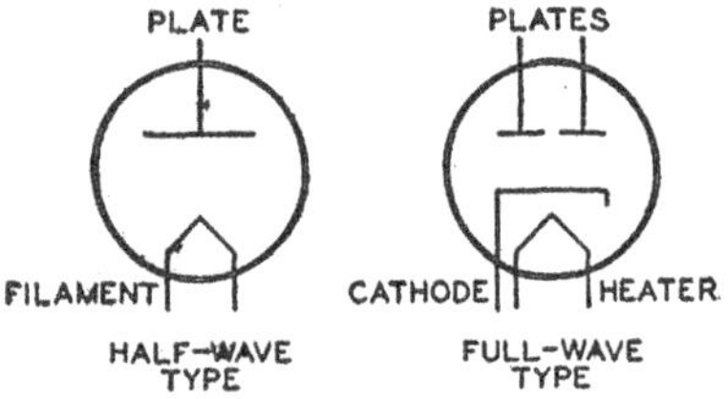

Fig. 3

Triodes

In triodes, or three-electrode tubes, an auxiliary control electrode, called a grid, is placed between the cathode and the plate, as shown in Fig. 4. The grid is usually a cylindrical or oval-shaped spiral of fine wire surrounding the cathode, although wire-mesh and grating-type grids may also be used.

Because of its open construction, the grid does not appreciably obstruct the movement of electrons from cathode to plate. When the grid is made positive or negative with respect to the cathode, however, its electric field can increase or decrease the rate of electron flow. This effect makes it possible for a triode to be used as an amplifier. In a typical amplifier circuit, such as that shown in Fig. 5, the energy required to attract electrons to the plate is obtained from a high-voltage dc plate supply and the electrical impulse to be amplified, the **input signal**, is applied between grid and cathode. Because the plate current of the tube flows through the load, variation of the grid-cathode voltage causes the dc power drawn from the plate supply to appear as ac power in the load. The power required by the grid for complete control is ordinarily only a fraction of the power developed in the load circuit. The ac power in the load circuit is always less than 100 per cent of the dc input power, however, because some power is dissipated at the plate of the tube and in the resistance of the load circuit. In addition to their use as audio-frequency and radio-frequency amplifiers, power triodes may be used in suitable circuit arrangements for oscillation, frequency multiplication, modulation, and various special purposes.

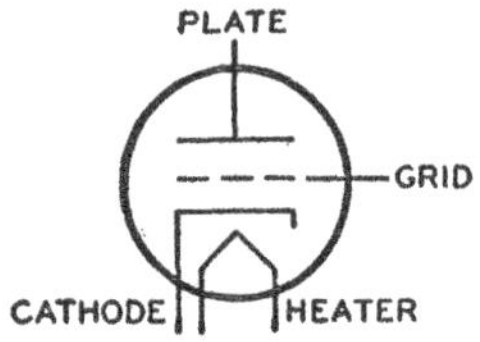

Fig. 4

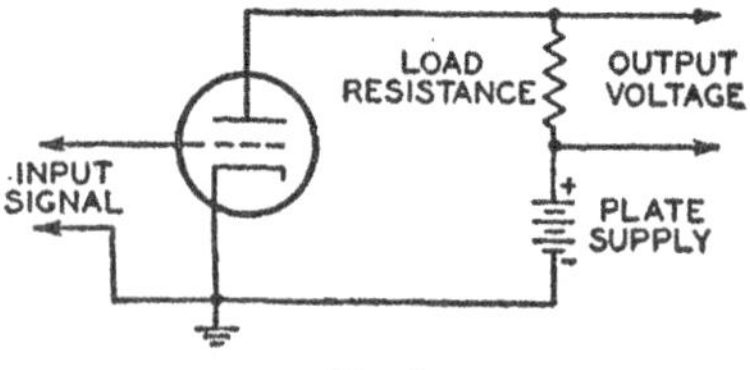

Fig. 5

The plate, cathode, and other electrodes of a tube form an electrostatic system, each electrode acting as one plate of a small capacitor. In a triode, capacitances exist between grid and cathode, grid and plate, and plate and cathode, as shown in Fig. 6. Although these interelectrode capacitances do not have values of more than a few micromicrofarads, they may have substantial

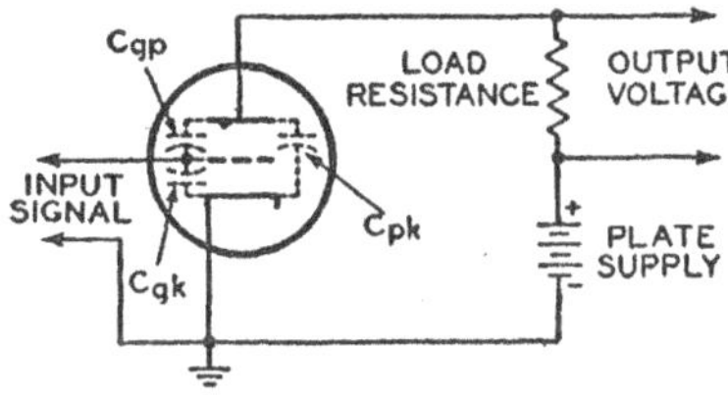

Fig. 6

effects on tube operation, especially at radio frequencies. For example, the grid-plate capacitance, C_{gp}, provides an internal path between the output and input circuits. When a triode is used as an amplifier at radio frequencies, sufficient energy may be fed back through this path to cause uncontrolled regeneration or oscillation. Although this type of internal feedback is frequently employed in oscillator circuits, it is undesirable in amplifier applications. Triode radio-frequency amplifiers, therefore, require either special circuit arrangements or the use of a feedback-cancelling technique known as **neutralization.** These special considerations are discussed at length in the *Power-Tube Applications* Section.

Tetrodes

Internal feedback between plate and grid, and the resulting need for neutralization in triode radio-frequency amplifiers, can be minimized by incorporation of a second grid (the **screen grid**) between the grid No.1 (the control grid) and the plate, as shown in Fig. 7. Tubes which employ a grid No.2 or screen grid, cathode, control grid, and plate are known generically as tetrodes.

When a tetrode is used as an amplifier, the screen grid is operated at a fixed positive potential (usually somewhat lower than the plate voltage), and is bypassed to the cathode through a capacitor having a very low impedance at the operating frequency. This capacitor diverts signal-frequency alternating currents from the screen grid to ground, and effectively short-circuits the capacitive feedback path between plate and control grid. The screen grid acts as an electrostatic shield between the control grid and the plate, and reduces the grid-plate capacitance to such a small value that internal feedback is usually negligible over the range of frequencies for which the tube is designed.

Because the screen grid is operated at a positive potential with respect to the cathode, it collects a substantial number of the available electrons and, therefore, reduces the plate current which can flow at a given plate voltage. The addition of a screen grid thus increases the internal resistance or plate resistance of a tube. However, it also gives the grid No.1 a greater degree of control over the plate resistance, and thus increases the voltage-amplification factor.

The voltage at which the screen grid is operated has a substantial effect on the plate current of a tetrode. This characteristic makes it practicable to control the gain of a tetrode by variation of the dc screen-grid potential, or to modulate the tube output economically by the application of signal voltage to the screen grid, as well as to the

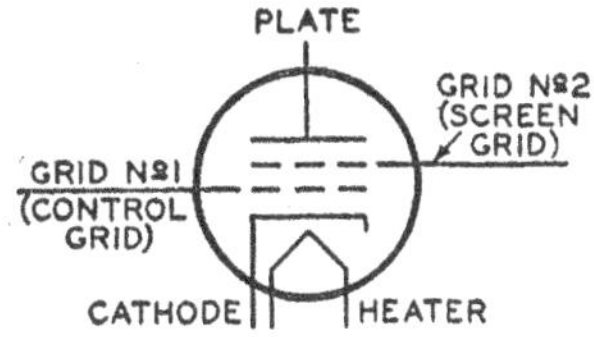

Fig. 7

control grid. It is usually necessary, therefore, to remove ripple and other fluctuations from the screen-grid supply voltage to prevent undesired modulation of the tube output.

Because the use of a grid No.2 or screen grid reduces internal coupling between the output and input circuits, tetrodes can furnish a high degree of stable amplification in relatively simple circuits. Some residual grid-plate capacitance is unavoidable, however, and internal feedback may be a problem. The amount of internal feedback that can be tolerated in any amplifier tube depends on the frequency at which the tube is operated, the effective gain of the stage, the characteristics of the tube input and output circuits, and the mechanical layout employed. Because of their high power sensitivity, tetrodes used in rf applications generally require shielding from external fields and careful circuit layout to minimize external feedback between the input and output circuits of the tubes. In certain amplifier applications involving high radio frequencies and high stage gains, tetrodes, as well as triodes, may require neutralization. Further information on this subject is given in the *Power-Tube Circuit-Design Considerations* Section.

If the negative excursion of the output signal swings the plate to a voltage less positive than that of the screen grid, electrons moving from the screen grid to the plate tend to reverse their direction and return to the screen grid. The resulting decrease in plate current causes a corresponding rise in plate voltage, which terminates the negative swing of the output signal before it completes a full excursion. This effect, which tends to reduce the power output of a tetrode below that obtainable from a triode having equivalent plate-input rating, is emphasized considerably when there is secondary emission from the plate.

The loss of a portion of the output energy which occurs in a tetrode under these conditions reduces the power-handling capabilities of the tube, and causes serious distortion of the signal waveform. The output of the tube, therefore, contains harmonics of the signal frequency and other spurious frequencies which may cause considerable interference to communications service. Such distortion may also be highly objectionable to the ear or to the eye when a tetrode is used as an audio or video amplifier. Although this effect can be minimized by reducing the amplitude of the plate-voltage swing so that the plate voltage never swings negative with respect to the screen-grid voltage, this expedient imposes further limitations on the tube output.

The abrupt rise in the plate voltage of a tetrode caused by the reversal of electron flow tends to draw both primary and secondary electrons back to the plate. Collection of these electrons then makes the plate less positive than the screen grid so that the tube current tends to reverse again. This interchange of electrons between plate and screen grid, called **dynatron action**, may continue for several cycles, and is equivalent to an oscillatory current. Although dynatron action forms the basis of certain tetrode oscillator circuits, it is highly objectionable when a tube is used solely as an amplifier.

Pentodes

The limitation imposed on the plate-voltage swing of a tetrode by "dynatron action" can be overcome by the use of a grid No.3, or **suppressor grid**, between the screen grid (grid No.2) and the plate, as shown in Fig. 8. Tubes which employ five-electrode structures of this type are called pentodes.

When a pentode is used as an amplifier, the grid No.3 or suppressor grid is generally operated at a fixed negative potential with respect to both the screen grid and the plate and thus establishes a negative electrostatic field between them. Although this field is not strong enough to prevent the desired movement of high-velocity primary electrons from screen grid to plate, it effectively prevents both primary and secondary electrons from flowing backward to the screen grid. Consequently, the plate voltage of a pentode may swing negative with respect to the screen-grid voltage without the loss of

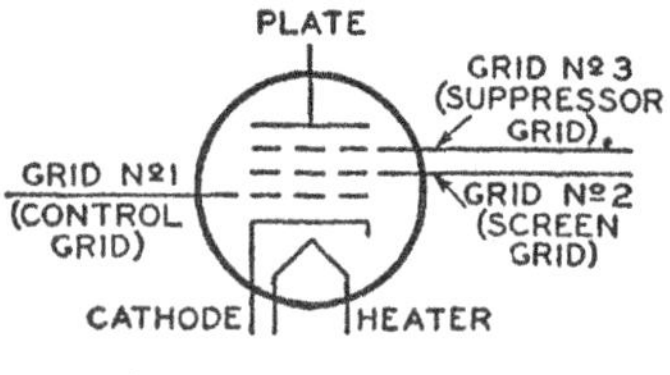

Fig. 8

output power and the waveform distortion that occur under the same conditions in a tetrode.

The grid No.3 or suppressor grid may be connected internally to the cathode, as in the 1613, so that it is automatically maintained at a negative potential with respect to the plate and screen grid. In most power pentodes, however, the suppressor grid is an independent electrode which can either be connected externally to the cathode or operated at a positive or negative potential with respect to the cathode to meet various application requirements. The use of an independent suppressor grid permits the introduction of an auxiliary signal or control voltage into the tube circuit. Although the screen grid can also be used for this purpose, a suppressor grid is generally a more effective control electrode because it requires much less signal power for full modulation of the tube output. In addition, the shielding action of the screen grid minimizes undesirable coupling between the suppressor grid and the control grid when signals are applied simultaneously to these electrodes.

Beam Power Tubes

The power-handling ability of a tetrode or pentode is limited to some extent because some of the available electrons are collected by the screen grid and, therefore, do not contribute to the plate current. In beam power tubes, however, the lateral wires of the screen grid are aligned with the control-grid wires to direct the flow of electrons through the screen grid to the plate. A sectional view of a typical beam power tube is shown in Fig. 9. As indicated by the dashed lines in the figure, the stream of electrons is divided into sheets or "beams" which tend to pass between the wires of the screen grid. Because relatively few electrons impinge on the screen grid, a substantial portion of the electron energy that would otherwise be absorbed by the screen grid and dissipated as heat is diverted to the plate, where it can be converted into useful output power.

In beam power tubes of the type illustrated in Fig. 9, dynatron action and other undesirable effects of secondary emission from the plate can be minimized by spacing the electrodes so that a space-charge effect is created in the heavily shaded region. The negative electrostatic field produced by the dense concentration of electrons in this region blocks the escape of secondary electrons from the plate, and also prevents the return of primary electrons to the screen grid when the plate swings negative with respect to the screen grid. Stray secondary electrons may be prevented from reaching the screen grid by paths outside the effective field of the space

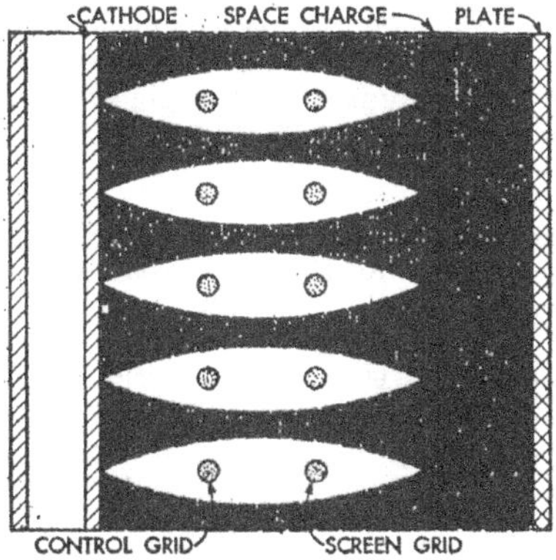

Fig. 9

charge by the incorporation of special beam-confining electrodes operated at cathode potential.

Beam power tubes may also employ suppressor grids rather than space-charge effects to prevent the reversal of electron flow when the plate swings negative with respect to the screen grid. Because beam power tubes are generally used in the same applications as power pentodes, they are represented in this Manual by a pentode tube symbol.

In general, pentodes and beam power tubes have higher power sensitivity than other generic types, *i.e.*, they require very little driving power in relation to obtainable power output. The use of pentodes and beam power tubes in multistage equipment, therefore, minimizes the number of stages required to obtain a specific power gain.

These tube types are especially useful as buffer-amplifier tubes and frequency-multiplier tubes in transmitters and other types of radio-frequency power equipment. Pentodes and beam power tubes are also widely used as audio-frequency power-amplifier tubes and modulator tubes, and in certain types of oscillator circuits.

Construction and Materials

Although power tubes may vary widely with respect to physical form, size, and terminal arrangement, they utilize two general forms of electrode assembly. In unit-type assemblies, such as that shown in Fig. 10(a), the various electrodes are assembled in a rigid framework formed of supports and insulating spacers, and are installed and supported in the envelope as a unit. This type of assembly is used in vacuum rectifiers such as the 5R4-GY and the 836, and in low- and medium-frequency power amplifiers such as the 805 and 813. Because the various electrodes are held in the desired spatial relationship by the common supporting framework, vibration and shock are received by the assembly as a unit, and the relative positions of individual electrodes are not appreciably affected.

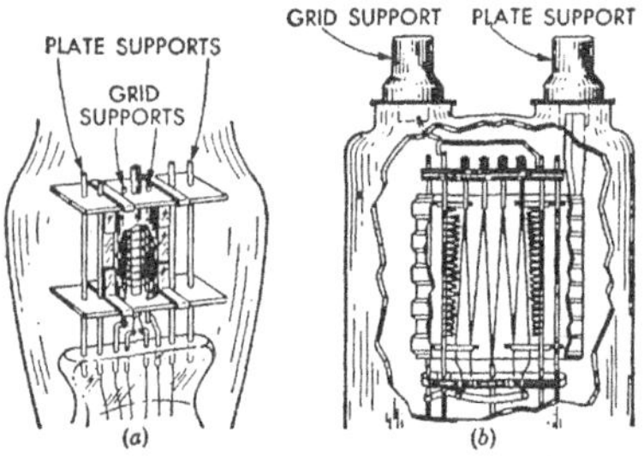

Fig. 10

Electrodes may also be suspended individually from various parts of the tube envelope, as shown in Fig. 10(b). Individually supported electrodes are used in mercury-vapor rectifiers such as the 866-A to eliminate metal framework members which might amalgamate or combine chemically with the mercury or affect the internal temperature distribution. They are also used in high-voltage vacuum tubes such as the 808 to eliminate possible leakage paths and thus provide maximum insulation between the various electrodes, and in very-high-frequency and ultra-high-frequency tubes such as the 826 and 833-A to minimize interelectrode capacitances and to eliminate the large energy losses which occur in most insulating materials at these frequencies.

Cathodes

The most efficient practical cathodes for power tubes utilize **thermionic emission.** Because such emission varies exponentially with temperature, a power-tube cathode must be operated at a constant temperature if substantial variations in emission are to be avoided. Because of the practical difficulties involved in measuring the cathode temperature of a tube, proper operating conditions are usually expressed in terms of a specific voltage and a specific current. Specific values of heating voltage and current for each tube type are given in the *Tube Types* Section.

A **directly heated** cathode, or **filamentary cathode,** is a metallic conductor drawn into wire or ribbon form, as shown in Fig. 11. The conductor is heated to emitting temperature by its own resistance to a flow of electric current. Emission may be obtained either from the conductor itself or from a coating of thermoemissive material bonded to its surface. Filamentary cathodes have the basic advantages of mechanical simplicity, high emission efficiency, and rapid heating. A single continuous filament can be wound or folded to provide uniform emission distribution over large areas, or to expose a minimum of surface to destructive positive-ion bombardment. Because of their high efficiency and quick heating, filamentary cathodes are especially suitable for portable and mobile equipment, in which economy of operating power is an important consideration.

Early filamentary cathodes were made of pure tungsten, a dense, tough metal having an extremely high melting point. Because tungsten must be heated to very high temperatures to emit electrons in useful quantities, such filaments require considerable electrical power for excitation. Much higher emission efficiencies can be obtained with thoriated-tungsten filaments, which are drawn from tungsten slugs impregnated with thoria (thorium oxide). During tube processing, some of the thorium oxide is driven to the surface of the filament and reduced to pure metallic thorium, which emits useful quantities of electrons when heated to a relatively low temperature. This surface thorium evaporates during tube operation, but is continuously replenished from the internal supply of thorium oxide.

Filamentary cathodes may also be made of inexpensive nickel alloys, rather than highly refractory metals, and coated with "alkaline-earth" oxides, which emit electrons freely at much lower temperatures than either pure tungsten or thoriated tungsten. The coating is applied to the filament in the form of a carbonate of the basic element (generally barium carbonate or a mixture of barium, calcium, and strontium carbonates), and is converted to the highly emissive oxide form during tube processing. Oxide-coated filaments are especially suitable for use in gas rectifiers, which require low-temperature cathodes capable of delivering high emission currents and withstanding intense positive-ion bombardment.

An **indirectly heated** cathode, or **heater-cathode,** is a hollow metal cylinder or sleeve having a coating of thermoemissive material bonded to its outer surface, as shown in Fig. 12. The cathode is heated by radiation from a metal filament, called the heater, which is mounted inside the sleeve. The cathode sleeve is usually electrically insulated from the heater. The emissive material employed is generally the same as that used on coated filamentary cathodes and operates at substantially the same temperature.

The electrical insulation between the heating and emitting elements in a heater-cathode provides several advantages from the standpoints of tube operation and circuit design. Because the current through the heater wire produces no voltage drop in its associated cathode, all points of the emitting surface are at the same dc potential with respect to the other electrodes of the tube. Because of this feature, this type of cathode is often called a **unipotential cathode.** The emission is substantially uniform over the entire cathode. An indirectly heated cathode may generally be operated at a fixed or variable potential of either polarity with respect to its heater, provided this potential does not exceed the maximum heater-cathode-voltage rating of the tube.

The heater of a heater-cathode is usually a folded or helically wound filament of very fine tungsten or tungsten-alloy wire. The actual form of a heater is determined by the application requirements of the tube, the amount of insulation required between heater and cathode, and the internal dimensions of the cathode sleeve. A refractory metal is required because the heater has very small effective area and, therefore, must be operated at a high temperature to supply the thermal energy required by the

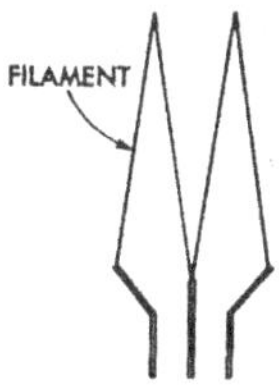

Fig. 11

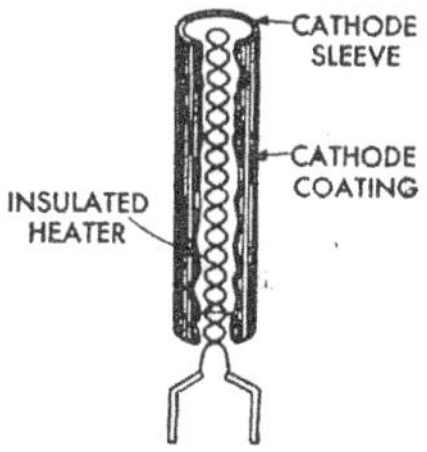

Fig. 12

cathode. The insulation must be capable of withstanding these high temperatures and, in addition, must possess sufficient flexibility to accommodate bends of very small radius because the heaters must be folded or wound into forms compact enough to fit inside the cathode sleeve. The insulation generally used is aluminum oxide, or a similar material known commercially as "alundum." The insulation is first applied to the heater as a suspension of fine particles in a nitrocellulose binder, and is then sintered into a solid coating by operation of the heater for a carefully controlled period of time at a temperature slightly above its normal operating value.

Heater-cathodes have excellent rigidity and dimensional stability, and permit the use of simpler, more compact, and more rugged electrode structures. They can also be placed very close to other tube electrodes, and thus make possible the reduction of internal losses caused by space-charge effects and electron transit time. Because tubes using these cathodes can usually be operated in any position, the equipment designer has greater freedom in locating tubes and components to provide maximum circuit efficiency or accessibility.

Plates

Plates or anodes of power tubes are designed to collect as many as possible of the electrons made available by the cathode. They must also be capable of dissipating heat. Typical plate designs are shown in Fig. 13. The plates shown at (a) and (b) are inherently rigid cylinders which surround the cathode and other electrodes. The plate at (a) is simple and extremely rugged. Plates of this type are used principally in low- and medium-frequency power tubes such as the 810 and 813.

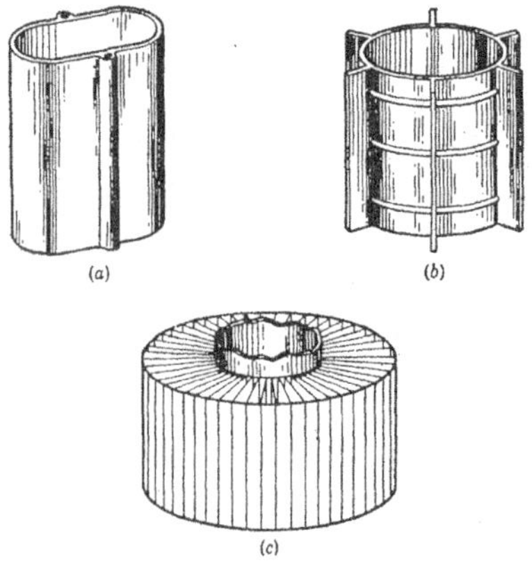

Fig. 13

The plate shown at (b) has radial fins to provide increased heat-radiating surface without appreciably increasing the capacitances between the plate and other electrodes. Plates of this type are used in tubes such as the 826.

The radiator design shown at (c) makes it possible to obtain substantial heat dissipation from plates of limited area by the use of forced-air cooling. This type of plate is used in tubes such as the 827-R.

Plates may be made of many materials, depending on the tube requirements. Nickel is often used for the plates of power tubes which operate at moderate temperatures because it can be formed readily into complex shapes and has the advantage of light weight, so that elaborate support structures are not needed. The heat-radiating ability of nickel plates can be substantially improved by means of a surface treatment called "carbonizing," in which a closely adhering layer of amorphous carbon is deposited on the surface of the nickel.

The thermal advantage of nickel is combined with high mechanical strength in a comparatively new material developed for the plates of small power tubes, which can be roughly described as carbonized nickel-plated steel.

Pure copper is now used extensively in so-called "external-plate" designs for tubes in various power ranges and physical sizes. In tubes of this type, the copper plate forms part of the envelope, and forced-air or water cooling is used to maintain the temperatures of the copper and of the copper-to-glass seal at safe values. With the aid of these cooling methods, tubes of relatively small physical size can handle very large amounts of power.

Other metals used for tube plates include materials such as tungsten, molybdenum, tantalum, and graphite. Zirconium is sometimes applied as a coating. The use of graphite, tantalum, or zirconium provides "getter" action which helps to maintain a high vacuum within a tube by cleaning up residual gases or those which may be given off by parts of the tube during operation. Graphite and molybdenum are usually subjected to some form of surface treatment during processing to improve their thermal efficiency.

Grids

In general, tube grids are constructed of individual wires arranged in parallel and swaged or welded to metal supporting rods. Fig. 14 shows typical grid structures used in power tubes. The grid at (a) is a cylindrical type consisting of individual parallel wires welded to side-rods. The grid at (b) is a cylindrical type consisting of a single wire wound in spiral form and swaged to the side-rods. The "cage" grid structures shown at (c) may be formed from single cylindrical metal blanks or of individual metal rods.

Tube grids may be made of pure metals such as tungsten, molybdenum, or tantalum, of various alloys of tungsten and molybdenum, or of a nickel-manganese alloy. Because of its physical position between the cathode and the plate,

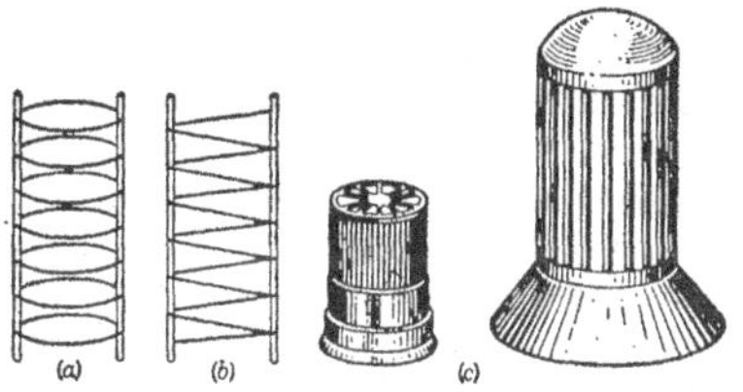

Fig. 14

the grid is subjected to heat radiated from both of these electrodes, and, if gas is present in the tube, may also undergo heavy positive-ion bombardment. As a result, the grid may emit primary electrons. Its tendency to emit electrons is further increased if it becomes contaminated with emissive material evaporated from the cathode. The grids are often coated with gold or platinum to reduce the possibility of primary emission. In the case of power tubes, platinum coatings are usually preferred to gold coatings because platinum can withstand higher temperatures than gold without vaporizing.

Because power tubes are often operated under conditions in which the grid is driven positive with respect to the cathode, the grid can attract electrons which may possess sufficient kinetic energy to liberate large numbers of secondary electrons from the grid. A carbon coating is sometimes applied to the grid to reduce its tendency to secondary emission.

Internal Insulation

Aside from the insulating materials employed in envelopes and bases, insulation is used in tube construction for electrode spacers. Spacers must be made of material which is unaffected by heat and can be formed with extreme accuracy. In small, low-power tubes, spacers are generally disks or wafers of high-quality mica; in larger tubes, they are usually bars or cross-arms of a low-loss refractory insulating material.

In many cases, insulating spacers are also used for centering the electrode assembly within the envelope. The mica wafers used for this purpose in smaller tubes usually incorporate special structural features which absorb vibration and mechanical shocks transmitted through the envelope. Refractory spacers are usually equipped with shock-absorbing metal springs at the points of contact with the envelope.

Getters

A chemical "getter" is used in electron tubes to absorb residual gases. The getter is usually a mixture of barium oxide and a reducing agent which frees the barium when the getter is "flashed." The getter material is usually concentrated in a small capsule, ribbon, or "tab," and is "flashed" or vaporized after the tube is sealed off. This tab is installed in the tube far enough from the main electrode structure to assure that the getter will not be flashed by the heat developed during the exhaust process, and that getter material will not be deposited on the tube electrodes during flashing.

Envelopes

Most small- and medium-sized low-frequency power tubes use simple cylindrical "soft"-glass envelopes and have the low-voltage electrode leads brought out through the base. "Hard" glasses of the borosilicate type are used for the envelopes of practically all medium- and high-power radiation-cooled tubes, particularly where compact construction is necessary to meet electrical-design requirements or equipment-space limitations. These glasses have relatively high softening temperatures, low rates of expansion, high electrical resistance, and excellent resistance to abrasion and "weathering."

In some high-power tubes and tubes designed for operation at very-high and ultra-high frequencies, parts of the electrode structure are utilized in the tube envelope. For example, in metal-glass types such as the 6161, the metal sections of the envelopes are extensions of the internal electrodes, while the intermediate glass sections provide the required interelectrode spacing and insulation. This type of envelope structure permits realization of good tube efficiency at ultra-high frequencies by the virtual elimination of objectionable lead reactances and losses in internal insulation. The metal sections of these envelopes are also used as electrode terminals, mounting facilities, heat-radiating surfaces, and often interelectrode shields. Pure copper is used for most of these envelope sections because of its high thermal and electrical conductivity and its high ductility, which readily permits the fabrication of special shapes.

In several metal-glass tubes, the plate sections of the envelopes are fitted with special radiators which make it possible to obtain substantially increased heat dissipation by the use of forced-air cooling and thus permit the use of relatively small tubes in high-power circuits. The grid-No.2 or screen-grid sections of the envelopes of some ultra-high-frequency metal-glass tubes provide external shielding between the grid-No.1 and plate sections. In the 5675 and other "pencil"-type tubes, the flange-type grid sections of the envelopes act as shields between the plate and cathode sections and thus minimize feedback when these tubes are used as amplifiers in ultra-high-frequency cathode-drive circuits.

Power-Tube Applications

The power tubes listed in this Manual represent the RCA types most frequently used in transmitters and other radio-frequency (rf) power equipment operating at power-input levels up to approximately 4 kilowatts and at frequencies up to approximately 3000 megacycles per second. These tubes may in general be used as audio-frequency (af) or video-frequency power amplifiers or modulators, as modulated or unmodulated rf power amplifiers, as frequency multipliers, or as oscillators. The variety of designs represented includes types suitable for use in practically all forms of communications and industrial or scientific service.

Amplification

Although power-tube applications may involve different circuit arrangements and operating conditions, they may all be considered forms of amplifier service in which the control voltage is applied between the grid (grid No.1 in a multigrid tube) and the cathode, and the output is taken from the plate circuit. (Oscillator service may be considered a form of amplifier service in which the output is fed back to the input.) Consequently, it is convenient to define tube operation in terms of the relationship between grid voltage and plate current when all other electrode voltages are held constant. This relationship, called the "mutual" or "transfer" characteristic of the tube, has the general form shown in Fig. 15. A system of classification based on this relationship is universally recognized by tube manufacturers and equipment designers.

In this system of classification, a portion of the generalized mutual characteristic is divided, as shown in Fig. 15, into three regions, A, B, and C, representing respectively the "linear" region, the region in the immediate vicinity of plate-current cutoff, and the region beyond cutoff. Tube operation may also be considered in three major categories—class A, class B, and Class C—each of which represents the type of response obtained when the operating point is in the corresponding region of the characteristic.

In **class A** operation, the operating point is centered in region A so that the tube can respond to both positive and negative excursions of grid voltage. In this type of operation, plate current flows at all times.

In **class B** operation, the operating point is in the vicinity of cutoff so that the tube can respond to positive excursions of grid voltage. In this type of operation, plate current flows for approximately one half (180 degrees) of each cycle of an alternating grid voltage.

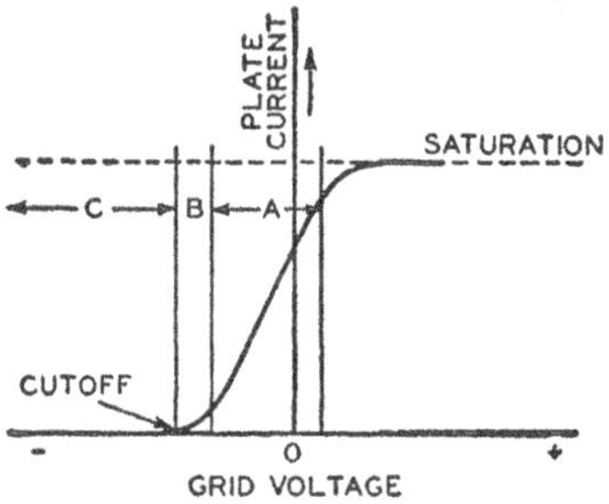

Fig. 15

In **class C** operation, the operating point is in the region beyond cutoff so that the tube can respond only to those portions of positive grid-voltage excursions which are positive with respect to the cutoff point. In this type of operation, plate current flows for less than one half (less than 180 degrees) of each cycle of an alternating grid voltage.

A fourth class of operation, **class AB**, is also used. In this class of operation, the operating point is in the lower portion of region A so that the tube responds unequally to positive and negative grid-voltage excursions above a certain amplitude. Consequently, the duration of plate-current flow on each cycle varies with the amplitude of the alternating grid voltage. In this service, plate current flows for more than one half

(180 degrees) of each cycle, but for less than the entire cycle.

The suffix 1 may be added to the letter or letters of a class identification to denote that grid current does not flow during any part of the grid-voltage cycle. The suffix 2 may be used to denote that grid current flows during some part of the cycle. In most cases, these suffixes are used only for class A_1 or class AB_1 and AB_2 operation.

Class A Amplifiers

The basic circuit and operating characteristics of a class A amplifier are shown in Fig. 16. The operating point is centered in region A of the mutual characteristic by the use of a suitable negative grid bias. The amplitude of the driving signal (alternating grid voltage) is controlled so that the grid is never driven sufficiently negative with respect to the cathode to cut off the plate current of the tube. Plate current, therefore, flows during the entire signal cycle (360-degree conduction). Although the general terms of class A operation permit the use of the grid-current region (**class** A_2 operation), the driving voltage is usually kept smaller than the grid bias so that the grid is not driven positive with respect to the cathode and, consequently, does not draw current. Under these conditions (**class** A_1 operation), waveform distortion (variation of output-signal waveshape from that of input signal) consists principally of even-order harmonics and can easily be limited to less than 5 per cent of full output in triodes and less than 7 per cent of full output in multigrid tubes by a proper choice of operating conditions. For symmetrical driving voltages, the dc plate current remains substantially constant at the quiescent (zero-signal) value.

Because operation of a class A amplifier is restricted to the linear region of the characteristics, the maximum plate-current swing available between cutoff and saturation is not fully utilized. Consequently, the power output, which is proportional to the square of the plate-current swing, is somewhat limited. The highest theoretical plate-circuit efficiency (ratio of output power to input power) obtainable under class A conditions is 50 per cent. Efficiencies in the order of 40 to 45 per cent can be achieved in certain beam power tubes and pentodes, and efficiencies of 25 to 30 per cent in triodes.

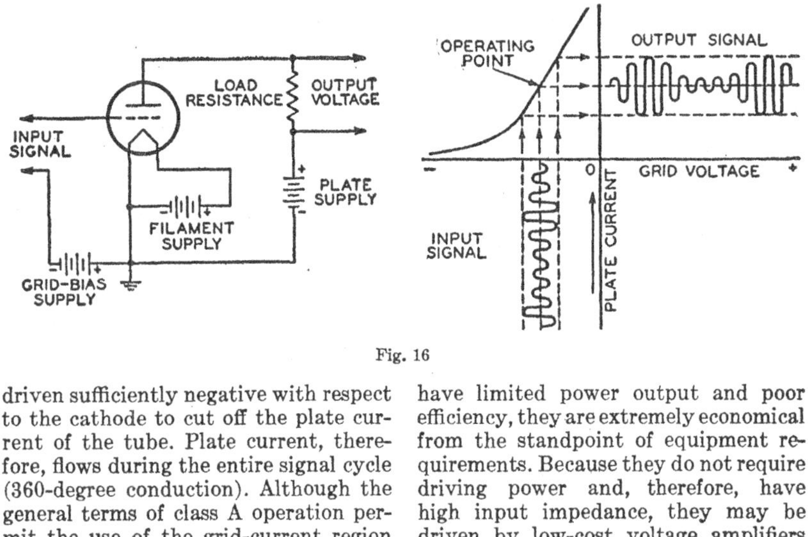

Fig. 16

Although class A power amplifiers have limited power output and poor efficiency, they are extremely economical from the standpoint of equipment requirements. Because they do not require driving power and, therefore, have high input impedance, they may be driven by low-cost voltage amplifiers employing direct coupling or simple resistance-capacitance coupling networks. Because the average plate currents remain substantially constant, plate supplies need not be designed for good regulation. The constant average plate current and moderate grid-bias voltage requirements also make it practicable to use self-bias without danger of excessive distortion, thus eliminating the expense of special bias supplies.

The power output required for a particular application may be obtained either from a single tube having suitable ratings, or from two or more tubes operated in parallel, push-pull, or push-pull-parallel. Although single-tube stages are usually the most efficient electrically and the simplest mechanically, parallel and push-pull stages can provide substantial amounts of power output from relatively small and inexpensive tubes operating at low plate voltages.

In general, the power output that can be obtained from a given number of tubes is the same in parallel and in push-pull operation. Each method, however, has advantages. **Parallel** operation improves stability and output regulation because it reduces plate resistance in direct proportion to the number of tubes employed. In addition, it is usually the simplest and most convenient method of adding tubes to an existing stage because it does not require a change in circuit configuration or an increase in driving voltage. It does not, however, reduce harmonic distortion in relation to total power output, and may actually result in an increase in the total harmonic output unless certain precautions discussed in the *Power-Tube Circuit-Design Considerations* Section are observed.

A **push-pull** stage requires a driving circuit supplying two signal voltages 180 degrees out of phase (each equal to the voltage required by a single tube) and a center-tapped output transformer or load. Because push-pull operation increases effective plate resistance, it results in poorer output regulation. However, it provides a number of very important advantages.

Even-order harmonics generated in the opposite sides of a push-pull stage develop voltages of opposite polarity and substantially equal amplitude in the load, and are thus cancelled or substantially reduced in relation to the total power output. Consequently, a push-pull stage can deliver output of substantially better quality than a parallel stage using the same tubes and operating under the same conditions, or it can deliver higher output for the same amount of even-harmonic distortion. Higher power output per tube can also be obtained without an increase in plate voltage by the use of a plate-to-plate load resistance only slightly larger than that recommended for single-tube operation. Although odd-order harmonic distortion is not cancelled or reduced by push-pull operation, this type of distortion is usually negligible in class A amplifiers, and may be minimized by the proper choice of operating conditions or by the use of inverse-feedback circuit arrangements.

Hum caused by the presence of ripple in dc plate, screen-grid (grid-No.2), or bias (grid-No.1) supply voltages, or by the use of ac filament or heater voltages, is also cancelled or substantially reduced in a push-pull stage. Push-pull operation thus simplifies power-supply filter requirements. Furthermore, it frequently eliminates the necessity for attenuating the low-frequency response of an audio or video amplifier to reduce interference from power-supply hum.

Push-pull af power amplifier stages can employ substantially smaller and less expensive output transformers than those required for equivalent single-ended stages. They are also inherently capable of better high-frequency response because corresponding tube and circuit capacitances are in series rather than in parallel, and thus cause substantially less shunting of the input and output circuits.

Class B Amplifiers

The highest efficiencies and power outputs attainable in linear amplifiers

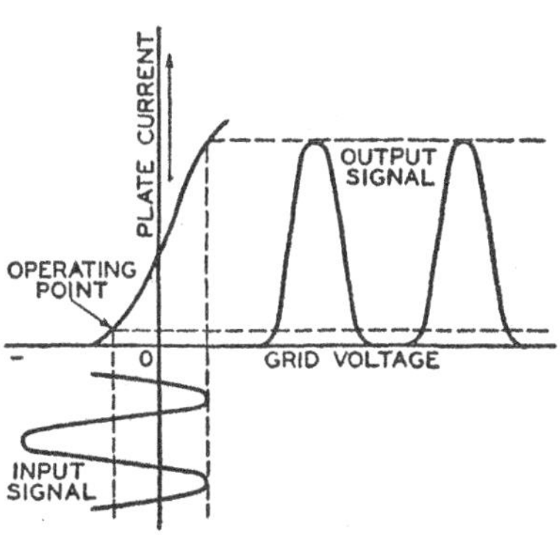

Fig. 17

are obtained under class B conditions. As shown graphically in Fig. 17, a class B amplifier is biased so that its operating point is just above plate-current cutoff. The tube, therefore, draws a very small zero-signal plate current, and responds only to the positive portions of an ac input signal. Because the operating characteristic is highly asymmetrical, the plate-current waveform contains a large amount of even-harmonic distortion and is similar to that of a half-wave rectifier.

In class B af amplifiers, push-pull circuits such as that shown in Fig. 18 are used to obtain cancellation of the

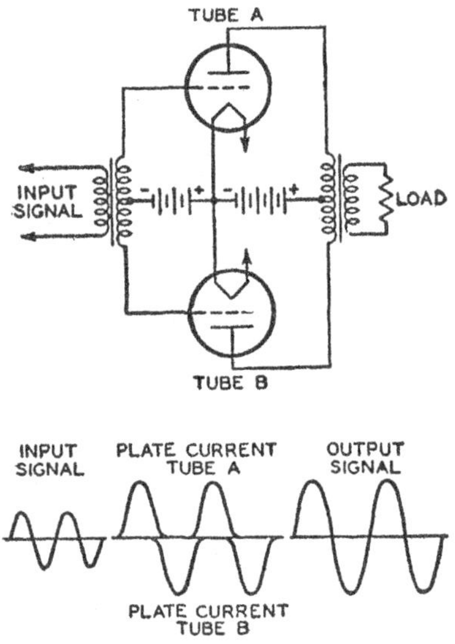

Fig. 18

even-harmonic distortion and amplification of both positive and negative portions of the signal waveform. In class B rf amplifiers, on the other hand, complete oscillations can be obtained from pulses of plate current in single-ended stages by the use of a tuned plate-tank circuit.

Because of the small zero-signal plate current, class B amplifiers may use higher plate voltages than are permissible for class A operation without danger of exceeding maximum plate-input ratings. The use of higher plate voltage and operation in the positive-grid region results in power outputs of four to six times the class A output.

Theoretically, the highest plate-circuit efficiency that can be achieved under class B conditions is 78.5 per cent. This value may be closely approached in well-designed class B audio amplifiers. To achieve maximum power output and efficiency in a class B stage, however, it is necessary to supply driving power to the grids. Because the average plate current and grid current vary with the amplitude of the driving signal, the plate supply must have very good voltage regulation so that serious distortion and loss of power output will not occur on large input signals. For the same reasons, bias must be obtained from a separate, stable, fixed supply, and not from a grid resistor or cathode resistor.

As a result of the discontinuity in the composite characteristic of a push-pull class B audio amplifier, shown in Fig. 18, the plate current never falls to zero, but transfers abruptly from one tube to the other each time the driving voltage swings through the operating point. This "switching" action results in the generation of an odd-harmonic component which cannot be cancelled by push-pull operation and, because of its steep waveform, may cause spurious oscillations in the output transformer. The amplitude of this harmonic can be minimized by moving the operating point toward the linear region of the tube characteristic, *i.e.*, by increasing the zero-signal plate current and thereby reducing the plate-circuit efficiency. The most desirable tubes for class B audio service, therefore, are those having very steep mutual characteristics and very short "lower bends" so that the discontinuity in the composite characteristic will be small even when the operating point is very close to cutoff.

Because of their linearity and relatively high efficiency, class B amplifiers are particularly suitable for use as output amplifiers in rf transmitters employing **"low-level"** amplitude modulation. Modulation applied to the final or output stage of a transmitter is called **"high-level"** modulation; that applied to any stages preceding the final stage is called "low-level" modulation. When "low-level" amplitude modulation is employed, any stages following the modulated amplifier must be linear amplifiers to avoid distortion of the modulated rf waveform. The circuit of a typical class

B linear rf output stage is shown in Fig. 19.

The quiescent plate current of a class B rf amplifier, unlike that of its af counterpart, is not approximately zero but is proportional to the amplitude of the unmodulated rf driving signal or carrier. Consequently, the maximum efficiency is lower than that obtainable in af service, and varies from approximately 33 per cent for an unmodulated carrier to approximately 66 per cent for a fully modulated carrier. With symmetrical modulating voltages, the average plate current remains constant, and it is not necessary to employ a regulated plate supply.

The high degree of linearity required for the reproduction of complex modulated rf waveforms may be obtained by careful control of the position of the operating point and the maximum and minimum amplitudes of the modulated driving signal. Consequently, bias, tuning, and other operating adjustments for class B linear rf amplifiers are usually

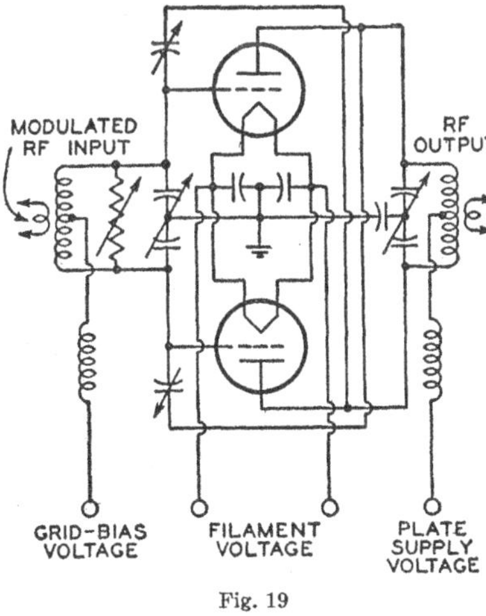

Fig. 19

much more critical than those for other types of rf power amplifiers.

Class B linear amplifiers are finding increased use as output amplifiers in **single-sideband, suppressed-carrier** radiotelephone transmitters. In amplitude modulation, the additional power obtained from the modulator at each modulating frequency appears in the rf output as a pair of "sideband" signals, as shown in Fig. 20. Each of these signals is separated from the carrier by a frequency f equal to the modulating frequency, and contains one-half the modulating power at that frequency. The output of the modulated amplifier, therefore, occupies a frequency band 2f

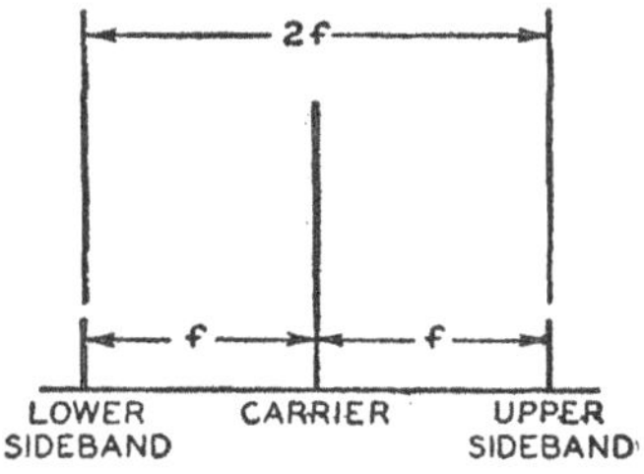

Fig. 20

wide, where f is the highest modulating frequency employed.

Because all the information represented by the modulation is present in either the upper or lower sideband group, the carrier and one group of sidebands are in a sense superfluous once modulation has been accomplished. Although transmission of the carrier and both sidebands is uneconomical of transmitter power and channel space, it is employed in standard radio broadcasting and in many radiotelephone communications services because it permits the use of simple transmitter and receiver circuit designs.

In single-sideband, suppressed-carrier radiotelephony, both the carrier and one sideband group are eliminated by the use of a special low-level modulator circuit. Because low-level modulation is employed, the output stage must be linear, and, for maximum efficiency, is usually a class B amplifier.

Class AB Amplifiers

Multigrid tubes and low-mu triodes are not usually recommended or rated for use as class B audio-frequency amplifiers. Multigrid types generate large amounts of odd-harmonic distortion when operated in the vicinity of plate-current cutoff, and low-mu triodes require uneconomically large fixed-bias voltages and relatively high driving power. These types can, however, deliver relatively high output with low distortion and good efficiency when operated under class AB conditions.

Class AB operation is an intermediate classification combining certain characteristics of both class A and class B operation, as shown in Fig. 21. Like class B operation, it results in severe

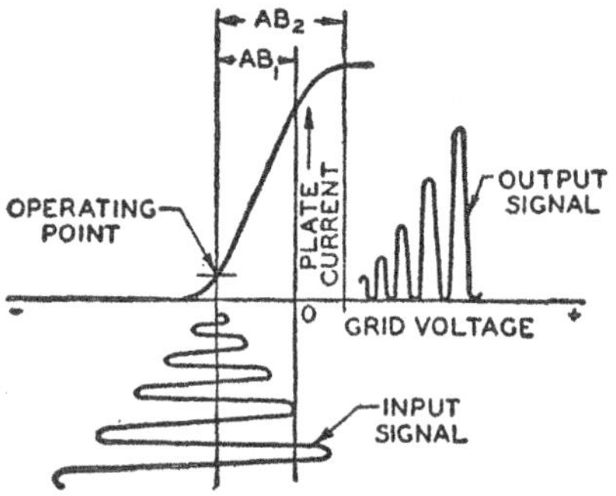

Fig. 21

even-harmonic distortion and, consequently, requires the use of a push-pull circuit when used in audio or video service. The bias is adjusted so that the operating point is in the lower portion of the linear region of the characteristic. Because of the relatively small quiescent plate current, the tube can be operated at a higher plate voltage than would be permissible under class A conditions, and can thus deliver a higher maximum power output.

On small input signals, operation takes place over a substantially linear region of the characteristic, and the tube operates as a class A amplifier. On large input signals, however, the negative grid-voltage excursions extend into the region beyond cutoff, and the tube operates as a class B amplifier.

In **class AB_1** operation, the grid is never driven sufficiently positive to draw current. Because no driving power is required under these conditions, class AB_1 amplifiers, like class A amplifiers, may be driven by voltage amplifiers using direct or resistance-capacitance coupling. In **class AB_2** operation, the grid is driven positive by the larger input signals and, therefore, draws current. Class AB_2 amplifiers thus require driving power, but can deliver substantially higher power outputs than class AB_1 amplifiers because of the larger plate-current swings that can be achieved.

The average plate current of a class AB amplifier varies with the amplitude of the driving signal, although this variation is smaller under class AB_1 than under AB_2 conditions. Consequently, plate and screen-grid (grid-No.2) supplies for these amplifiers must have good voltage regulation to assure that the full output capabilities of the tubes can be realized and the harmonic distortion kept low. Cathode-resistor bias can be employed for class AB_1 amplifiers, although higher power output and lower distortion can usually be obtained by the use of fixed bias. Fixed bias must be used for class AB_2 amplifiers.

The plate-circuit efficiencies that can be attained in class AB_1 amplifiers range from about 30 to 40 per cent for triodes to as high as 50 to 60 per cent for multigrid tubes. Efficiencies of 60 to 70 per cent can be attained in beam power tubes used as class AB_2 amplifiers.

Class C Amplifiers

Maximum power output and plate-circuit efficiency can be obtained from triodes or multigrid tubes under class C conditions. Because these advantages are obtained at the expense of linearity, class C amplifiers cannot be used if it is necessary to reproduce variations in the waveform of the driving signal. Class C amplifiers can be modulated linearly, however, and are extremely useful as rf

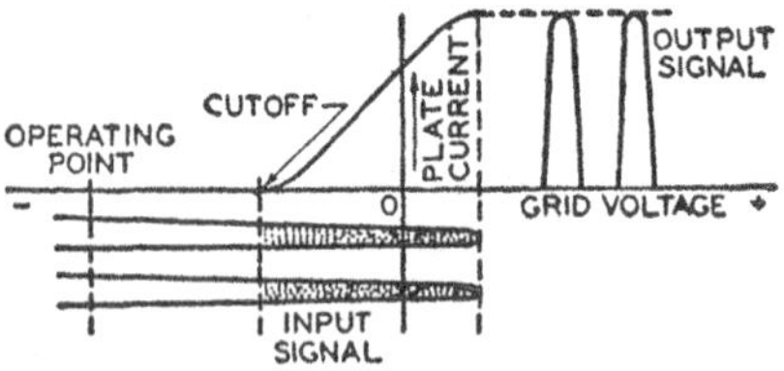

Fig. 22

power amplifiers, frequency multipliers, and oscillators.

A class C amplifier is operated with a negative control-grid (grid-No.1) bias substantially higher than that required for plate-current cutoff, as shown in Fig. 22. The quiescent plate current, therefore, is zero, and the tube responds only to those portions of positive grid-voltage excursions which are positive with respect to the cutoff voltage (indicated by the shaded areas of the input-signal waveform in Fig. 22). In practice, the grid is excited by an rf voltage having constant amplitude, and the plate-current waveform consists of relatively narrow pulses of equal height which have the same frequency as the excitation voltage but contain very strong odd- and even-order harmonic components. The height of these pulses (the **peak plate current**) is determined by the point on the transfer characteristic to which the tube is driven by the rf driving voltage. For a given pulse height, the average or dc value of the plate current is determined by the pulse width (*i.e.*, the conduction angle employed) and, therefore, varies inversely with the magnitude of the negative voltage for constant peak driving voltage.

The power output of a class C amplifier is proportional to the square of the plate voltage. Maximum power output is achieved when the excitation swings the plate current between zero and the saturation value during each conduction interval. To achieve this swing, it is necessary to drive the grid highly positive and, consequently, supply it with a substantial amount of driving power. The plate-circuit efficiency increases as the conduction angle is reduced, and theoretically may reach 100 per cent when the conduction angle is made infinitely small. Very small conduction angles usually cannot be obtained, however, without increasing the bias and excitation voltages to such high values that they exceed the maximum grid-voltage ratings of the tube. Driving-power requirements, which increase as the square of the excitation voltage, are also a limiting factor. However, plate-circuit efficiencies of 75 to 80 per cent are easily achieved.

The large grid-bias voltages required by class C amplifiers are conveniently and economically obtained by grid-rectification of the driving voltage (grid-resistor bias). This type of bias automatically adjusts itself to the amplitude of the excitation voltage to maintain the desired conduction angle, and allows the full plate-supply voltage to be applied between the plate and cathode of the tube. (Because grid-resistor bias depends on the presence of excitation, it is also necessary to employ some means for protecting the tube against damage by excessive plate current in the event that excitation fails or is accidentally removed.)

Class C Telegraphy

The term "Class C Telegraphy" applies to applications in which power tubes may be operated at their highest ratings. It includes "straight-through" rf power amplifiers which are not "keyed" or modulated as well as those which are actually "keyed" for telegraphy service, oscillators, and amplifiers for frequency-modulated rf carriers.

The circuit of a typical "straight-through" class C rf amplifier employing a beam power tube is shown in Fig. 23.

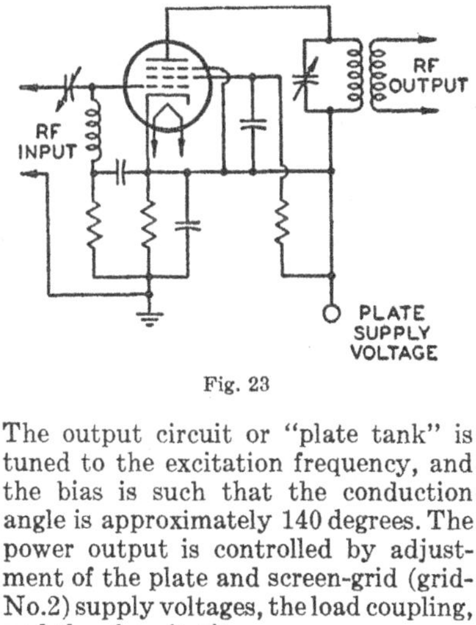

Fig. 23

The output circuit or "plate tank" is tuned to the excitation frequency, and the bias is such that the conduction angle is approximately 140 degrees. The power output is controlled by adjustment of the plate and screen-grid (grid-No.2) supply voltages, the load coupling, and the rf excitation.

Triode "straight-through" rf amplifiers must be neutralized to prevent self-oscillation resulting from internal feedback through the grid-plate capacitance. Multigrid-tube "straight-through" amplifiers may also require neutralization to assure stability at the higher radio frequencies.

The circuit of a "keyed" class C rf amplifier is essentially the same as the one shown in Fig. 23 except that a "key" (a manually or automatically operated switch) is inserted in the plate, screen-grid, or cathode circuit.

The circuit and operating conditions of a class C amplifier for frequency-modulated signals are the same as those shown in Fig. 23 and described above. The only special consideration involved in the operation of such an amplifier is that the plate-tank circuit must be designed to have constant impedance over the entire frequency band covered by the carrier at maximum deviation.

Modulated Class C Amplifiers

The plate current of a class C amplifier is proportional to plate voltage and, in the case of a multigrid tube, to screen-grid (grid-No.2) voltage. Within certain limits it is also proportional to control-grid (grid-No.1) bias and, in the case of certain pentodes and beam power tubes, to suppressor-grid (grid-No.3) voltage. Consequently, the output of a class C rf power amplifier can be modulated in amplitude by varying one or more of its dc electrode voltages in accordance with the amplitude variations of an audio or video signal.

Distortionless modulation requires that the relationship between the dc control voltage and the plate current be linear, and that both vary between zero and twice their unmodulated values on the peaks of the modulating signal. Under these ideal conditions, the peak power output of the class C amplifier at full (100-per-cent) modulation is 4 times the unmodulated output, and the average power output 1.5 times the unmodulated output.

Plate input and plate dissipation also increase 50 per cent when a class C amplifier is fully modulated. For plate modulation, therefore, the plate input and dissipation under carrier conditions must not exceed two-thirds the maximum values for class C telegraphy. For control-grid, screen-grid, suppressor-grid, or cathode modulation, the permissible dc plate input is even smaller. Maximum dc plate-voltage and plate-current ratings for modulated class C amplifiers are usually not more than 80 per cent of the class C telegraphy values.

The audio or video power required for 100-per-cent modulation of a class C amplifier is equal to one-half the dc power input to the modulated circuit. For symmetrical modulating voltages, the dc plate current of the modulated amplifier and the dc supply voltage and current of the modulated-electrode circuit remain constant. The additional power output obtained by amplitude modulation does not increase the carrier power, but is equally divided between two symmetrical "sideband" signals.

The method of modulation that provides the greatest plate-circuit efficiency and linearity is **plate modulation.** In this method, the modulating voltage is connected in series with the dc plate supply for the class C amplifier, as shown in Fig. 24. In a beam power

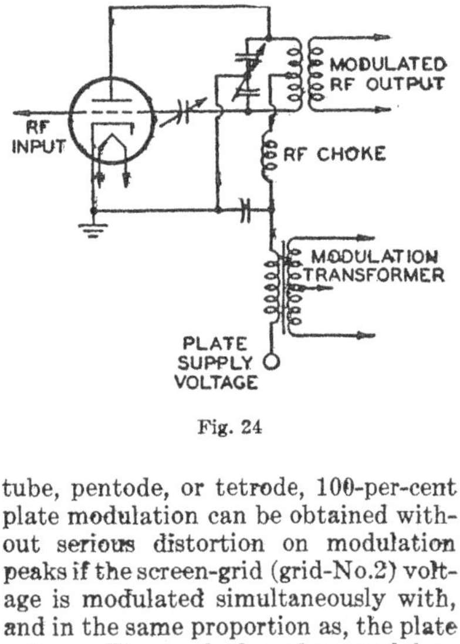

Fig. 24

tube, pentode, or tetrode, 100-per-cent plate modulation can be obtained without serious distortion on modulation peaks if the screen-grid (grid-No.2) voltage is modulated simultaneously with, and in the same proportion as, the plate voltage. The method used to modulate the screen grid depends on the type of screen-grid-supply circuit used. If screen-grid voltage is obtained from a separate supply, the method shown in Fig. 25(a) may be used. If screen-grid voltage is obtained from the plate supply through a series resistor, the resistor should be connected to the modulated side of the plate supply circuit, as shown in Fig. 25(b). In all such cases, the modulator must be capable of supplying af power at least equal to one-half the combined dc inputs to the plate and screen-grid circuits.

A circuit in which modulation power is applied only to the plate of a beam power tube is shown in Fig. 25(c). The reactance of the af choke at the lowest modulating frequency should be at least equal to the dc screen-grid voltage divided by the dc screen-grid current.

The plate-circuit efficiency of a plate-modulated class C amplifier is usually in the order of 65 to 70 per cent.

Control-grid (grid-No.1) or "grid-bias" modulation requires very little modulating power and can provide good linearity. However, the power output obtainable is only one-third to one-half that obtainable with plate modulation, and plate-circuit efficiency is not usually greater than 33 per cent.

In control-grid modulation, the audio or video modulating voltage is connected in series with the bias supply for the class C amplifier. Consequently, the operating point of the modulated amplifier varies with the modulation. In order to obtain 100-per-cent modulation with good linearity, the plate current and effective plate voltage must swing between zero and twice their unmodulated values on the peaks of the modulating signal. The dc plate voltage, therefore, can only be about one-half that for plate modulation. Operating conditions, plate-circuit efficiency, and power output are almost identical with those for class B rf service.

The modulator must be capable of supplying the power required by the grid of the modulated amplifier on the positive peaks of the modulating signal. It must also have good output regulation because of the wide variation in the load impedance presented by the grid-circuit over the entire modulation cycle. The driver supplying the unmodulated carrier and the bias supply for the modulated amplifier must also have very good regulation to avoid serious distortion. Bias must be obtained from a separate low-impedance, fixed supply, and not from a grid resistor or cathode resistor.

Because pentodes and beam power tubes are substantially free from the secondary-emission effects which occur in

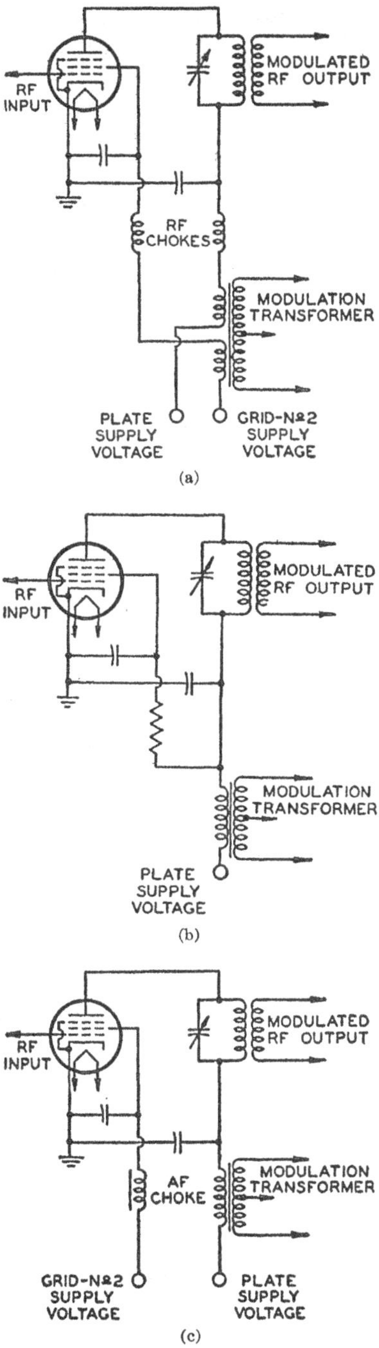

Fig. 25

other multigrid types when the screen grid (grid No.2) becomes more positive than the plate, they may use screen-grid modulation without danger of serious distortion. **Screen-grid modulation** is similar to grid-bias modulation in that it requires relatively little af power, and provides substantially the same power output and efficiency. Unlike grid-bias modulation, however, it does not require the use of fixed bias or good driver regulation.

When screen-grid voltage is obtained from a separate supply, the modulating voltage may be connected directly in series with the supply circuit, as shown in Fig. 26(a). When screen-grid voltage is obtained by the series-resistor method, it is generally necessary to use the "clamp-tube" method of modulation shown in Fig. 26(b).

Suppressor-grid (grid-No.3) modulation can be used with certain beam power tubes and pentodes. Operating conditions are similar to those used in screen-grid modulation, except that the suppressor grid is supplied with a fixed negative dc bias voltage in addition to the modulating voltage. This bias voltage is adjusted so that the plate current and rf output current of the modulated amplifier under carrier conditions are one-half those obtained in class C telegraphy service with zero voltage on the suppressor grid. Under these conditions, the modulator is required to supply only a peak voltage equal to the suppressor-grid bias, and does not have to supply power because the suppressor-grid is not driven positive. Suppressor-grid modulation has only limited application, however, because relatively few beam power tubes and pentodes have the neccessary linear relation between suppressor-grid voltage and plate current.

Cathode modulation combines the characteristics of plate and grid-bias modulation. The modulating voltage is introduced in the common dc cathode-return circuit of the class C amplifier and, therefore, varies the plate voltage and grid bias simultaneously. This method requires less modulating power than plate modulation, and permits the modulated amplifier to be operated with a plate-circuit efficiency proportional to the amount of modulating power available. However, the power output obtainable is less than that obtainable with plate modulation.

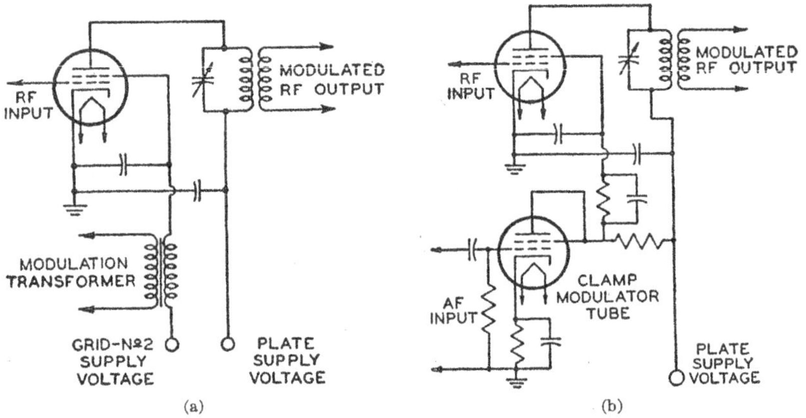

Fig. 26

The type of coupling used between a modulator and the modulated circuit of a class C rf amplifier depends primarily on the amount of modulating power required. In suppressor-grid modulation or "clamp-tube" screen-grid modulation, it is usually practicable to use resistance-capacitance or impedance coupling because little or no modulating power is required. In other cases, it is usually necessary to employ transformer coupling to obtain proper impedance matching and most efficient use of the available modulator power.

The bypass capacitors shown in

Figs. 24 through 26 should have very low reactance at the rf carrier and sideband frequencies and high reactance at the highest modulating frequency. The modulation transformer must convert the equivalent resistance of the modulated dc supply circuit into the proper plate or plate-to-plate load resistance, Z, for the modulator output tubes and, consequently, should have a primary-to-secondary turns ratio, N_1/N_2, equal to $\sqrt{ZI/E}$, where I and E are the average current and dc input voltage of the modulated circuit, respectively.

The value used for I in this calculation is the current under carrier conditions (no modulation). In the case of plate modulation it is the total dc plate current; in the case of combined plate and screen-grid modulation using series-resistor screen-grid supply, it is the sum of the dc plate and screen-grid currents. In the case of grid-bias modulation, I is the dc grid current and E the grid-bias voltage.

Frequency Multiplication

Any amplifier which generates harmonics can be used as a frequency multiplier provided the desired harmonic of the excitation frequency is present in the plate-current pulse. The fundamental and other harmonics may then be eliminated by means of a plate-tank circuit tuned to the desired harmonic. This procedure can be repeated in successive stages as often as desired.

By frequency multiplication, high-frequency carriers having a very high degree of frequency stability can be obtained. Frequency multiplication also makes it possible to obtain output in several harmonically related frequency bands (such as those assigned for amateur service) from a single oscillator circuit. For example, an oscillator operating in the 80-meter band (at a frequency between 3.5 and 3.58 megacycles per second) can be used with a series of frequency-doubler stages to obtain output in the 40-, 20-, and 10-meter bands.

Frequency multipliers are almost invariably class C amplifiers because maximum harmonic output can be achieved under class C conditions. When a class C amplifier is operated under the conditions normally employed for "straight-through" amplifier service, however, its efficiency as a frequency multiplier is relatively poor because even the strongest harmonics represent only a small fraction of the total power output. To obtain good efficiency in multiplier service, it is necessary to select a plate-conduction angle which has high harmonic content at the desired harmonic frequency. Consequently, frequency multipliers require substantially higher bias and excitation voltages and more driving power than "straight-through" class C amplifiers. The plate-circuit efficiency that can be achieved is usually not more than 60 per cent (doubler operation), and decreases rapidly as the degree of multiplication is increased.

Frequency multiplication of more than four is seldom practicable in a single stage because of the relatively small output at the high harmonics and the large amounts of driving power required. Although a triode frequency multiplier does not require neutralization because the grid and plate circuits are not tuned to the same frequency, neutralization can be used to reduce the amplitude of undesired frequency components in the plate-current waveform and thus increase the output at the desired harmonic frequency.

Because of its smaller conduction angle, a frequency multiplier is more sensitive to small changes in excitation voltage and loading than an equivalent "straight-through" class C amplifier and, therefore, has poorer output regulation. From the excitation standpoint, this difficulty can be minimized by the use of beam power tubes or pentodes rather than triodes. Improved regulation can also be obtained by the use of tubes in parallel. Very good output regulation can be obtained in doubler service by the use of a **"push-push"** circuit such as that shown in Fig. 27. In this type of circuit, the grids are excited in push-pull so that the tubes conduct alternately on successive half-cycles of the excitation voltage. Because the plates are connected in parallel, two pulses of plate current flow in the common plate-tank circuit for each excitation cycle, doubling the power output and reducing the output impedance to one-half the value for one tube.

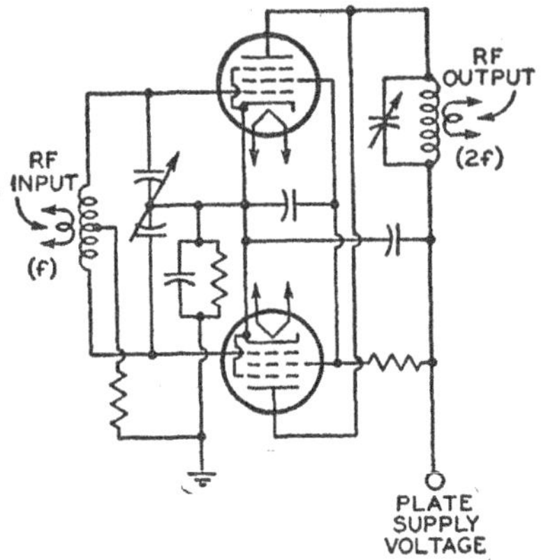

Fig. 27

Additional information on the characteristics of frequency multipliers and the efficiencies obtainable for various degrees of multiplication is given in the *Power-Tube Circuit-Design Considerations* Section.

Oscillators

RF power oscillators are usually class C amplifiers which obtain excitation from their own output circuits and employ either quartz crystals or inductance-capacitance tuned circuits as frequency-determining elements. Crystal-controlled oscillators can provide the highest degree of frequency stability, and are used in equipment which operates entirely or predominantly on fixed frequencies or on fixed harmonically related frequencies. In general, mechanical considerations make it impracticable to cut crystals for fundamental frequencies higher than about 20 megacycles per second. A technique known as "overtone operation," however, permits crystals to be used for the control of oscillators operating at frequencies up to 100 megacycles per second and higher. Representative crystal oscillators are shown in the *Circuits* Section.

Inductance-capacitance frequency-determining elements are used for oscillators which must be capable of operating at any frequency within a specific band. They are also used for oscillators which must operate at frequencies above and below those for which crystals can be cut. The mechanical form of the LC tank and the type of oscillator circuit employed are usually determined by the operating frequencies involved. At the lower radio frequencies, well-designed electron-coupled oscillators employing conventional coils and tuning capacitors can provide stabilities comparable to those obtained in crystal oscillators. When followed by suitable frequency-multiplier stages, such oscillators can be used to control equipment operating at frequencies up to about 30 megacycles per second. Tuned-line oscillators of the type shown in the *Circuits* Section are usually employed in very-high-frequency (vhf) equipment. Ultra-high-frequency (uhf) oscillators usually require the use of coaxial- or cavity-type circuits as frequency-determining elements.

Circuit Configuration

The amplifier applications discussed in this chapter have been illustrated by "**grid-drive**" circuits of the type shown in Fig. 16. In this type of circuit, the grid is employed as the "drive" electrode, the plate as the "output" electrode, and the cathode as the "ground" or reference electrode common to the input and output circuits of the tube.

As mentioned previously, a grid-drive triode rf amplifier must be neutralized to cancel the regenerative feedback which takes place through the grid-plate capacitance of the tube. Neutralization, however, becomes less effective and more difficult to achieve as the operating frequency is increased because of unavoidable resonance effects in the components of the neutralizing circuit. These effects alter the phase of the neutralizing voltage and, in most cases, make it impossible to obtain neutralization at frequencies of more than a few hundred megacycles. Although multigrid tubes capable of operating as grid-drive uhf amplifiers are available, triodes are generally preferable for uhf service because of their lower noise and shorter electron-transit time, and because their simpler electrode structures and power-supply requirements make them more readily adaptable to installation in coaxial and cavity-type uhf tank-circuit components.

In many cases, this difficulty may be overcome by the use of **"cathode-drive"** circuits such as that shown in Fig. 28. In this method of operation, the cathode is the "drive" electrode and the grid is the "ground" electrode common to the input and output circuits. The grid thus acts as an electrostatic shield between the input and output terminals, and reduces internal feedback in the same manner and to approximately the same degree as the screen grid (grid No.2) of a multigrid tube.

A cathode-drive amplifier requires more driving power than a grid-drive amplifier because its input is shunted not only by the grid-cathode capacitance but also by the plate resistance, r_p, and load resistance, R_L, in series. This additional power is not wasted, however, but

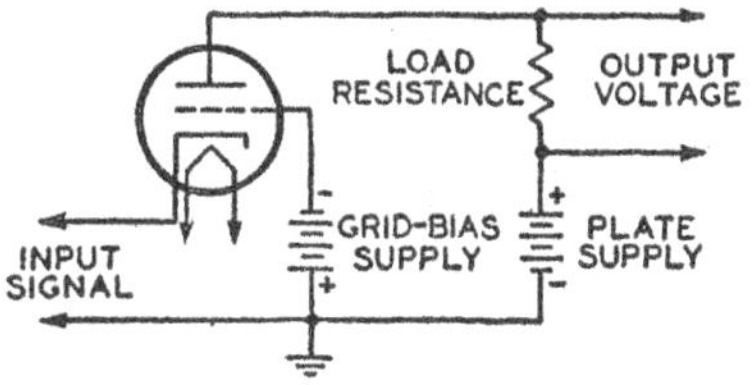

Fig. 28

is added to the output because the driving voltage and plate-supply voltage are effectively in series across the load. The input of a cathode-drive amplifier is also shunted by the heater-cathode capacitance or by the capacitance to ground of the filament-supply circuit. This capacitance, however, may be neutralized by the use of suitable rf chokes in the heater or filament circuit.

A **"cathode follower,"** shown in Fig. 29, is a grid-drive amplifier in which the cathode is used as the output electrode and the plate as the ground or common terminal of the input and output circuits. Because the grid-cathode

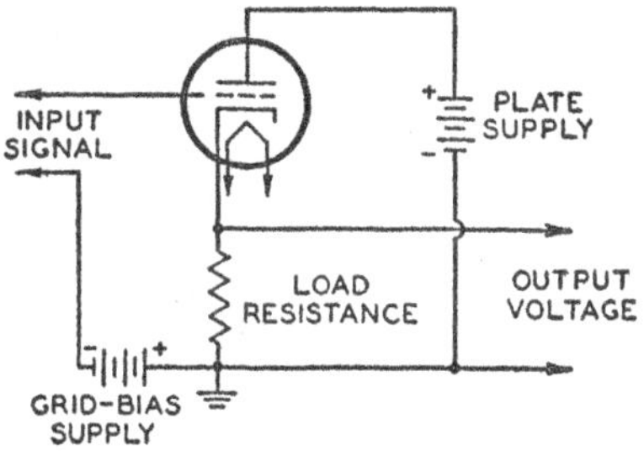

Fig. 29

capacitance of the tube does not shunt the driving circuit, the cathode follower has higher input impedance than a conventional grid-drive amplifier and, consequently, requires less driving power for the same power output. The output impedance, which is composed of the external cathode resistance, R_k, and the plate resistance, r_p, of the tube in parallel, can be made as low as desired by the use of a suitable cathode resistor. Because the driving voltage and output are both developed across R_k, the voltage gain cannot exceed unity. Substantial power gains can be achieved, however, by the transformation from a high to a low impedance.

Because the voltage gain of a cathode follower is always less than unity, this type of amplifier cannot oscillate and, therefore, does not require neutralization, regardless of the operating frequency.

Power-Tube Circuit-Design Considerations

The performance of a power tube depends not only on the conditions under which the tube is operated but also on the design of the associated circuits.

Proper circuit design assures economical and effective use of tubes and other components, simplifies equipment adjustment, provides for stable operation, thereby minimizing the likelihood of interference with other services, and provides a substantial measure of protection for the equipment, as well as greater personal safety.

In the production of moderate to large amounts of power at audio or radio frequencies, a signal or voltage having suitable characteristics is usually generated at a low power level. This signal is then amplified in one or more stages until the desired power level is achieved. In rf equipment, one or more amplifier stages may also be used to modify some characteristic of the signal, such as frequency, phase, or instantaneous amplitude. Consequently, the individual stages usually operate under substantially different conditions. Power-tube equipment, therefore, is designed one stage at a time, the usual procedure being to start with the output stage and work backward through preceding stages to the oscillator or input stage of the equipment. The design of a stage involves selection of the most suitable tube type; design of input and output coupling circuits; design of power-supply circuits; design of circuits for controlling gain or power output, or for varying the instantaneous amplitude, frequency, or phase of the output signal; and provision of means for stabilization against self-oscillation or other conditions which may result in interference, unauthorized radiations, distortion, or other undesirable effects.

In af equipment, all stages usually operate into non-resonant loads and have substantially the same frequency-response characteristics. The dc input to the tubes is constant, and power output is controlled by attenuation of the signal at a relatively low-level point in the system and/or by the use of remote-cutoff tubes. Input, interstage, and output coupling is fixed, and control of over-all frequency response, where required, is usually accomplished by fixed or adjustable filters in one or more stages. Stabilization seldom involves procedures other than those necessary to prevent self-oscillation or minimize distortion.

In rf power-tube equipment, all stages usually operate into resonant loads. In a transmitter, individual stages may operate at different frequencies and, in many cases, each stage must also be capable of operating at any frequency within one or more bands. The power output of an rf stage is controlled by adjustment of the dc input, rf excitation, and loading. In transmitters, consideration must also be given to the design of "keying" or modulating circuits. Because the input and output impedances of rf amplifier stages vary considerably with changes in operating frequency, excitation, and loading, interstage and output coupling circuits are generally made adjustable.

Stabilization of rf equipment usually involves the elimination not only of self-oscillation, but also of undesired harmonics, and may also involve the isolation and elimination of parasitic oscillations in circuit components and wiring.

Tube Selection

The selection of the most suitable tube type for a particular application depends to a large extent upon the type of primary power available and the desired power sensitivity. Tubes having the same filament voltage or current ratings should be used throughout the equipment wherever possible to simplify power-supply requirements. Driving-power requirements vary widely with application, operating frequency, type of circuit employed, and other factors. Because of its importance in circuit design, driving power is discussed at greater length later in this section. Mechanical considerations such as equipment space limitations, layout, and ventilation, as well as economic considerations, also affect tube selection.

An initial selection of types having suitable filament-voltage, plate-voltage, plate-input, and plate-dissipation ratings for a particular application can be made from the Power-Tube Selection Guides in the *Charts* Section. The final selection is then made by comparison of the technical data for the individual types.

In the selection of a tube for use as an unmodulated rf amplifier, frequency multiplier, or oscillator, the maximum plate-input and plate-dissipation ratings and the relative plate-circuit efficiency of the tube at the highest frequency at which the equipment is to operate must be considered. When ability to change frequency quickly is an important consideration in the design of a transmitter, it is desirable to select types which require few or relatively minor changes in operating conditions with changes in frequency. In this respect beam power tubes and other multigrid types are generally superior to triodes.

Additional factors which must be considered in the selection of tubes for use as modulated rf amplifiers depend on the type and degree of modulation to be employed.These factors are discussed in the *Power-Tube Applications* Section and in the *Tube Types* Section.

Multiple-Tube Stages

Most satisfactory operation of parallel, push-pull, or push-pull-parallel stages is obtained when the plate currents of the individual tubes are equal. Equalization of average plate currents minimizes the danger of excessive plate dissipation in one or more tubes, particularly in stages which obtain bias from a common fixed supply or a common grid resistor. Equalization of zero-signal plate currents in push-pull af amplifier stages substantially aids the cancellation of even-order harmonic distortion. For complete cancellation of even-order harmonics, the plate-current excursions in the two sides of a push-pull stage must also be equal. This type of equalization (dynamic balance) is difficult to achieve, however, because of the large number of tube and circuit variables involved.

Zero-signal or average plate currents in multiple-tube stages are most easily equalized by means of individual grid-bias adjustments. The particular method used in any case depends on the type of cathode employed in the tubes and on the circuit configuration. Two methods in general use are shown in Fig. 30.

Multiple-tube stages employing beam power tubes and other multigrid

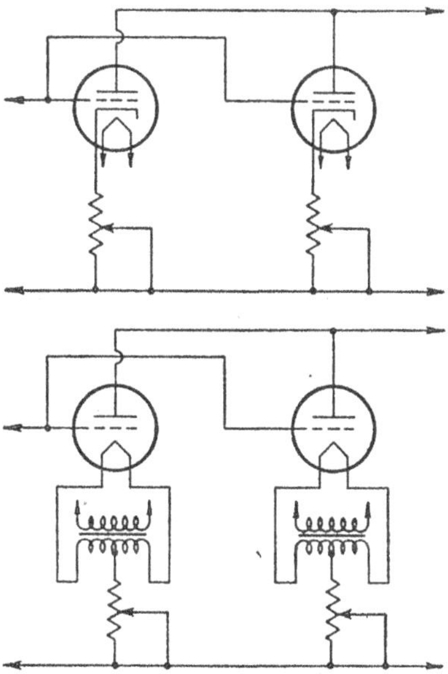

Fig. 30

types should be provided with individual adjustments for screen-grid (grid-No.2) voltage as well as for control-grid (grid-No.1) bias. Such adjustments make it possible to avoid excessive screen-grid dissipation in individual tubes and are frequently of considerable aid in obtaining plate-current equalization.

AF Power Amplifiers

Class A af power amplifiers do not normally draw grid current or require driving power. Furthermore, they draw substantially constant plate and screen-grid currents and, therefore, can employ simple cathode-resistor (self) bias. After the most suitable tube type has been selected and the tube operating conditions determined, the principal considerations in the design of a class A amplifier are: (1) the selection of a driver capable of supplying the required peak driving voltage; (2) the selection of input and output coupling devices having the desired frequency and impedance characteristics; (3) the selection of bypassing and decoupling components necessary to minimize hum, assure stability, or improve the over-all frequency response.

For this class of amplifier, the driver may be a class A voltage amplifier and the input-coupling device a simple resistance-capacitance network. Resistance-capacitance coupling provides good frequency-response characteristics economically and permits the use of simple phase-inverter circuits for driving push-pull stages. Transformer coupling can also be used between the driver and the class A power amplifier. Interstage transformers having wide frequency response are relatively expensive, however, and are seldom used unless a substantial voltage step-up must be obtained between driver and class A power amplifier.

Plate- and screen-grid-supply circuits for single-ended class A power amplifiers must be well filtered to minimize hum and undesired coupling with other stages in the equipment. These circuits, as well as the cathode-bias resistor, must also be adequately bypassed to the cathode at the lowest frequency to be reproduced to assure full output from a single-ended stage. When particularly good response at low audio frequencies is required in a single-ended stage, it may be necessary to use parallel feed, as shown in Fig. 31, to eliminate unbalanced dc from the output transformer and the driver transformer.

Circuit-design considerations for **class AB_1 af power amplifiers** are substantially the same as those for class A amplifiers, except that special consideration must be given to the characteristics of plate and screen-grid (grid-No.2) supply circuits, and to the method used for obtaining grid bias. Because the average plate and screen-grid currents of a class AB_1 amplifier vary with the amplitude of the driving signal, serious distortion and inadequate power output may result on large input signals unless plate and screen-grid supply voltages are well regulated and the bias is extremely stable. For optimum performance, plate-supply regulation should be within 10 per cent, screen-grid-supply regulation within 5 per cent, and grid-bias-supply regulation within 3 per cent.

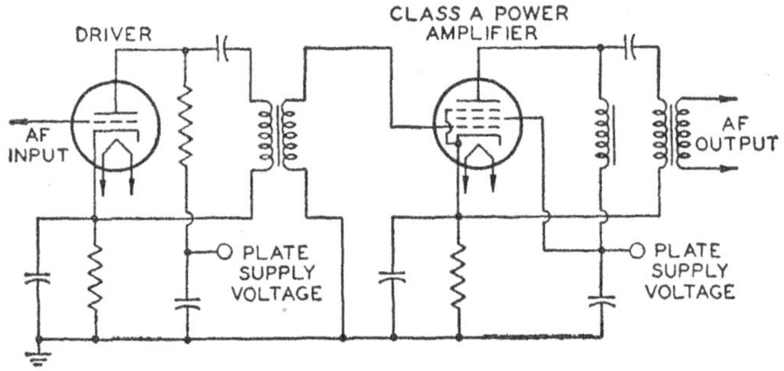

Fig. 31

Class B and class AB_2 af power amplifiers normally draw grid current on large input signals and, therefore, require appreciable driving power. Power output, frequency response, and harmonic distortion are critically dependent on the circuit constants employed in the amplifier and in the driving circuit. Consequently, the design of a class B or class AB_2 amplifier involves the design of a complete system, including the driver stage, the interstage coupling circuit, the output (class B or class AB_2) stage, and the power-supply and bias circuits for both stages.

The driver must be capable of supplying both the signal power required to drive the class B or class AB_2 stage to full output and the power lost in the interstage coupling circuit.

The driving circuit must also have very good regulation characteristics because the input impedance of a class B stage varies from a very high value on small input signals (open-circuit value when no grid current is drawn) to a very low value on large input signals (when maximum grid current is drawn). Consequently, it is usually necessary to use an amplifier having very low output impedance as the driver, and an efficient transformer as the interstage coupling device. For minimum over-all harmonic distortion, the driver should be a push-pull class A or class AB_1 amplifier. If the driver stage uses triodes, it may be operated into a load impedance higher than that normally used for the tube type employed to minimize distortion at some reduction of available output power.

The interstage or "driver" transformer must provide the proper load for the driver under maximum-drive conditions (*i.e.*, when the input impedance of the output stage is minimum) and, therefore, is usually designed as a step-down transformer. The step-down ratio required will depend on the specific tube types used in the driver and output stages, the load resistance used for the output stage, the peak power efficiency of the driver transformer, and the amount of harmonic distortion that can be tolerated in the output.

The driver transformer must also have the desired frequency-response characteristics when operated into a very high load impedance (or even an open circuit) such as that presented by the grid circuit of the class B or class AB_2 stage on very small driving signals. To assure good response at the higher audio frequencies, the transformer must also be designed to have low leakage reactance. In addition, the resistance of the secondary windings must be kept low to minimize dc voltage drops which might affect the operating bias during grid-current flow.

For maximum power output and minimum harmonic distortion, the operating point of a class B or class AB_2 amplifier must not be affected by the normal variations in average plate, screen-grid, and control-grid currents. Consequently, bias must be obtained from a separate fixed supply, such as a battery or a rectifier having very low internal resistance, and plate and screen-grid supplies must have exceptionally good regulation characteristics. For optimum performance, plate-supply regulation for class B and class AB_2 amplifiers should be within 5 per cent, and screen-grid-supply and grid-bias-supply regulation should be within 3 per cent.

Output transformers for class B and class AB_2 amplifiers should have low-resistance windings to minimize power losses at the large plate currents which flow under maximum-signal conditions. They should also have very low leakage inductance to assure good response at the higher audio frequencies and to minimize the danger of parasitic oscillations and "ringing."

Modulators

An af power amplifier used to modulate a class C rf amplifier must be capable of delivering an undistorted power output equal to one-half the average power in the modulated circuit to permit 100-per-cent modulation. In addition, the modulation transformer must convert the equivalent resistance of the modulated circuit into the proper plate-load resistance for the modulator stage.

The average power, Wa, in watts in the modulated circuit is equal to EI, and the effective resistance, R_2, is equal to E/I, where E is the dc potential across the modulated circuit in volts and I is the total direct current in amperes. The proper turns ratio (primary to secondary), N_1/N_2, for the modulation transformer is then given by

$$\frac{N_1}{N_2} = \sqrt{\frac{R_1}{R_2}}$$

where R_1 is the effective plate (or plate-to-plate) load resistance required for the af amplifier and R_2 is the effective resistance of the modulated circuit in ohms.

Example (1): Determine the amount of af power, Wo, required for 100-per-cent plate modulation of push-pull class C 812-A triodes operating under ICAS conditions. (Values are given in the technical data for the 812-A under Plate-Modulated RF Power Amplifier—Class C Telephony, Typical Operation.)

$$W_0 = \frac{W_a}{2} = \frac{(1250)(2 \times 0.140)}{2} = 175 \text{ watts.}$$

This amount of af power can be obtained from a push-pull 811-A class B amplifier operating under CCS conditions at a dc plate potential of 750 volts. (Values are given in the technical data for the 811-A under AF Power Amplifier and Modulator—Class B, Typical Operation.) The effective plate-to-plate load resistance required for the 811-A's is 5100 ohms. The equivalent resistance of the 812-A plate circuit is

$$R_2 = \frac{1250}{2 \times 0.140} = 4464$$

or approximately 4500 ohms.

Consequently, the turns ratio (primary to secondary) required for the modulation transformer is

$$\frac{N_1}{N_2} = \sqrt{\frac{5100}{4500}} = \frac{1.1}{1} \; (approx.)$$

Example (2): Determine the amount of af power, Wo, required for 100-per-cent simultaneous plate and screen-grid modulation of a single 813 class C amplifier operating under ICAS conditions. (Values are given in the technical data for the 813 under Plate-Modulated RF Power Amplifier—Class C Telephony, Typical Operation.) Screen-grid voltage for the 813 is obtained through a series voltage-dropping resistor from the plate supply, as shown in Fig. 25(c).

$$W_0 = \frac{W_a}{2} = \frac{(2000)(0.200+0.040)}{2} = 240 \text{watts}$$

This amount of power can be obtained from a push-pull 811-A class B amplifier operating under ICAS conditions at a dc plate potential of 1000 volts. (Values are given in the technical data for the 811-A under AF Power Amplifier and Modulator—Class B, Typical Operation.) The effective plate-to-plate load required for the 811-A's is 7400 ohms. The equivalent resistance of the 813 plate and screen-grid circuit is

$$R_2 = \frac{2000}{0.200+0.040} = 8333$$

or approximately 8400 ohms.

Consequently, the turns ratio (primary to secondary) required for the modulation transformer is

$$\frac{N_1}{N_2} = \sqrt{\frac{7400}{8400}} = \frac{0.94}{1} \; (approx.)$$

In the design of af power amplifiers for modulator service, consideration should also be given to the magnetizing effect of the unbalanced dc current flowing in the secondary windings of the modulation transformer. If this current is large enough to cause a decrease in low-frequency response, a suitable blocking capacitor and af choke should be used to isolate the unbalanced dc current from the secondary winding.

RF Power Amplifiers

Class B and class C rf power amplifiers normally operate into resonant load circuits which can be designed to filter out undesired harmonics of any order. Consequently, push-pull circuits do not have to be used to minimize even-order harmonics. Push-pull operation is sometimes used for "straight-through" class B and class C amplifier stages, however, as a means of obtaining increased output or improved operation at the higher radio frequencies. It is also used in frequency-multiplier service as a means of emphasizing odd-order harmonic frequencies.

Driving Power

One of the most important considerations in the design of a class B or class C rf power-amplifier stage is the provision of adequate driving power. "Typical" driving-power figures given in the technical data for tubes rated for use in class B and class C rf service indicate only the signal power dissipated in the internal grid-cathode circuit of the tube and in the resistance of the bias circuit. These figures do not normally include driving power that may be lost in tube sockets or in the components and wiring of driving circuits, or tube losses due to electron-transit-time phenomena, internal lead impedances, or other factors.

The driver stage must be capable of delivering sufficient signal power to supply all the tube and circuit losses. Although these losses vary with frequency, tube operating conditions, circuit configuration, and the components and layout of the circuit, they can be estimated with reasonable accuracy for "straight-through" amplifiers. At frequencies up to about 30 megacycles per second, total tube and circuit losses are

ponents to the peak value of the plate-current pulse is a maximum. These maxima occur at conduction angles of about

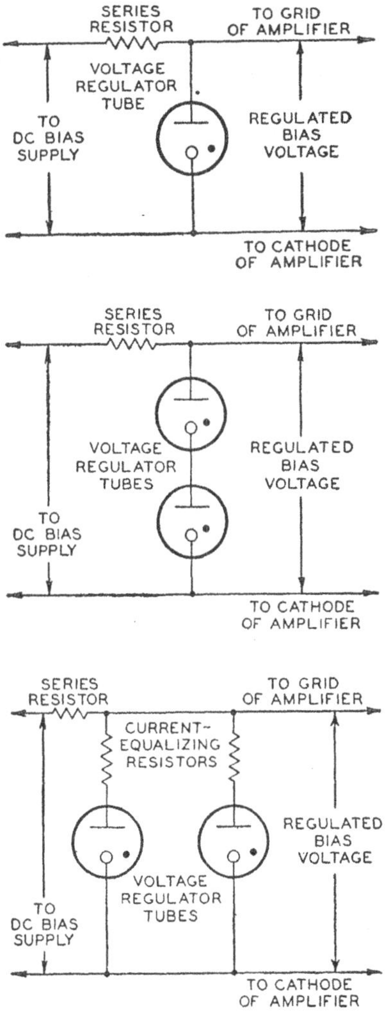

Fig. 33

120 degrees for frequency doublers, 80 degrees for triplers, and 60 degrees for quadruplers.

Because the use of small conduction angles usually requires the use of large values of negative bias, power output and plate-circuit efficiency at the higher harmonics are limited by the grid-bias rating of the tube, as well as by the peak-emission capabilities of the cathode. The over-all efficiencies obtainable in frequency-multiplier service are also limited by driving-power requirements, which increase as the square of the grid-driving voltage. Tube types for use in frequency-multiplier stages should have high-wattage filaments or cathodes capable of supplying the very high peak-emission currents required, and high transconductance or high amplification factors to provide high power sensitivity.

Oscillators

The principal consideration in the design of an oscillator is usually frequency stability, rather than high efficiency or high power output. The frequency stability of an oscillator is determined partly by the mechanical characteristics of a crystal or an inductance-capacitance tuned circuit, and partly by the conditions under which the tube is operated.

It is usually necessary to employ one or more of the following measures to obtain a high degree of frequency stability:

(1) Minimize mechanical vibration and variations in ambient temperature which might alter the characteristics of the frequency-determining crystal or tuned circuit.

(2) Limit the amplitude of oscillation to minimize internal heating in the frequency-determining crystal or tuned circuit which might alter its characteristics.

(3) Minimize variations in supply voltages by the use of regulated plate and screen-grid (grid-No.2) supplies.

(4) Minimize variations in loading, or isolate the oscillator from a varying load by means of a "buffer" stage (usually a class A or class AB_1 amplifier).

(5) Use special components or circuit arrangements to compensate for variations in temperature, load, or supply voltage.

The frequency stability of a crystal oscillator is determined principally by the temperature coefficient and mounting of the crystal, and only to a limited extent by tube operating conditions and loading. Consequently, it is not usually necessary to use regulated plate and screen supplies for such oscillators, or to isolate them from varying loads by means of buffer stages. When extremely high stability is required, however, (*e.g.*, in frequency standards and commercial transmitters), it is usually necessary to employ all of the stabilizing measures described above and to maintain the crystal at a constant temperature in a thermostatically controlled oven.

Crystals, particularly those which are ground, "grown," or otherwise dimensioned for the higher radio frequencies, are extremely fragile and may be destroyed by overloading or the use of excessive feedback. Triodes used in crystal oscillators should, therefore, be low-power types, or be operated at substantially reduced plate voltages to minimize crystal loading and limit the amplitude of oscillation. Beam power tubes, pentodes, and tetrodes cause relatively little crystal loading because of their small driving-power requirements, and provide limited feedback even when operated at full plate voltage because of their internal shielding. Consequently, these types are especially suitable for use in crystal oscillators. They can also deliver substantially higher power outputs than triodes of comparable size, and thus permit the use of fewer stages in achieving a desired final power output.

When multigrid tubes having very good internal shielding are used in crystal-oscillator circuits, it may be necessary to use external capacitive feedback to obtain oscillation. This feedback may be provided by a small adjustable capacitor (usually not more than 2 or 3 micromicrofarads) connected between the grid-No.1 terminal and the plate terminal of the tube. Under no circumstances should the external feedback capacitance be larger than necessary for oscillation, because even small excess values may provide sufficient feedback to destroy the crystal.

To obtain good frequency stability in a variable-frequency oscillator, it is usually necessary to use all the stabilizing measures described above. It is particularly important to employ good components and sturdy mechanical construction, and generally desirable to enclose the entire oscillator tank circuit in a heavy metal shield having good thermal stability. Good isolation from load variations can be obtained without a buffer stage by the use of an electron-coupled circuit. In this type of oscillator circuit, the control grid (grid No.1) and screen grid (grid No.2) of a multigrid tube are the actual oscillator terminals, the screen grid acting as the anode. Power output is taken from the plate circuit, which is coupled to the oscillator only by the internal electron stream.

Crystal oscillators and variable-frequency oscillators can also be used as harmonic generators and frequency multipliers. Electron-coupled oscillators are particularly suitable for use as frequency multipliers because selection of desired harmonics can be accomplished in the plate circuit without affecting the oscillator frequency.

Parallel-Tuned Tank Circuits

The performance of an rf power amplifier, frequency multiplier, or oscillator is critically dependent on the characteristics of the circuit which forms its plate load. The characteristics of the load circuit affect the power output, harmonic output, plate dissipation, and driving-power requirements of the stage.

The plate-circuit load of a class B or class C rf amplifier is usually a parallel-tuned resonant tank of the type shown schematically in Fig. 34. The resonant

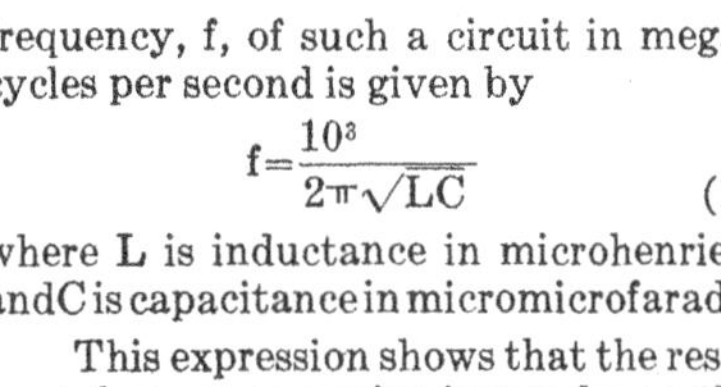

Fig. 34

frequency, f, of such a circuit in megacycles per second is given by

$$f = \frac{10^3}{2\pi\sqrt{LC}} \qquad (1)$$

where L is inductance in microhenries, and C is capacitance in micromicrofarads.

This expression shows that the resonant frequency varies inversely as the square root of the product LC. Doubling both L and C halves the resonant frequency. For any given frequency, f, the product of L and C is a constant.

Except in circuits operating at ultra-high and higher frequencies, L is usually "lumped" or concentrated in a coil or specially formed conductor, and C is a combination of lumped and distributed capacitance. The lumped capacitance component is usually a variable capacitor, and the distributed component is composed of the self-capacitance of the tank, tube capacitances, and the stray capacitance of the circuit. Consequently, distributed capacitance should always be taken into account, particularly in calculations for the higher radio frequencies, at which it is usually either the principal component or the entire tank capacitance.

The plate-tank circuit of a class B or class C rf amplifier must resonate at the desired output frequency, and must also convert relatively short, unidirectional pulses of plate current into complete oscillations at this frequency. In other words, it must act as an electrical "flywheel." The plate tank must also have sufficient impedance at resonance to limit the no-load plate current of the stage to a safe value.

The effectiveness of a tank circuit's flywheel action is indicated by the ratio of the "wattless" power (in volt-amperes) developed in the tank to the actual power (in watts) delivered by the tube. This ratio is known as the "operating Q" of the tank, and is proportional to the tank capacitance. Its approximate value in terms of tube operating conditions is given by

$$Q = \frac{C \times f \times E_b}{300 \times I_b} \qquad (2)$$

where C is the total capacitance across the tank in micromicrofarads, f is the frequency in megacycles per second, Eb is the dc plate potential in volts, and Ib is the total dc plate current of the stage in milliamperes.

The impedance of a parallel-tuned circuit at resonance (its equivalent resistance, Req) is proportional to the tank inductance and inversely proportional to the tank capacitance and the tank-coil resistance. The approximate value Req in ohms is given by

$$R_{eq} = \frac{L}{Cr} \qquad (3)$$

where L is the tank inductance in microhenries, C is the tank capacitance in microfarads, and r is the ac resistance of the tank-circuit inductor in ohms.

Because there is a conflict between the characteristics required for high operating Q and those required for high equivalent resistance, determination of proper values for plate-tank circuits is one of the most important considerations in rf amplifier design.

The first step in the design of a plate-tank circuit is the determination of the most suitable operating Q for the type of service in which the stage is to be used. The use of too low a Q results in a distorted waveform containing very strong harmonics and, therefore, is wasteful of power and likely to result in serious interference. The use of too high a Q, on the other hand, usually results in large circulating currents and, therefore, in substantial tank-circuit losses. A value between 10 and 15 is generally recommended for rf telegraphy or telephony service. A value of 12 is most frequently used in the design of amateur and industrial equipment.

The next step is the determination of the tank capacitance, C, for the Q value and tube operating conditions selected. This value is obtained from equation (2) transposed to the form

$$C = \frac{300 \times Q \times I_b}{f \times E_b} \qquad (4)$$

Fig. 35 shows C as a function of the ratio Eb/Ib for a Q value of 12. The curves in Fig. 35 can be used to determine values of tank-circuit capacitance suitable for use in equipment operating in the amateur bands. Values of C obtained from this chart or calculated by the use of Equation (4) apply only for single-ended tank circuits which are not split for neutralization or other purposes, such as that shown in Fig. 36 (a). These values represent the total capacitance required for resonance at the corresponding frequencies, and include tube and stray circuit capacitance. Values slightly higher than those indicated can generally be used without appreciable reduction of power output.

When a split tank circuit is employed for a single-ended stage, as shown in Fig. 36 (b), the total tank capacitance should be one-fourth that indicated by Fig. 35 or Equation (4). The corresponding tank inductance, therefore, is 4 times that required for a tank circuit which is not split. If the tank tuning capacitor is a split-stator type, such as that shown in Fig. 36 (c), each section should have one-half the capacitance indicated by Fig. 35 or Equation (4).

A push-pull stage operating at the same dc plate voltage and total dc plate current as a single-ended stage also requires one-fourth the tank-circuit capacitance indicated in Fig. 35 or Equation (4), or if the tuning capacitor is a split-stator type, each section should have one-half the capacitance indicated. A push-pull stage operated at the same plate voltage but drawing twice as much plate current as a single-ended stage requires one-half the tank-circuit capacitance indicated. In this case, each section of a split-stator tank capacitor should have the capacitance indicated in Fig. 35 and in Equation (4).

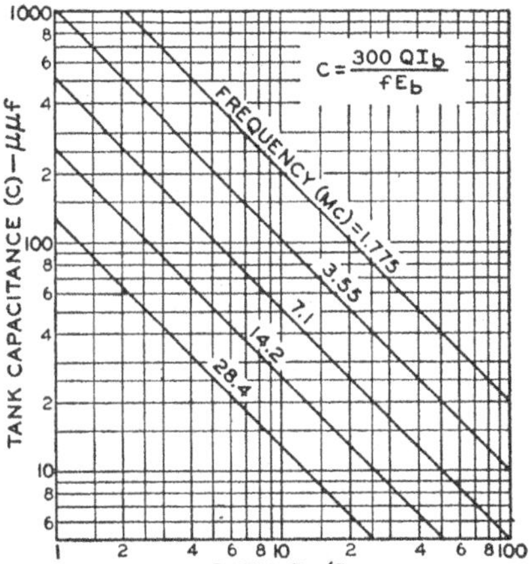

Fig. 35

When the required tank-circuit capacitance is known, the tank inductance required for resonance at the desired frequency can be determined by substitution of the value of C in Equation (1). Approximate winding data for single-layer coils, such as that shown in Fig. 37, suitable for use in amateur transmitters can then be obtained from the following formula:

$$L = \frac{R^2 \times N^2}{9R \times 10B}$$

where L is the inductance of the coil in microhenries, R is the mean radius in

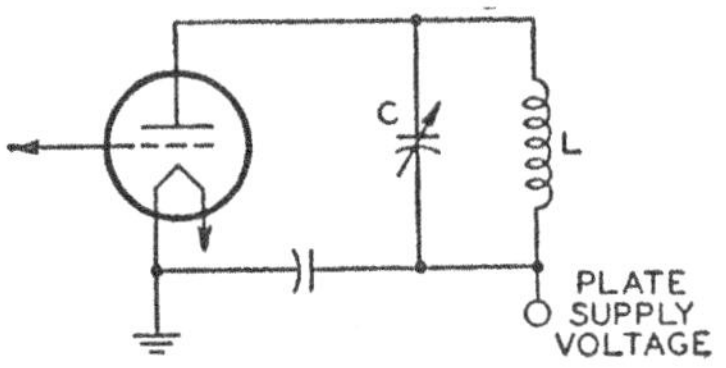

(a)

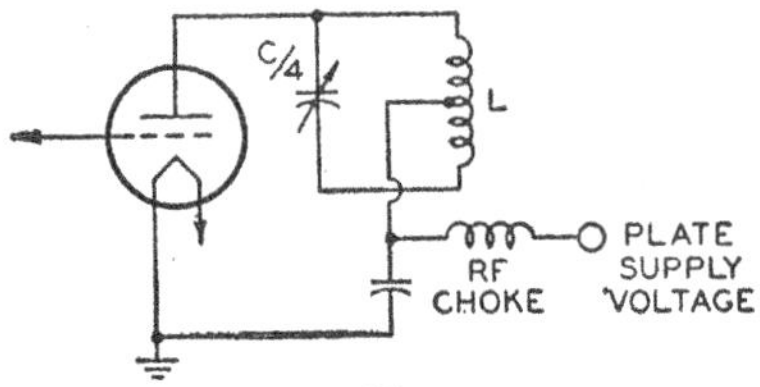

(b)

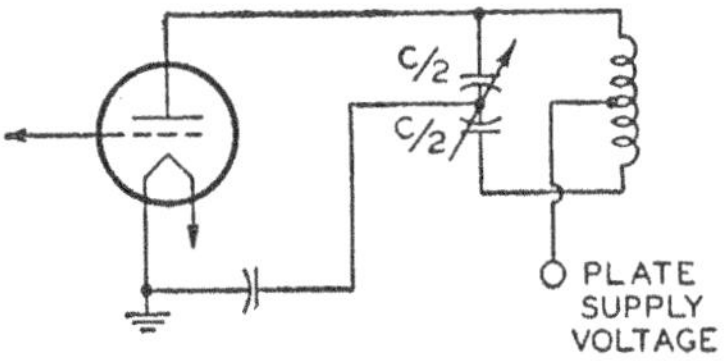

(c)

Fig. 36

inches, N is the number of turns, and B is the length in inches.

It is sometimes impracticable to limit the operating Q of a plate-tank circuit to the desired value under the proposed operating conditions. For example, in parallel-tube stages or stages operating at the higher radio frequencies, tube and stray circuit capacitance may be larger than the optimum total capacitance indicated in Equation (4). In such cases, the designer has a choice of the following procedures:

(1) Retain the proposed tube-operating conditions and design the plate-

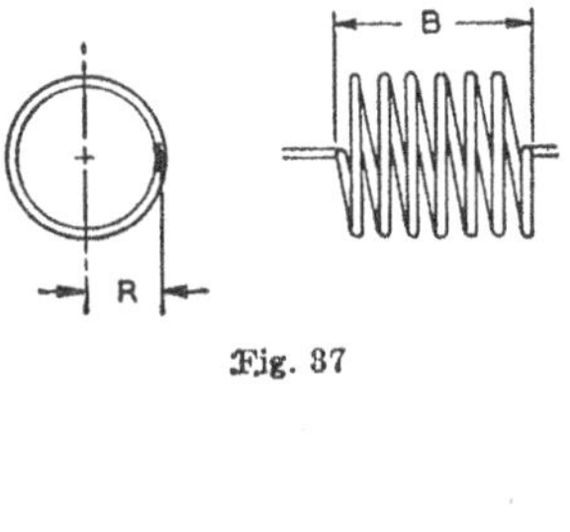

Fig. 37

tank circuit for the lowest Q value obtainable under these conditions;

(2) Modify the tube-operating conditions (provided the tube ratings are not exceeded) to obtain the proper Eb/Ib ratio for the desired operating Q;

(3) Design the stage for push-pull operation, thereby reducing tube output capacitance to one-half that of a single tube, or to one-fourth that of parallel tubes;

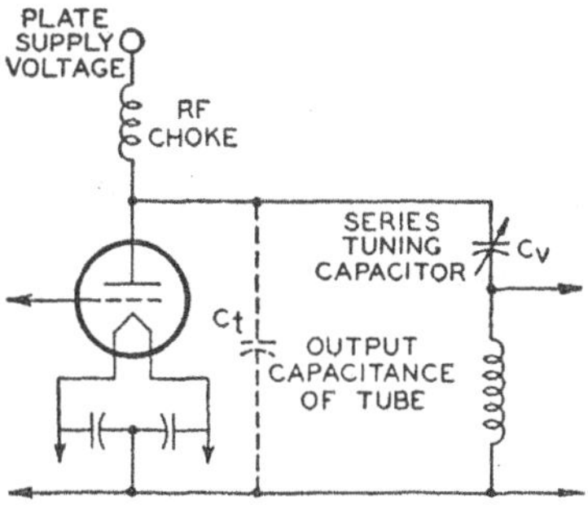

Fig. 38

(4) Employ a "series-tuned" tank circuit of the type shown in Fig. 38, in which the variable capacitance C_v is several times larger than the tube capacitance C_t.

Interstage Coupling

One of the most important considerations in rf circuit design is the method used for coupling the input of an amplifier or frequency multiplier to the output of the preceding stage. An interstage rf coupling circuit must permit efficient transfer of energy at the desired frequency; discriminate, if possible, against harmonics of the desired frequency; and, where necessary, provide dc isolation between the driver and the driven stage. It should also permit adjustment of the loading for the driver and the excitation supplied to the following stage. Three principal types of interstage coupling are employed in rf equipment: capacitive coupling, direct inductive coupling, and indirect inductive ("link") coupling.

In **capacitive coupling,** a capacitor having very low reactance at the desired frequency is connected between the plate-tank circuit of the driver stage and the grid of the following tube. This capacitor should be designed for use at radio frequencies, and should have a voltage-breakdown rating adequate to withstand the maximum potential difference developed between the driver plate circuit and the grid of the following tube. The input side of the coupling capacitor may be connected directly to the driver plate, as shown in Fig. 39 (a), or to a tap on the plate-tank coil, as shown in Fig. 39 (b).

A tapped plate-tank coil provides a convenient means for controlling loading and excitation, and generally makes it unnecessary to tune the grid circuit of the driven stage. Unused portions of tapped tank coils, however, frequently resonate with stray capacitances to form unloaded "parasitic" tank circuits which are readily shocked into oscillation and may interfere with the operation of the equipment. Consequently, it is usually preferable to use an untapped plate-tank

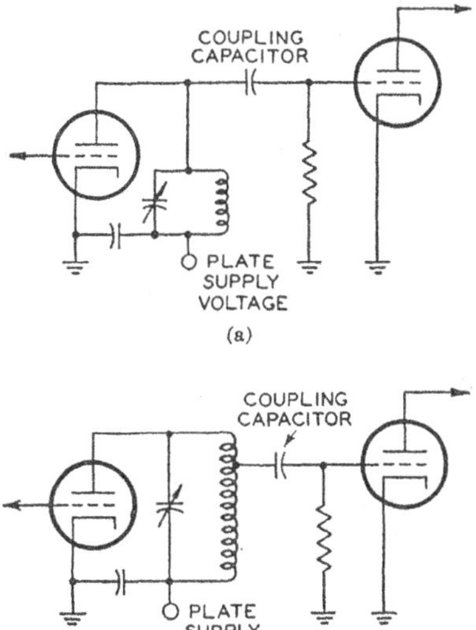

(a)

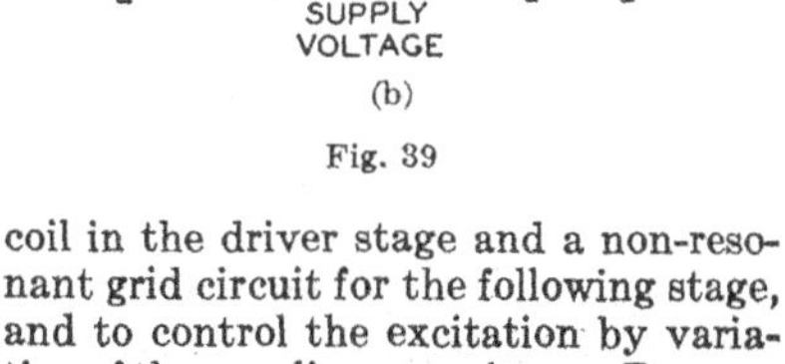

(b)

Fig. 39

coil in the driver stage and a non-resonant grid circuit for the following stage, and to control the excitation by variation of the coupling capacitance. Because of the relatively high impedances on both sides of the coupling capacitor, the driver and the driven stage should be in close proximity. Capacitive coupling tends to increase the transfer of harmonics because the reactance of the coupling capacitor decreases as the frequency increases.

Direct inductive coupling, shown in Fig. 40, is very efficient, but also involves high coupling impedances and, therefore, requires that the driver and driven stage be in close proximity. The

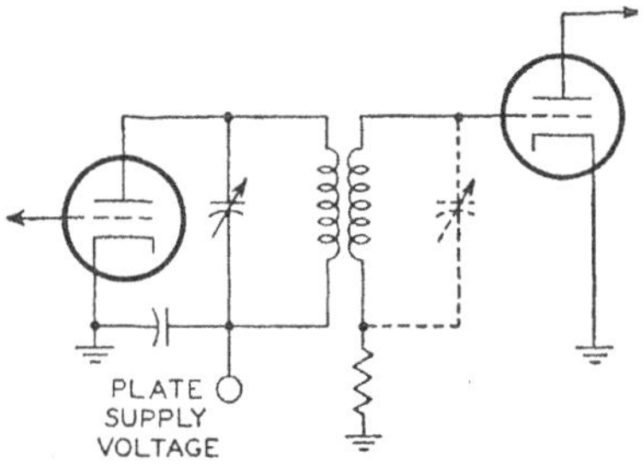

Fig. 40

coupling between the plate and grid windings may be fixed or adjustable. Adjustable coupling provides a convenient means for controlling loading and excitation. The grid winding may be either tuned or untuned. Although the tuned type provides maximum efficiency, the additional control complicates tuning and is rather critical of adjustment.

Indirect inductive coupling or **"link" coupling** is used extensively in rf power equipment. Although it does not provide the high efficiency obtainable with direct inductive coupling, it allows considerable flexibility in equipment design because it does not require close physical proximity between the coupled stages. "Link" coupling is especially useful for equipment which is frequently modified or which must be designed to permit concentration of principal control functions in a particular stage or unit of the equipment.

In this method of coupling, shown in Fig. 41, substantially identical "link" windings of a few turns each are inductively coupled to the plate-tank coil of the driver and to the grid-tank coil of the following stage. Because of their low impedance, these link windings may be connected together through suitable transmission lines of considerable length with little danger of excessive radiation or interference pickup. Because the links are inductively coupled to the plate and grid circuits, the transmission lines are not required to carry dc and, therefore, may be grounded. These interstage transmission lines may be any of the various types commercially available, such as twisted pair, ribbon line, open-wire line, or coaxial cable, depending on the requirements of the circuit.

The coupling between link windings and their respective tank coils may be either fixed or adjustable. Fixed links should be coupled as tightly as possible to their tank coils in order to assure maximum energy transfer. When variable coupling is desired, it is usually sufficient to have only one of the links adjustable. Link windings should always be coupled to their tank coils at points of minimum rf potential. In single-ended tank circuits (not split), the correct location for a link winding is at the end of the plate-tank coil connected to the plate-voltage supply or at the ground (or bias-supply) end of the grid-tank coil. In split single-ended circuits or push-pull circuits, link windings should

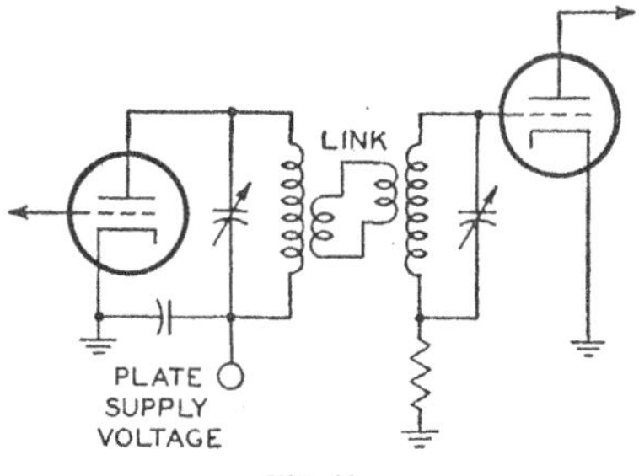

Fig. 41

be coupled to the centers of their respective tank coils.

Both direct inductive coupling and link coupling inherently provide better discrimination against harmonics than capacitive coupling.

Output Coupling

Output coupling circuits must deliver as much as possible of the power supplied to them because there is no subsequent amplification to make up for any losses. Because these circuits are usually required to work into low-impedance antennas, transmission lines, or other load devices, they must also deliver heavy output currents. Consequently, they must be designed to have the highest possible efficiency. In addition, any harmonics present in the output of the final stage must be eliminated in the output coupling circuit so that they will not enter the antenna or output transmission line.

Safety considerations usually require that the load side of an output coupling circuit be completely insulated from the ac and dc power-supply circuits of the equipment, and particularly from the plate-supply voltage of the output stage. In some cases the antenna, transmission line, or load device must also be insulated from ground.

Capacitive output coupling has the advantage of simplicity. It also permits matching to loads of substantially different impedance by the selection of a suitable feed point on the plate-tank coil of the output stage. However, it does not discriminate against harmonics which may be present in the output of the final stage, and may create serious safety hazards if leakage or voltage breakdown occurs in the coupling capacitor.

Probably the simplest and most convenient type of output coupling is **inductive coupling.** This type permits accurate impedance matching to high- or low-impedance antennas, transmission lines, or other loads, and inherently tends to discriminate against harmonics. Because it does not involve the use of series capacitors, it also minimizes the possibility of breakdowns which might place the plate voltage of the output stage across the rf output terminals and load.

When the load winding of an inductively coupled output circuit is untuned, the turns ratio between the input and output windings must be such that the proper load impedance is reflected in the plate circuit of the final amplifier. This turns ratio (primary to secondary) is equal to Zp/Zs, where Zp is the plate-load impedance desired for the final amplifier, and Zs is the impedance of the antenna, transmission line, or other load device. The plate-load impedance, Zp, in ohms can be determined approximately from the following relations:

For unmodulated or plate-modulated class C amplifiers, Zp=Eb/2Ib; for class B amplifiers and grid- or suppressor-grid-modulated class C amplifiers, Zp=Eb/(4 Ib); where Eb is the dc plate potential in volts and Ib is the dc plate current in amperes. These values of Zp are for unbalanced, single-ended output circuits. For split-tank or push-pull circuits, the values of Zp determined from these relations should be multiplied by four.

Stabilization

Any amplifier will oscillate if sufficient energy having the same frequency and the same phase as the grid voltage is fed back from the plate circuit to the grid circuit. Feedback of the proper phase for oscillation (regenerative feedback) may take place through the grid-plate capacitance of the tube, or through external capacitive or inductive coupling between plate and grid circuits. The amount of feedback necessary to cause self-oscillation is inversely proportional to the power sensitivity of the amplifier and, therefore, is much smaller for beam power tubes and other multigrid types than for triodes. In most multigrid types, however, the internal shielding provided by the screen grid (grid No.2) is so effective that any tendency to self-oscillation is usually the result of external, rather than internal, feedback. To assure stability in a multigrid rf amplifier stage, therefore, it is essential that the input and output circuits be completely shielded from each other. In some cases, it may also be necessary to shield these circuits from the tube.

In a triode, the relatively large grid-plate capacitance provides a low-impedance path for regenerative feedback which cannot be eliminated by the use of external shielding. The effect of this capacitance can be nullified, however, by taking voltage from the plate circuit and feeding it back to the grid in the proper phase and amplitude to cancel the regenerative feedback. This technique, known as "**neutralization,**" can also be employed with multigrid tubes to improve their stability at the higher radio frequencies.

The method of neutralization most frequently used, **plate neutralization,** is shown in Fig. 42. This method employs a balanced plate-tank circuit having its mid-point effectively at rf ground potential, so that rf voltages of substantially equal amplitude and opposite phase are developed across the two halves of the tank. The neutralizing voltage is taken from the bottom end of the tank and applied to the grid through the neutralizing capacitor, C_n. Although the theoretical value of C_n is exactly equal to the grid-plate capacitance of the tube, the value actually required may vary because of stray capacitances.

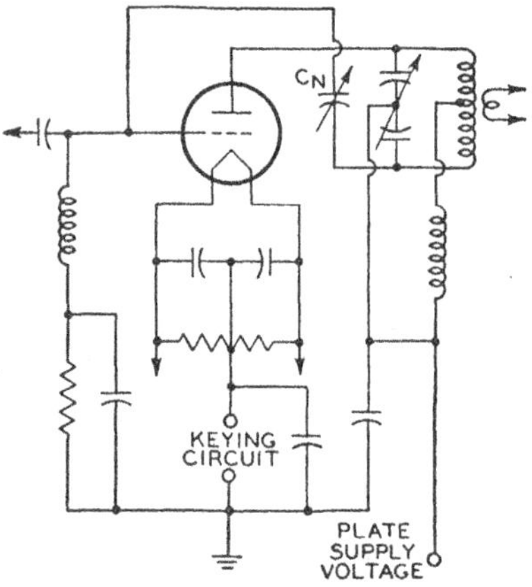

Fig. 42

Consequently, C_n is usually made adjustable over a small range on either side of the theoretical value.

Another method of neutralization for single-ended stages, **grid neutralization,** is similar to plate neutralization except that the split tank circuit which provides the neutralizing voltage is located in the grid circuit.

Parasitic Oscillations

Parasitic oscillations are oscillations which occur in a circuit at frequencies other than the desired signal frequency, its harmonics, or its subharmonics. They may be continuous, or occur only during keying, modulation, or surges in the power-supply circuits of the equipment. Because they absorb power from the circuits in which they occur, parasitics reduce efficiency and performance at the desired operating frequency. They may also be responsible for voltage flashover, instability, or premature failure of tubes and other circuit components, and may create serious interference by causing radiation of spurious carrier and sideband frequencies.

Parasitics are generated when resonance at some frequency other than the normal operating frequency occurs simultaneously in the input and output circuits of a tube. Under these conditions the stage functions as a "tuned-grid-tuned-plate" oscillator, the grid-plate capacitance of the tube providing the feedback path. These simultaneous resonance conditions may be created by the use of similar circuit constants in the plate and grid circuits (*e.g.*, the use of identical rf chokes in both circuits) or by the "secondary" characteristics (small amounts of capacitance and inductance) of the tubes, circuit components, or circuit conductors.

Parasitics in multistage equipment must be eliminated on a stage-by-stage basis. Identification of the particular components forming a parasitic circuit often requires considerable study and "cut-and-try" experimentation. The first step is to distinguish true parasitics from self-oscillation in the stage in question, and to determine the frequency or frequencies of the parasitics. For this step, excitation is removed from the offending stage, and also from the preceding stage to minimize the possibility of feed-through at the normal operating frequency or a subharmonic. The stage is then operated at about one-half normal plate and screen-grid (grid-No.2) voltage and checked for oscillations.

When the presence of parasitics has been verified, and their frequency or frequencies determined, vhf parasitics should be eliminated first. VHF parasitics can usually be traced to one or more of the following sources:

(1) Long connecting leads between grid and plate terminals of tubes and the corresponding tank circuits.

(2) Push-pull tank circuits employing split-stator tank capacitors in which

the common terminals of the tank capacitors are not at rf ground potential.

(3) Inadequate bypassing, or the use of long connecting leads to bypass capacitors, particularly in the screen-grid-to-cathode circuits of multigrid tubes.

(4) Long leads in neutralizing circuits.

(5) Tapped tank-circuit coils. (Unused portions of tapped tank coils are particularly troublesome in this respect because they are not loaded and, therefore, can form resonant circuits of very high Q.)

(6) Inadequate separation between components in the input and output circuits of the stage.

Two methods can be used to minimize parasitics in resonant circuits. In one method, the constants of one of the circuits involved are changed to shift its resonant frequency. The lengths of the leads to the circuit may be reduced (preferably to a minimum), or the position of a connecting lead or component may be shifted to reduce its capacitance. When such a change is made, however, the new resonant frequency of the circuit may be the same as that of another combination of circuit elements, with the result that a new parasitic oscillation is created.

The second method is the insertion in one of the tube circuits (grid, plate, or cathode circuit) of a special load which will rapidly dissipate parasitic oscillations but will not appreciably affect the performance of the stage at the desired frequency. In a low-current circuit, this load may be a non-inductive resistor having a value between 10 and 100 ohms inserted directly at the tube socket. In a high-current circuit, a small rf choke (5 to 10 turns of wire) should be connected in parallel with the resistor.

Fig. 43 shows a beam power tube in an rf amplifier which has been stabilized to eliminate parasitics. L_g, L_k, and L_p represent the distributed inductance of the grid, cathode, and plate leads, respectively. C_{gp} and C_{gk} are the grid-plate and plate-cathode capacitances of the tube. L_1, C_1, L_2, and C_2 are the normal grid and plate tank-circuit components. The following stabilization measures are shown in the circuit:

(1) The screen grid (grid No.2) is bypassed to the cathode directly at the tube socket with a mica or ceramic capacitor of not less than 0.002 microfarad having extremely short leads.

(2) Because the tube has an indirectly heated cathode, an unbypassed

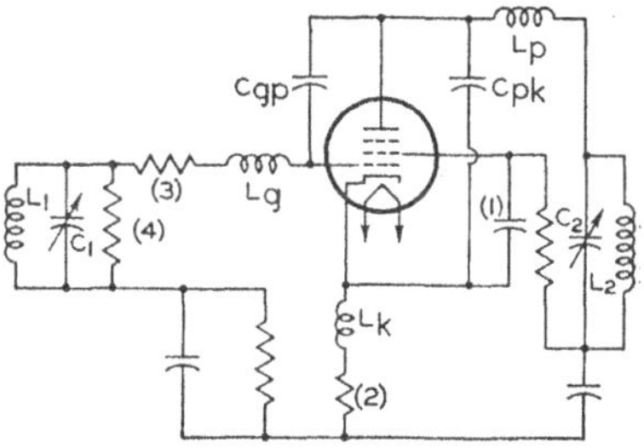

Fig. 43

non-inductive resistor having a value of 25 ohms or less is installed in the cathode-return lead directly at the tube socket.

(3) A non-inductive resistor having a value of 50 ohms or less is installed in series with the grid-tank circuit directly at the grid terminal of the tube socket.

(4) The grid-tank circuit is loaded with a non-inductive resistor having a value between 5000 and 50000 ohms.

Besides the measures shown in the circuit, the screen-grid voltage is reduced proportionally when the tube is operated at less than the maximum rated value of plate current. In addition, ample driving power is provided. If necessary, the grid current and bias are increased to provide ample driving power, but the maximum ratings for grid current and grid voltage should not be exceeded. A "saturated" tube (*i.e.*, one supplied with ample driving power) is relatively immune to parasitics.

When all vhf parasitics have been eliminated, attention should be directed to the elimination of low-frequency parasitics. Low-frequency parasitics are frequently caused by:

(1) The use of rf chokes in series with both the plate and grid circuits of the amplifier, particularly when identical chokes are used in both circuits.

(2) Resonance conditions in power-supply filter circuits.

(3) Resonance conditions in modulation-circuit components.

(4) The use of high-impedance RC circuits in screen-grid-supply circuits for multigrid tubes.

(5) The use of parallel feed in both the grid and plate circuits of a tube.

In addition to the stabilization of individual stages in power-tube equipment, it is also necessary to prevent undesired coupling and feedback between stages operating at the same frequency. Over-all stabilization of multistage equipment may require shielding of individual tubes or entire stages, the use of filtering and decoupling networks in power-supply leads and in grid-, plate-, or other circuit-return leads, or combinations of such measures.

Power-Supply Considerations

Because class B and class C rf amplifiers may be operated without plate, screen-grid, or bias voltages (or at voltages substantially below normal values) during certain tuning adjustments, they should incorporate means for reducing or completely removing these voltages independently in each stage. It is also desirable that plate, screen-grid, and fixed-bias voltages for individual rf amplifier stages be adjustable up to the maximum values for the tubes employed so that maximum operating efficiency is attainable at a particular power output or frequency.

Calculation of Operating Conditions

The only restrictions on tube operating values are those imposed by the published maximum ratings. When it is necessary or desirable to operate tubes under conditions other than those shown under "Typical Operation" in published data, suitable values may be approximated by simple calculations. These approximate values may then be used in a tentative operating setup, and adjustments made, if necessary, to assure that desired output and efficiency are obtained without any of the maximum ratings for the tube being exceeded.

Simple calculations can be used to determine operating conditions for any type of service in which plate current flows for less than the entire signal cycle. They can be used for triode and multi-grid-tube class C amplifiers (both modulated and unmodulated), for push-pull class AB and class B audio amplifiers and for class B linear amplifiers.

The basic factors used in these calculations are the peak plate current of the tube, and the corresponding instantaneous plate voltage, grid voltages, and grid currents. The peak plate current is determined by the average or dc plate current and by the plate-conduction angle (*i.e.*, the fraction of the signal cycle during which plate current flows). For a given dc plate current, peak plate current varies inversely with conduction angle and is equal to the dc value times a conversion factor K_1, given in Table I. The corresponding instantaneous values of the other tube currents and voltages are obtained from the "Average Characteristics" curves for the tube.

Table I

Conduction Angle (degrees)	K_1	K_2	K_3	K_4	K_5
180	3.14	0.785	—	—	0.250
160	3.50	0.825	0.210	1.210	0.224
150	3.75	0.844	0.350	1.350	0.213
140	4.00	0.862	0.520	1.520	0.200
130	4.25	0.880	0.732	1.732	0.187
120	4.60	0.897	1.000	2.000	0.174
110	5.00	0.913	1.345	2.345	0.160
100	5.50	0.927	1.800	2.800	0.145
90	6.10	0.940	2.410	3.410	0.130

Table I also gives four other conversion factors or constants (K_2, K_3, K_4, and K_5) used in these calculations. A sixth factor, K_6, which is a function of grid bias and driving voltage, is given in Table II. The values given for constants K_1, K_2, K_3, K_4, K_5 are based on the use of sinusoidal signal waveforms and conduction angles between 90 and 180 degrees. Angles between 100 and 160 degrees are generally used in "straight-through" class C amplifiers. Angles of 90 degrees are usually employed only in frequency multipliers, and angles of 180 degrees in class AB and class B amplifiers.

Experience has shown that the most satisfactory relation between power output and power gain in "straight-through" class C amplifier service is achieved at a conduction angle of about 140 degrees. The use of larger conduction angles reduces driving-power requirements, but results in substantially reduced plate-circuit efficiency. The use of smaller conduction angles, on the other hand, tends to increase plate-circuit efficiency, but makes it necessary to provide substantially higher driving power.

Table II

E_{c1}/E_{g1}	K_6	E_{c1}/E_{g1}	K_6
0.25	4.67	0.65	6.95
0.30	4.84	0.70	7.52
0.35	5.04	0.75	8.25
0.40	5.26	0.80	9.25
0.45	5.50	0.85	10.70
0.50	5.78	0.90	13.12
0.55	6.10	0.95	18.63
0.60	6.49		

Use of Curves

Average characteristics of power tubes are usually given in the form of sets or "families" of curves, such as those shown in the *Tube Types* Section. The separate "plate," "grid-No.1," and "grid-No.2" families given for the RCA-6146 beam power tube are typical of curves furnished for multigrid types. Combined "plate" and "grid" families such as those given for the RCA-812-A are usually furnished for triodes.

Plate families show the simultaneous relationships between plate voltage, control-grid voltage, and plate current. Consequently, they may be used for determining effective minimum plate voltages and peak positive control-grid voltages corresponding to desired or calculated values of peak plate current. They also may be used for determination of the grid-bias voltages required to obtain desired values of quiescent (zero-signal) plate current in class A, class AB, and class B amplifiers. In addition, they permit such factors as plate-load resistance, power output, plate dissipation, and harmonic distortion to be determined graphically.

Grid families are used in determining the peak currents in the corresponding grid circuits. Like peak plate current, these peak grid currents flow at the instant control-grid voltage is at positive peak value, and plate voltage is minimum.

A single set of curve families for a multigrid tube shows the characteristics of the tube at a particular grid-No.2 (or screen-grid) voltage. If a different grid-No.2 voltage is to be used, appropriate "Average Characteristics" curves must be obtained, or values shown in the available curves must be converted mathematically. A simple method of conversion is given later.

Class C Telegraphy Service

Multigrid Tubes

(1) Choose a plate voltage (E_b), a dc grid-No.2 (screen-grid) voltage (E_{c2}), and a dc plate current (I_b) which provide a plate input (P_i) within the maximum rating for the tube. Also select a conduction angle smaller than 180 degrees (preferably 140 degrees).

(2) Using the value of K_1 given in Table I for the conduction angle selected, calculate the peak plate current (i_{bmax}) as follows:

$$i_{bmax} = K_1 \times I_b$$

(3) Determine the effective minimum plate voltage (e_{bmin}) and peak positive grid-No.1 voltage (e_{c1max}) from the plate-family curves for the chosen value of E_{c2} and the calculated value of i_{bmax}. For maximum plate-circuit efficiency and maximum power gain, both e_{bmin} and e_{c1max} should be as small as possible. Because of other considerations, however, e_{bmin} should be slightly above and to the right of the "knee" in the appropriate grid-No.1 voltage curve. The use of e_{bmin} and e_{c1max} values below the knee causes excessive grid-No.1 and grid-No.2 current; the use of values too far to the right of the knee reduces power output and may result in excessive plate dissipation.

(4) Using the value of K_2 given in Table I for the conduction angle selected, calculate power output (P_o) as follows:

$$P_o = K_2 \times (E_b - e_{bmin}) \times I_b$$

(5) Plate dissipation or plate loss (P_p) is then given by

$$P_p = (E_b \times I_b) - P_o$$

If this value exceeds the maximum plate-dissipation rating for the tube, it will be necessary to recalculate steps (1) through (5) using a smaller conduction angle.

(6) Using the values of K_3 and K_4 given in Table I, calculate the dc grid-No.1 voltage or bias (E_{c1}) as follows:

$$E_{c1} = -(K_3 \times e_{c1max}) - \frac{K_4 \times E_{c2}}{\mu_{g2g1}}$$

where μ_{g2g1} is the mu-factor (grid No.2 to grid No.1) of the tube.

(7) The peak rf grid-No.1 voltage (E_{g1}) required to drive the tube to full output is given by

$$E_{g1} = -E_{c1} + e_{c1max}$$

(8) Determine peak grid-No.1 current (i_{c1max}) from the grid-current characteristics curves for the appropriate value of E_{c2}. (Like peak plate current, peak grid-No.1 current flows at the instant that plate voltage is equal to e_{bmin} and grid-No. 1 voltage is equal to e_{c1max}). Then, using the value of K_6 given in Table II for the calculated values of E_{c1} and E_{g1}, determine the dc grid current (I_{c1}) as follows:

$$I_{c1} = i_{c1max}/K_6$$

(9) The approximate driving power (P_d) required by the grid-cathode circuit of the tube is then given by

$$P_d = 0.9 \times E_{g1} \times I_{c1}$$

(It should be noted that this value of P_d does not represent the total power that must be delivered by the driver stage, which must be sufficient to supply the various tube and circuit losses described previously.)

(10) It is now necessary to calculate the dc grid-No.2 current (I_{c2}) and grid-No.2 input (W_{c2}). First determine the peak grid-No.2 current (i_{c2max}) from the screen-grid-current characteristics curves for the appropriate value of E_{c2}. (The value of i_{c2max} is determined at the intersection of the plate-voltage coordinate corresponding to e_{bmin} with the grid-No.1 voltage coordinate corresponding to e_{c1max}). Then, using the value of K_5 given in Table I for the conduction angle employed, calculate the dc grid-No.2 current (I_{c2}) as follows:

$$I_{c2} = K_5 \times i_{c2max}$$

Grid-No.2 input (W_{c2}) is then given by

$$W_{c2} = E_{c2} \times I_{c2}$$

If this value of W_{c2} exceeds the maximum rating for grid-No.2 input given in the tube data, it will be necessary either to reduce E_{c2} or to employ a smaller conduction angle.

Example:

Calculate operating values for the RCA-6146 in Class C Telegraphy Service under CCS conditions. The basic operating values are selected to be: E_b=600 volts; I_b=112 milliamperes, E_{c2}=150 volts; plate-conduction angle=140 degrees.

(1) Plate input (P_i) = 600 volts × 0.112 ampere=67.2 watts. This value is just within the maximum CCS rating of 67.5 watts.

(2) From Table I, K_1 for a conduction angle of 140 degrees is 4. Therefore, peak plate current (i_{bmax})=0.112 ampere × 4 =0.448 ampere, or 448 milliamperes.

(3) From the plate family for the 6146 given in Fig. 44 (E_{c2}=150 volts), a suitable value for effective minimum plate voltage (e_{bmin}) to the right of the "knee" is 70 volts. The corresponding peak positive grid-No.1 voltage (e_{c1max}, determined from E_{c1} curves) for a peak plate current of 448 milliamperes is approximately +16 volts.

(4) From Table I, K_2 for a conduction angle of 140 degrees is 0.862. Therefore, power output (P_o)=0.862 × (600−70) × 0.112=51 watts.

(5) Plate dissipation (P_p) = (600 × 0.112)−51 = 16.2 watts. This value is well within the maximum plate-dissipation rating of the 6146 for class C telegraphy under CCS conditions (20 watts).

(6) The dc grid-No.1 or bias voltage (E_{c1}) and peak rf grid-No.1 voltage (E_{g1}) are calculated next. (Note that bias voltage E_{c1} is not the E_{c1} shown in the characteristics curves, which represents total grid voltage, *i.e.*, the algebraic sum of the bias E_{c1} and peak rf grid-No.1 voltage e_{c1max}). From table I, K_3 and K_4 for a conduction angle of 140 degrees are, respectively, 0.520 and 1.520. From the technical data for the 6146, mu-factor (μ_{g2g1}) is 4.5. Therefore,

$$E_{c1} = -(0.520 \times 16) - \frac{1.520 \times 150}{4.5} = -8.3 - 50.6 = -58.9$$, or approximately −59 volts.

(7) Peak rf grid-No.1 voltage (E_{g1}) = −(−59) + 16 = 75 volts.

(8) The next step is to determine dc grid-No.1 current (I_{c1}). From the grid-No.1 average characteristics curves

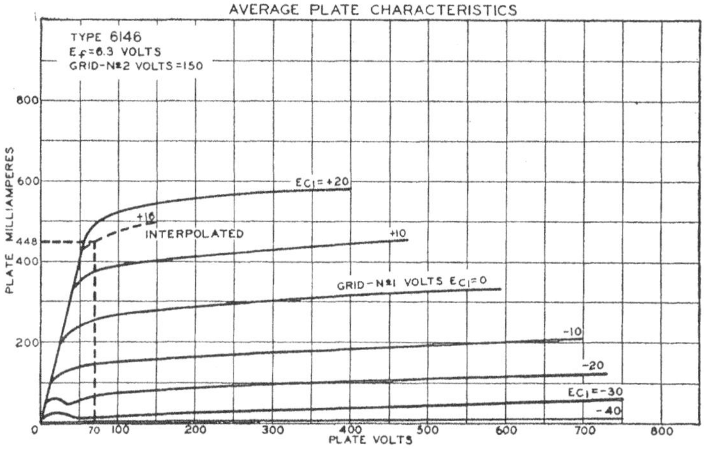

Fig. 44

shown in the tube data (E_{c2} = 150 volts), for e_{bmin} of 70 volts and e_{c1max} of +16 volts, peak grid-No.1 current (i_{c1max}) = 28 milliamperes.

From Table II, K_6 for the ratio E_{c1}/E_{g1} = 59/75 = 0.787 is between the values given for ratios of 0.75 and 0.80, and is approximately 9. Consequently, I_{c1} = 0.028/9 = 0.0031 ampere, or approximately 3 milliamperes.

(9) The driving power required by the grid (P_d) = 0.9 × 75 × 0.003 = 0.203, or approximately 0.2 watt.

(10) From the grid-No.2 characteristics curves shown in the tube data (E_{c2} = 150 volts), for E_b = 70 volts and E_{c1} = +16 volts, peak grid-No.2 current (i_{c2max}) = 59 milliamperes (approx.)

From Table I, K_5 for a conduction angle of 140 degrees is 0.200. Consequently, dc grid-No.2 current (I_{c2}) = 0.200 × 0.059 = 0.0118 ampere, or 11.8 milliamperes. Grid-No.2 input (W_{c2}) = 150 × 0.0118 = 1.77 or approximately 1.8 watts. This value is well within the maximum rating for the 6146 (3 watts).

These calculated values are compared below with the "Typical Operation" values given in the published data for the 6146 in Class C Telegraphy Service, CCS conditions, as amplifier up to 60 Mc:

	Calculated	*Published*	
DC Plate Voltage (E_b)	600	600	volts
DC Grid-No.2 Voltage (E_{c2})	150	150	volts
DC Grid-No.1 Voltage (E_{c1})	-59	-58	volts
Peak RF Grid-No.1 Voltage (e_{g1max})	75	73	volts
DC Plate Current (I_b)	112	112	ma
DC Grid-No.2 Current (I_{c2})	11.8	9	ma
DC Grid-No.1 Current (I_{c1})	3	2.8	ma
Driving Power (Approx., P_d)	0.2	0.2	watt
Power Output (Approx., P_o)	51	52	watts

Class C Telegraphy Service

Triodes

Calculations for triode class C amplifiers are similar to those described for multigrid tubes except that somewhat different considerations are involved in the determination of effective minimum plate voltage (e_{bmin}) and peak positive grid voltage (e_{cmax}), and that calculations for grid-No.2 current and input are not required.

(1) Choose a plate voltage (E_b) and a dc plate current (I_b) which provide a plate input (P_i) within the maximum rating for the tube. Also select a suitable conduction angle (preferably 140 degrees).

(2) Using the value of K_1 given in

Table I for the conduction angle selected, calculate the peak plate current (i_{bmax}) as follows:

$$i_{bmax} = I_b \times K_1$$

(3) Determine peak positive grid voltage (e_{cmax}) and effective minimum plate voltage (e_{bmin}) for this value of i_{bmax} from the plate-family curves for the tube.

The maximum permissible value of e_{cmax} and the minimum permissible value of e_{bmin} are determined at the point where the horizontal coordinate representing the peak current intersects the "$E_c = E_b$" line (sometimes called "Diode Line"). It is generally desirable that e_{bmin} be slightly more positive than e_{cmax}. If e_{bmin} is smaller than e_{cmax}, the grid will be driven more positive than the plate and will draw excessive current, and the peak plate current will be reduced. In addition, the harmonic output of the stage will be greatly increased.

(4) Using the value of K_2 given in Table I, calculate the power output (P_o) as follows:

$$P_o = K_2 \times (E_b - e_{bmin}) \times I_b$$

(5) Plate dissipation or plate loss (P_p) is then given by

$$P_p = (E_b \times I_b) - P_o$$

If this value exceeds the maximum plate-dissipation rating of the tube, it will be necessary to recalculate steps (1) through (5) using a smaller conduction angle.

(6) Using the value of K_3 given in Table I, calculate the grid bias required (E_c) as follows:

$$E_c = -[K_3 \times (e_{cmax} + e_{bmin}/\mu) + E_b/\mu]$$

where μ is the amplification factor shown in the published data for the tube.

(7) The peak rf grid voltage (E_g) required to drive the grid from bias level to the peak positive value determined in step (3) is given by

$$E_g = -E_c + e_{cmax}$$

(8) Determine peak grid current (i_{cmax}) from the grid-current characteristics curves. (The value of i_{cmax} is shown at the intersection of the plate-voltage coordinate corresponding to e_{bmin} with the grid-voltage curve corresponding to e_{cmax}). Then, using the value of K_6 given in Table II for the calculated values of E_c and E_g, determine the dc grid current (I_c) as follows:

$$I_c = i_{cmax}/K_6$$

If this value of Ic is greater than the maximum grid-current rating for the tube, or is undesirably large, it will be necessary to recalculate using a higher value for e_{bmin}.

(9) The approximate driving power (P_d) required by the tube is then given by

$$P_d = 0.9 \times E_g \times I_c$$

Example:

Calculate operating values for the RCA-812-A for Class C Telegraphy Service under ICAS conditions. The plate voltage is selected to be 1500 volts; the plate input, the maximum rated value for the tube; and the plate-conduction angle, 140 degrees.

(1) From the published data for the 812-A, the maximum plate-input rating is 260 watts. The dc plate current (I_b) required to provide this input at a plate voltage, (E_b) of 1500 volts is $I_b = 260/1500 = 0.173$ ampere, or 173 milliamperes.

(2) From Table I, K_1 for a conduction angle of 140 degrees is 4. Therefore, peak plate current (i_{bmax}) $= 0.173 \times 4.00 = 0.692$ ampere, or 692 milliamperes.

(3) The average characteristics curves given in Fig. 45 show that a peak plate current of 692 milliamperes is obtained at a peak positive grid voltage (e_{cmax}) of 118 volts and an effective minimum plate voltage (e_{bmin}) of 140 volts.

(4) From Table I, K_2 for a conduction angle of 140 degrees is 0.862. Therefore, power output (P_o) $= 0.862 \times (1500 - 140) \times 0.173 = 203$ watts (approx.).

(5) Plate dissipation (P_p) $= (1500 \times 0.173) - 203 = 57$ watts (approx.)

This value is well within the 65-watt maximum rating for the 812-A for class C telegraphy under ICAS conditions.

(6) From Table I, K_3 and K_4 are 0.520 and 1.520, respectively. From the published data, the amplification factor μ is 29. Therefore, the dc grid voltage or bias(E_c) $= -[0.520 \times (118 + 140/29) + 1500/29] = -[0.520 \times (118 + 4.8) + 52] = -(64 + 52) = -116$ volts.

(7) Peak rf grid voltage (E_g) $= -(-116) + 118 = 234$ volts.

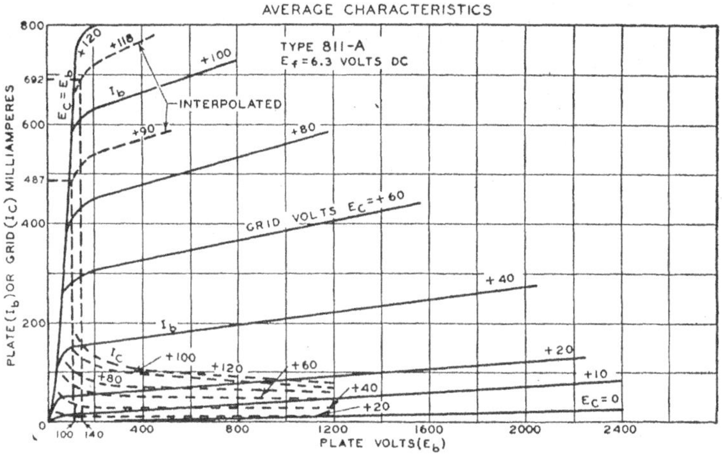

Fig. 45

(8) From the average characteristics curves shown in Fig. 45, for e_{cmax} of + 118 volts and e_{bmin} of 140 volts, peak grid current (i_{cmax}) = 195 milliamperes (approx.).

From Table II, K_6 for the ratio $E_c/E_g = 116/234$, or approximately 0.5, is 5.78. Consequently, the dc grid current (I_c) = 0.195/5.78 = 0.0337 ampere, or 34 milliamperes (approx.).

(9) The driving power required at the grid (P_d) = 0.9 × 234 × 0.034 = 7.2 watts.

These calculated values are compared below with the "Typical Operation" values given in the published data for the RCA-812-A in Class C Telegraphy Service, ICAS conditions:

	Calculated	*Published*	
DC Plate Voltage (E_b)	1500	1500	volts
DC Grid Voltage (E_c)	-116	-120	volts
Peak RF Grid Voltage (E_g)	234	240	volts
DC Plate Current (I_b)	173	173	ma
DC Grid Current, (Approx., I_c)	34	30	ma
Driving Power (Approx., P_d)	7.2	6.5	watts
Power Output (Approx., P_o)	203	190	watts

Plate-Modulated Class C Telephony Service

Operating values for plate-modulated class C amplifiers may also be calculated by the procedure described above. As mentioned previously, however, dc plate-voltage and dc plate-input values selected for plate-modulated amplifiers must be within the maximum ratings given in the tube data for this type of service.

In general, adequate protection against excessive dc plate input is obtained when the dc plate voltage and plate current do not exceed 80 per cent of the maximum class C telegraphy values. It is also usually desirable to employ a conduction angle smaller than that used in telegraphy service to assist in obtaining linear modulation, as discussed previously.

Frequency Multipliers

Operating values for frequency multipliers are also calculated as described above, except that values for the constants K_1, K_2, K_3, K_4, and K_5 are obtained from Table III instead of Table I, and the following equation is used to determine the value of grid-bias voltage for triodes:

$$E_c = -(K_3 \times E_{gmax}) + \frac{K_4}{2\mu}(3\,E_b - e_{bmin})$$

Table III

	K_1	K_2	K_3	K_4	K_5
Doubler	4.60	0.63	1.00	2.00	0.174
Tripler	6.90	0.63	3.27	4.27	0.116
Quadrupler	9.00	0.63	6.46	7.46	0.089

Class AB and Class B AF Amplifier Service

Push-pull class AB and class B af amplifiers are assumed to have a conduction angle of 180 degrees.

This assumption is permissible (even though the actual conduction angle per tube is slightly greater than 180 degrees) because any plate currents drawn simultaneously by the two sides of the circuit are effectively cancelled in the output transformer and do not appear in the composite plate-current waveform. DC voltage, current, input, and dissipation values for af amplifiers are calculated on a per-tube basis; ac values such as power output, driving voltage, and driving power are calculated for the entire stage.

The plate-circuit loads for af amplifiers are usually iron-core transformers, which are not adjustable to the same degree as the resonant tank circuits used as loads for rf amplifiers. To assure proper loading for a class AB or B stage, therefore, it is necessary to calculate the plate-to-plate load resistance required, and to provide an output transformer or coupling device which presents this resistance to the plate circuit of the amplifier when connected to the external load. Because the dc plate current of a class AB or class B af amplifier is small under zero-signal conditions and increases with amplitude of the driving signal, it is also necessary to calculate both the zero-signal plate current (I_{b0}) and the maximum-signal plate current (I_{bmax}). The maximum-signal value should not be confused with the peak plate current (i_{bmax}), which is the highest instantaneous value and, at the assumed conduction angle of 180 degrees, is equal to $3.14 \times I_{bmax}$.

Class AB₂ Amplifiers

Multigrid Tubes

(1) Choose a plate voltage (E_b), a dc grid-No.2 (screen-grid) voltage (E_{c2}), and a maximum-signal dc plate current (I_{bmax}) which provide a maximum-signal plate input within the maximum ratings for the tube. Assume a plate-conduction angle of 180 degrees.

(2) Using the value $K_1 = 3.14$ given in Table I for a conduction angle of 180 degrees, calculate the peak plate current (i_{bmax}) per tube as follows:

$$i_{bmax} = K_1 \times I_{bmax} = 3.14\ I_{bmax}$$

(3) Determine peak positive grid-No.1 voltage (e_{c1max}) and effective minimum plate voltage (e_{bmin}) from the plate-family curves for the tube for the calculated value of i_{bmax} and the chosen value of E_{c2}. As mentioned earlier for class C amplifiers, the best compromise from the standpoints of plate-circuit efficiency and power sensitivity is obtained when e_{bmin} is slightly to the right of the "knee" in the appropriate grid-voltage curve.

(4) Using the value of $K_2 = 0.785$ given in Table I, calculate the power output (P_o) for the stage (two tubes in push-pull) as follows:

$$P_o = 2K_2 \times (E_b - e_{bmin}) \times I_{bmax}$$
$$= 1.57 \times (E_b - e_{bmin}) \times I_{bmax}$$

(5) The plate dissipation (P_p) per tube is then given by

$$P_p = (E_b \times I_{bmax}) - P_o/2$$

If this value exceeds the maximum plate dissipation rating per tube for class AB_2 service, it will be necessary to recalculate steps (1) through (5) using either a smaller peak plate current (and, consequently, a smaller maximum-signal dc plate current), or a lower value of e_{bmin}.

(6) The zero-signal dc plate current (I_{b0}) per tube is selected to provide a combination of high power output with low odd-harmonic distortion. A small value of I_{b0} is desirable for high power output, but a value above the "knee" of the tube characteristic must be used to minimize distortion.

In most cases, a suitable value for I_{b0} is one which results in a zero-signal plate dissipation per tube of one-third to one-half the maximum rated value (P_{pmax}). For one-third maximum dissipation, the zero-signal plate current (I_{b0}) per tube is given by

$$I_{b0} = P_{pmax}/(3 \times E_b)$$

(7) The dc grid-No.1 bias voltage (E_{c1}) required to obtain the desired value of I_{b0} can then be determined from the plate-family curves for the chosen value of E_{c2}.

(8) The peak af grid-No.1 (driving) voltage (E_{g1}) required for each tube is given by

$$E_{g1} = -E_{c1} + e_{c1max}$$

The total driving voltage (E_{g1-g1}) required for the stage, therefore, is given by

$E_{g1-g1} = 2 \times (E_{g1}) = 2 \times (-E_{c1} + e_{c1max})$

(9) The plate-to-plate load resistance (R_{Lp-p}) required for a push-pull class AB_2 or class B af amplifier is given by

$$R_{Lp-p} = 1.27 \times (E_b - e_{bmin})/I_{bmax}$$

This value is four times the resistance represented by a load line drawn on the appropriate plate-family curves for the tube from the i_{bmax}, e_{bmin} point to the intersection of the plate-voltage (E_b) coordinate with the $I_b = 0$ axis.

(10) Determine the peak grid-No.1 current (i_{c1max}) per tube from the grid-No.1-current curves given for the tube. The value of i_{c1max} is shown at the intersection of the e_{bmin} coordinate with the e_{c1max} curve.

(11) The maximum-signal driving power (P_d) required by the push-pull stage is given by

$$P_d = i_{c1max} \times E_{g1}/2$$

(12) The peak grid-No.2 current per tube (i_{c2max}) is obtained from the grid-No.2 characteristics curves for the chosen grid-No.2 voltage.

(13) Using the value $K_5 = 0.25$ given in Table I for a conduction angle of 180 degrees, calculate the maximum-signal grid-No.2 current (I_{c2max}) per tube as follows:

$$I_{c2max} = K_5 \times i_{c2max} = 0.25\ i_{c2max}$$

(14) The maximum-signal grid-No.2 input (W_{c2}) per tube is then given by

$$W_{c2} = E_{c2} \times I_{c2max}$$

If this value of W_{c2} exceeds the maximum rating for the tube, it will be necessary to reduce either e_{bmin} or E_{c2}.

The zero-signal grid-No.2 current (I_{c2o}) is usually a small fraction of the maximum-signal current (I_{c2max}). Consequently, it has little or no effect on the maximum grid-No.2 input, and is not an important consideration.

Example:

Calculate operating values for a push-pull class AB_2 af amplifier stage using two RCA-6146 tubes operating under ICAS conditions. The basic operating values are E_b = 600 volts, E_{c2} = 200 volts, and I_{bmax} = 135 milliamperes per tube.

(1) Plate input per tube (P_i) = 600 $\times$ 0.135 = 81 watts. This value is well within the maximum rating of the 6146 for this type of service (90 watts).

(2) For a conduction angle of 180 degrees, peak plate current per tube (i_{bmax}) = 3.14 $\times$ 0.135 = 0.424 ampere, or 424 milliamperes.

(3) From the average plate characteristics curves for E_{c2} = 200 volts given in the data section, the peak positive grid-No.1 voltage per tube (e_{c1max}) = +5 volts (approx.) and the effective minimum plate voltage (e_{bmin}) = 65 volts (approx.).

(4) Power output for two tubes in push-pull (P_o) = 1.57 $\times$ (600-65) $\times$ 0.135 = 113.5 watts.

(5) Plate dissipation per tube (P_p) = (600 $\times$ 0.135) - 113.5/2 = 24.2 watts.

(6) For one-third maximum rated plate dissipation, zero-signal dc plate current (I_{bo}) = 25/(3 $\times$ 600) = 0.0139 ampere, or 14 milliamperes (approx.) per tube.

(7) From the plate-family curves for E_{c2} = 200 volts, the dc grid-No.1 voltage or bias (E_{c1}) required to produce a zero-signal plate current of 14 milliamperes per tube at a plate voltage of 600 volts is approximately −51 volts.

(8) The peak af grid-No.1-to-grid-No.1 (driving) voltage (E_{g1-g1}) = 2 [−(−51) +5] = 112 volts.

(9) The effective plate-to-plate load resistance (R_{Lp-p}) = $\frac{1.27 \times (600 - 65)}{0.135}$ = 5033, or approximately 5000 ohms.

(10) From the grid-No.1 curves given in the data section for E_{c2} = 200 volts, peak grid-No.1 current (i_{c1max}) is 8 milliamperes (approx.) for e_{c1max} = +5 volts and e_{bmin} = 65.

(11) The driving power required to produce maximum power output (P_d) = (0.008 $\times$ 56)/2 = 0.22 watt.

(12) From the grid-No.2 curves for E_{c2} = 200 volts given in the data section, for e_{c1max} = +5 volts and e_{bmin} = 65 volts, peak grid-No.2 current per tube (i_{c2max}) = 45 milliamperes.

(13) The dc maximum-signal grid-No.2 current per tube ($I_{c2\,max}$) = 0.25 $\times$ 45 = 11.2 milliamperes.

(14) Maximum-signal grid-No.2 input per tube (W_{c2}) = 200 × 0.0112 = 2.24 watts. This value is well within the maximum rating for the 6146 (3 watts per tube).

These calculated values are compared below with the nearest "Typical Operation" shown in the published data for the 6146 in Class AB_2 Operation, ICAS conditions.

Values are for two tubes	*Calculated*	*Published*	
DC Plate Voltage (E_b)..	600	600	volts
DC Grid-No.2 Voltage (E_{c2})........	200	190	volts
DC Grid-No.1 Voltage (Fixed Bias, E_{c1}).....	-51	-48	volts
Peak AF Grid-No.1-to-Grid-No.1 Voltage (E_{g1-g1})............	112	109	volts
Zero-Signal DC Plate Current ($2I_{bo}$).......	27	28	ma
Maximum-Signal DC Plate Current ($2I_{bmax}$)	270	270	ma
Zero-Signal DC Grid-No.2 Current ($2I_{c20}$)...	—	1.0	ma
Maximum-Signal DC Grid-No.2 Current ($2I_{c2max}$)	22.4	20	ma
Effective Load Resistance (Plate to plate, R_{Lp-p})	5000	5000	ohms
Maximum-Signal Driving Power, (Approx., P_d).	0.22	0.3	watt
Maximum-Signal Power Output, (Approx., P_o).	113.5	110	watts

Class B Amplifiers

Triodes

The procedure for calculating operating values for push-pull triode class B stages is substantially the same as that given above for multigrid-tube class AB_2 stages, but does not involve calculations for grid-No.2 voltage, current, input, or dissipation.

Example:

Calculate operating values for a class B modulator stage using two RCA-812-A's operating under ICAS conditions. The dc plate voltage (E_b) is 1500 volts, and the maximum-signal dc plate current (I_{bmax}) per tube is 155 milliamperes.

(1) Plate input per tube (P_i) = 1500 × 0.155 = 232.5 watts. This value is slightly less than the maximum plate-input rating of the 812-A for ICAS operation (235 watts).

(2) For a conduction angle of 180 degrees, the peak plate current per tube (i_{bmax}) = 3.14 × 0.155 = 0.487 ampere, or 487 milliamperes.

(3) From the average plate characteristics curves shown in Fig. 45, for i_{bmax} = 487 milliamperes, the peak positive grid voltage (e_{cmax}) = +90 volts (approx.) and the effective minimum plate voltage (e_{bmin}) = 100 volts.

(4) Power output for two tubes (P_o) = 157 × (1500−100) × 0.155 = 340 watts (approx.).

(5) Plate dissipation per tube (P_p) = (1500 × 0.155) −340/2 = 62.5 watts. This value is within the maximum rating for the 812-A (65 watts).

(6) For one-third maximum rated dissipation, zero-signal dc plate current per tube (I_{bo}) = 65/(3 × 1500) = 0.0145 ampere = 14.5 milliamperes.

(7) From the plate characteristics curves given in Fig. 45, dc grid voltage or bias (E_c) required to produce this value of plate current at a plate voltage of 1500 volts is approximately −45 volts.

(8) The peak af grid-to-grid driving voltage required for maximum power output (E_{g-g}) = $2E_g$ = 2[−(−45) +90] =270 volts.

(9) The effective plate-to-plate load resistance (R_{Lp-p}) = $\frac{1.27 \times (1500-100)}{0.155}$ = 11500 ohms (approx.).

(10) From the grid-current curves shown in Fig. 45, peak grid current (i_{cmax}) for e_{cmax} = + 90 volts and e_{bmin} = 100 volts is 140 milliamperes (approx.).

(11) The driving power required for maximum output (P_d) = (0.140 × 135)/2 = 9.45, or approximately 9.5 watts. These calculated values are compared below with the "Typical Operation" values for ICAS conditions shown in the published data for the RCA-812-A in Class B Modulator Service, ICAS conditions.

Values are for two tubes	*Calculated*	*Published*	
DC Plate Voltage (E_b)..	1500	1500	volts
DC Grid Voltage (E_c)..	-45	-48	volts
Peak AF Grid-to-Grid Voltage (E_{g-g}).......	270	270	volts
Zero-Signal DC Plate Current ($2I_{bo}$).......	29	28	ma
Maximum-Signal DC Plate Current ($2I_{bmax}$)....	310	310	ma
Effective Load Resistance (Plate-to-plate, R_{Lp-p})	11500	13200	ma
Maximum-Signal Driving Power (Approx., P_d)..	9.5	5	watts
Maximum-Signal Power Output (Approx., P_o)..	340	340	watts

Conversion Factors

Operating conditions for voltage values other than those shown in the published data can be obtained by the use of the **nomograph** shown in Fig. 46 when all electrode voltages are changed simultaneously in the same ratio. The nomograph includes conversion factors for current (F_i), power output (F_p), plate resistance or load resistance (F_r), and transconductance (F_{gm}) for voltage raitos between 0.5 and 2.0. These factors are expressed as functions of the ratio between the desired or new voltage for any electrode (E_{des}), and the published or original value of that voltage (E_{pub}). The relations shown are applicable to triodes and multigrid types in all classes of service.

To use the nomograph, simply place a straight-edge across the page so that it intersects the scales for E_{des} and E_{pub} at the desired values. The desired conversion factor may then be read directly or estimated at the point where the straight-edge intersects the F_i, F_p, F_r, or F_{gm} scale.

For example, the dashed lines on the nomograph show that for a ratio E_{des}/E_{pub} of 2/2.5 (all electrode voltages reduced 20 per cent), F_i is approximately 0.72, F_p is approximately 0.57, F_r is 1.12, and F_{gm} is approximately 0.892. These factors may be applied directly to operating values shown in the tube data, or to values calculated by the methods described previously.

When only one electrode voltage of

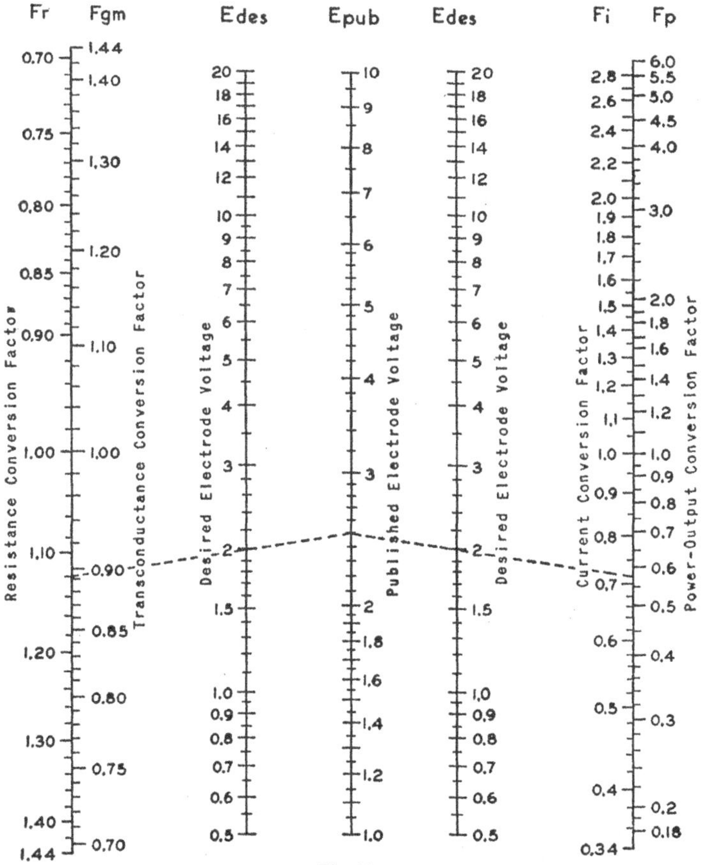

Fig. 46

a tube is changed, for example in the calculation of operating conditions for a multigrid tube operated at a grid-No.2 voltage for which curve families are not available, the nomograph is used twice. The procedure is shown in the following example:

Determine operating values for an RCA-6146 beam power tube in Class C Telegraphy Service at its maximum ICAS plate-voltage (E_b) and plate-input (P_i) ratings of 750 volts and 90 watts, and at a grid-No.2 voltage (E_{c2}) of 160 volts. (The dc plate current I_b of the tube under the desired conditions is 90 watts/750 volts, or 120 milliamperes.)

Because curve families are not available for an E_{c2} of 160 volts, operating conditions must first be calculated for the nearest value of E_{c2} for which curves are available (*i.e.*, 150 volts). For this calculation, the chosen values of E_b and I_b must be converted to the corresponding values for E_{c2} = 150. The plate voltage (E_b) becomes $\frac{750 \times 150,}{160}$ or approximately 703 volts. Using conversion-factor values obtained from the nomograph for the voltage ratio 150/160, the plate current (I_b) = $F_i \times I_b = 0.91 \times 120$, or approximately 109 milliamperes.

For a conduction angle of 140 degrees, K_1 = 4 and the peak plate current ($i_{b_{max}}$) = 4 × 109 = 436 milliamperes:

From the plate-family curves of the 6146 for E_{c2} = 150 volts shown in the tube data, the effective minimum plate voltage ($e_{b_{min}}$) = 75 volts and the peak positive grid voltage (e_{c1max}) = +15 volts.

From the corresponding grid-No.1 and grid-No.2 curve families, peak grid-No.1 current (i_{c1max}) = 24.5 milliamperes and peak grid-No.2 current (i_{c2max}) = 39.5 milliamperes.

These instantaneous voltages and currents can now be converted to corresponding values for the desired E_{c2} of 160 volts. For the voltage ratio 160/150, or 1.066, $e_{b_{min}} = 75 \times 1.066$, or approximately 80 volts, and $e_{c1max} = +15 \times 1.066$, or approximately 16 volts.

From the nomograph, the current conversion factor F_i for the ratio 160/150 is 1.1. Consequently, $i_{c1max} = 24.5 \times 1.1$, or approximately 27 milliamperes, and $i_{c2max} = 39.5 \times 1.1$, or approximately 43.5 milliamperes.

The remaining operating values can then be calculated: Power output (P_o) = $K_2 \times (E_b - e_{b_{min}}) \times I_b$ = 0.862 (750–80) × 0.120 = 69.3 watts.

The dc grid-No.1 voltage or bias (E_{c1}) = $-(K_3 \times e_{c1max}) - \frac{K_4 \times E_{c2}}{g_{2g1}}$ = –(0.52 × 16) –1.52 (160/4.5), o approximately –62 volts.

The peak rf grid-No.1 voltage (E_{g1}) = –(–62) +16 = 78 volts.

From Table II, the constant K_6 = 9.15 (approx.) for an E_{c1}/E_{g1} ratio of 62/78, or 0.795. Consequently, the dc grid-No.1 current (I_{c1}) = 27/9.15, or approximately 3 milliamperes.

The dc grid-No.2 current (I_{c2}) = $K_5 \times i_{c2max}$ = 0.2 × 43.5, or 8.7 milliamperes. The dc grid-No.2 input (W_{c2}) = 160 volts × 0.0087 amperes, or approximately 1.4 watts.

These calculated values are compared below with the published "Typical Operation" values for the 6146 in Class C Telegraphy, ICAS conditions:

	Calculated	*Published*	
DC Plate Voltage (E_b)	750	750	volts
DC Grid-No.2 Voltage (E_{c2})	160	160	volts
DC Grid-No.1 Voltage (E_{c1})	–62	–62	volts
Peak RF grid-No.1 Voltage (E_{g1})	78	79	volts
DC Plate Current (I_b)	120	120	ma
DC Grid-No.2 Current (I_{c2})	8.7	11	ma
DC Grid-No.1 Current (I_{c1})	3	3.1	ma
Driving Power, (Approx., P_d)	0.21	0.2	watt
Power Output, (Approx., P_o)	69.3	70	watts
Plate-input power (P_i)	90	90	watts
Plate dissipation (P_d)	21	20	watts
Grid-No.2 Input (W_{c2})	1.39	1.76	watts

Because this method for conversion of characteristics is necessarily an approximation, the accuracy of the nomograph decreases progressively as the ratio E_{des}/E_{pub} departs from unity. In general, results are substantially correct when the value of the ratio E_{des}/E_{pub} is between 0.7 and 1.5. Beyond these limits, the accuracy decreases rapidly, and the results obtained must be considered rough approximations.

The nomograph does not take into

consideration the effects of contact potential or secondary emission in tubes. Because contact-potential effects become noticeable only at very small dc grid-No.1 (bias) voltages, they are generally negligible in power tubes. Secondary emission may occur in conventional tetrodes, however, if the plate voltage swings below the grid-No.2 voltage. Consequently, the conversion factors shown in the nomograph apply to such tubes only when the plate voltage is greater than the grid-No.2 voltage. Because secondary emission may also occur in certain beam power tubes at very low values of plate current and plate voltage, the conversion factors shown in the nomograph do not apply when these tubes are operated under such conditions.

Adjustment and Tuning

AF equipment does not normally require tuning or preliminary adjustments other than those necessary for obtaining plate-current balance in push-pull stages. Subsequent operating adjustments of gain or input-signal level and "tone" or frequency response can usually be made without the aid of auxiliary equipment.

Tuning and operating adjustments in rf power equipment, however, are numerous and complex and require the use of instruments for accurate measurement of frequency, dc grid current, dc plate voltage and current, and dc screen-grid (grid-No.2) voltage and current of multigrid tubes. Other equipment which may be necessary or useful includes: a grid-dip oscillator for preliminary tuning of resonant tank circuits and for neutralization adjustments; a "dummy load" (an incandescent lamp or non-inductive resistor having suitable resistance and wattage rating) used to absorb the power output of the final stage so that unauthorized frequencies or other improper signals which may be produced during preliminary adjustments are not radiated by the antenna system or load; simple rf indicators, such as a neon lamp or a small flashlight bulb which is connected to a one- or two-turn loop of wire; and simple devices for measuring approximate frequency, such as absorption-type wavemeters. A cathode-ray oscilloscope is desirable for proper adjustment of radiotelephone, television, and facsimile transmitters.

Because a class C stage may draw excessive plate current if operated even momentarily into an improperly tuned plate-tank circuit, all plate-tank circuits should be tuned to their approximate operating frequencies (with the aid of a grid-dip oscillator) before actual operating adjustments are begun. During this preliminary tuning procedure, all plate, screen-grid, and grid-bias supplies should be turned off, but all tubes and circuit components should be in place and normal filament or heater voltages should be applied to the tubes to assure that the stray capacitance and inductance of each stage are substantially the same as those present during operation.

Tuning Procedure

Tuning and adjustment of rf power equipment starts in the oscillator or input stage, and continues through succeeding stages along the path followed by the rf signal. The procedure used in tuning class C stages is generally the same for all types of service, circuit configurations, and tube types. Consequently, the procedure given below for tuning a "straight-through" rf amplifier stage also applies to frequency multipliers. It is assumed that the amplifier has been properly neutralized, if required, by the method described later, and that the preceding stage or "driver" has been properly tuned and is delivering full output at the desired frequency.

(1) Make sure that all power to the equipment is *off*.

(2) Disconnect all positive plate, screen-grid, and suppressor-grid supply leads from the amplifier and from all following stages.

(3) If variable coupling is used between driver and amplifier, adjust the coupling to approximately one-half maximum.

(4) Apply only normal filament or heater voltage to the amplifier, and all normal operating voltages to the driver.

(5) Quickly tune the driver plate circuit to resonance, which is indicated by a dip in driver plate current, as shown in Fig. 47, and by maximum grid current in the amplifier stage. If the ampli-

fier has a tuned grid circuit, this circuit should also be tuned to resonance (indicated by an increase in the amplifier grid current).

(6) Increase the coupling between driver and amplifier, being careful not to exceed the maximum permissible grid current for the amplifier tube or tubes. It should be possible to obtain full rated grid current for the amplifier stage without overloading the driver (overload being indicated by excessive driver plate current at resonance).

(7) Retune the driver plate circuit (and the amplifier grid circuit) to resonance. This procedure should always be followed after a change is made in coupling or loading to compensate for the normal detuning effects of such changes.

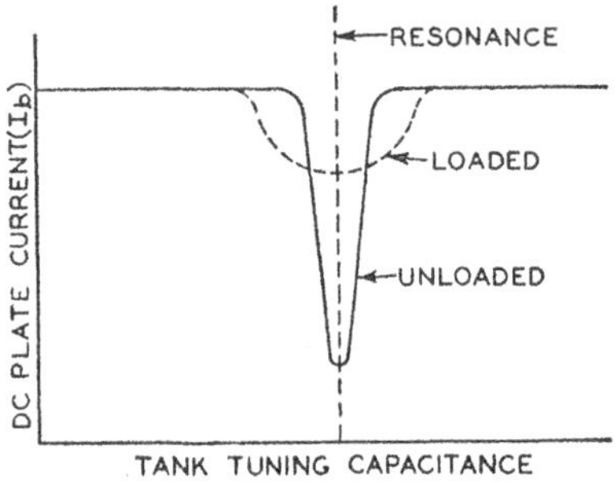

Fig. 47

(8) Turn on any fixed-bias supplies for the amplifier, and make any circuit changes or adjustments necessary to assure that the plate, screen-grid, and suppressor-grid voltages for the amplifier will not be more than 50 per cent of their normal values when applied. Disconnect the external load from the amplifier plate-tank circuit, or, if this change is not practicable, reduce the coupling between amplifier and external load to minimum. If the load for the amplifier is another tube, remove this tube from its socket.

(9) Apply plate, screen-grid, and suppressor-grid voltages (50 per cent of normal values) to the amplifier, but not to any following stages, and quickly tune the amplifier plate circuit to resonance. When an amplifier is operated without a load connected to its plate tank, its plate current will usually dip at resonance to between 10 and 20 per cent of the normal full-load value. The absolute value of the no-load plate current at resonance depends on the Q of the plate-tank circuit, the type of bias used, and the rf excitation voltage, and should not be considered an indication of the amplifier efficiency.

If the plate current of an unloaded triode does not dip in the normal manner, the trouble may be caused by inadequate grid excitation, excessive tank-circuit losses, or improper neutralization. If the plate-tank circuit of any class C amplifier cannot be tuned to resonance, the tank-circuit inductance or capacitance, or both, may have to be increased or decreased in value, depending on whether the circuit is found to tune higher or lower than the desired frequency. An absorption-type wavemeter is useful in such adjustments.

If flashover occurs in the plate-tank capacitor during tuning adjustments, reconnect the load to the amplifier output circuit and/or increase the coupling between amplifier and load until the rf voltage is reduced sufficiently to eliminate the flashover.

(10) Connect the external load to the amplifier plate tank. (If this step has already been taken to eliminate flashover, as described above, tighten the load coupling.) When the load is applied or the load coupling increased, the plate current of the amplifier should rise. Retune the amplifier plate tank to resonance after each change in coupling. The amplifier plate current should still dip at resonance, but its minimum value should be considerably higher than under no-load conditions, as shown by the dashed curve in Fig. 47.

(11) Apply full plate, screen-grid, and suppressor-grid voltages to the amplifier. Increase the coupling between amplifier and load, retuning the amplifier plate tank to resonance as often as necessary, until the plate current at the resonance dip has the desired value. In no case should the plate input (the product of the dc plate voltage and dc plate current) exceed the maximum value given in the tube ratings for the type of service involved.

Because the dc grid current of an amplifier decreases as the load on the amplifier is increased, grid current should be checked after each change in load or load coupling to make sure it has not dropped appreciably below the normal or desired value. If it has, the cause may be insufficient grid excitation or excessive grid bias.

Neutralizing Adjustments

The procedure used in neutralizing rf amplifiers is substantially the same regardless of the neutralizing circuits or tube types employed. The tube operating conditions used are similar to those employed for preliminary tuning of plate-tank circuits, except that excitation at the highest operating frequency is applied to the stage being neutralized.

(1) Make sure that all power to the equipment is *off*.

(2) Disconnect all positive plate, screen-grid, and suppressor-grid supply leads from the amplifier and from all following stages. Adjust the coupling between driver and amplifier to maximum, and loosely couple a fairly sensitive rf indicator to the amplifier plate-tank coil. Although a simple indicator is usually satisfactory, a sensitive rf meter connected to a one- or two-turn loop or a vacuum-tube voltmeter equipped with a suitable rectifier probe provides more exact indications, particularly for final adjustments.

(3) Apply normal filament or heater voltage to the amplifier, and all normal operating voltages to the driver, and tune the driver plate circuit to resonance.

(4) Tune the plate-tank circuit of the amplifier to resonance (shown by maximum brightness or maximum reading of the rf indicator). Adjust the neutralizing capacitor until the rf indicator shows minimum brightness reading.

(5) Carefully retune the amplifier plate-tank circuit to resonance. The rf indicator should now show a new maximum reading, but one having substantially smaller magnitude than the original reading. Again adjust the neutralizing capacitor for a minimum reading on the rf indicator. The driver plate-tank circuit should be checked and, if necessary, retuned to resonance during these adjustments.

Repeat step (5) until a setting for the neutralizing capacitor is found which produces no indication of rf voltage in the amplifier plate circuit. As this setting is approached, it will probably be necessary to increase the coupling between the rf indicator and amplifier plate tank to obtain useful indications. A stage may be considered properly neutralized when the rf indicator shows zero at maximum coupling.

In neutralizing a push-pull amplifier, both neutralizing capacitors should be adjusted simultaneously. However, both capacitors will seldom have the same setting at the point of complete neutralization because of slight differences in tube and stray circuit capacitance, and because split tank circuits are seldom electrically symmetrical.

A dc milliameter connected in the grid-return circuit of an amplifier can also be used as a very sensitive indicator for neutralizing adjustments. The amplifier is operated without plate, screen-grid, or suppressor-grid voltage, and sufficient rf excitation is applied to produce a normal value of grid current. If the amplifier is not properly neutralized, its grid current will vary when its plate-tank circuit is tuned through resonance. The neutralizing capacitor should then be adjusted slowly while the amplifier plate-tank circuit is tuned back and forth through resonance. As the point of neutralization is approached, the variations in grid current decrease. When the amplifier is perfectly neutralized, tuning of its plate-tank circuit through resonance does not cause even a slight change in the reading of the grid-current meter.

In some cases, it may not be possible to eliminate rf feedthrough entirely by adjustment of the neutralizing capacitor. This difficulty is usually an indication of stray coupling between the amplifier and driver plate tanks, or of stray capacitances in various portions of the amplifier which tend to unbalance the neutralizing circuit. Adequate shielding between the driver and amplifier and between the grid and plate circuits of the amplifier will usually eliminate this difficulty.

The difficulty may also arise in a stage employing a split-stator tank capacitor if the ground lead of the capacitor is not connected by the shortest possible path to the cathode-return point of the stage.

Power-Tube Installation

Because power tubes usually operate at high voltages and temperatures, draw heavy currents, and are used in high-efficiency circuits, terminal connections for such tubes should have large-area, low-resistance contacts capable of accommodating relatively large wire sizes and utilize high-quality insulation.

Sockets or **mountings** for power tubes having filamentary cathodes should be installed, as a general rule, so that the tubes are operated in a vertical position with the base or filament end down. Vertical operation minimizes the danger of internal short circuits which may be caused by thermal expansion or sagging of the filament. Certain filamentary-cathode vacuum types may be operated in other than vertical positions, provided precautions specified in the tube data are observed. Tubes having indirectly heated cathodes may generally be operated in any position.

If equipment is to be subjected to mechanical shock or vibration, the equipment housing, the tube mountings, or both should include some form of shock-absorbing suspension, and suitable means should be employed to lock the tubes in their sockets or mountings.

Ventilation

Power-tube equipment design should always permit the unimpeded circulation of air around all tubes and include provision for adequate ventilation of tube and equipment enclosures so that envelope temperatures will not become high enough to damage the tubes or their associated circuit components.

Most of the tubes listed in this Manual are designed for operation at maximum ratings with natural convection cooling. Certain types, however, such as the 6161, require forced-air cooling. Other types, such as the 826, 829-B, and 833-A, can be operated with natural convection cooling, but carry substantially higher ratings when forced-air cooling is employed. Maximum permissible bulb temperatures and forced-air flow and pressure requirements are given in the *Tube Types* Section for most types.

The glass portions of a tube envelope should not be exposed to the spray of any liquid or be permitted to come in contact with metal objects such as circuit wiring or grounded metal shields because excessive temperature differences may cause envelope fractures. Shields should not fit so closely as to impede the free circulation of air around the tubes. In many cases, they may be designed to produce a "chimney" effect which will increase the draft and improve tube ventilation.

The maximum permissible bulb temperature of a vacuum tube or inert-gas tube is determined principally by the softening point of the glass employed, or by the point at which gas may be released by the envelope. In the case of mercury-vapor tubes, both minimum and maximum bulb-temperature limits are specified to assure satisfactory vaporization of the mercury. Temperature considerations for mercury-vapor tubes are discussed in the *Rectifier Considerations* Section.

Wiring Considerations

Energy losses in power-tube circuit wiring limit operating efficiencies and may produce undesirable heat. These losses may be caused by conductor resistance (I^2R losses), leakage (E^2/R losses), radiation, or stray coupling.

Excessive I^2R losses in power-tube circuit wiring can be avoided by the use of conductors having adequate current-carrying capacity and the lowest possible resistance, and layouts which permit short, direct, connecting leads. Filament- and heater-circuit conductors are particularly susceptible to large I^2R losses because they carry currents of high average (dc) or rms (ac) value, and because their resistance is increased by heat received by direct thermal conduction from the tube filaments or heaters. When an installation requires the use of long filament-supply leads or operation of several high-current tubes from a common filament-supply line, these losses may cause filament voltages to decrease below the minimum values specified in the tube data and the tubes may be damaged. In such cases, conductors of adequate size should be used to avoid excessive losses or sufficient excess voltage should be provided at the supply to compensate for the resulting losses. In the latter case, means of adjusting the supply voltage and suitable metering facilities should be provided to assure that correct filament or heater voltage is received at all terminals.

Excessive I^2R losses in signal conductors may also cause improper operation and tube damage, particularly in driving circuits where the signal provides the required operating bias as well as protection of the tube. In the selection of signal conductors, consideration must be given to "skin effect," which causes current to concentrate nearer the surface of a conductor as the frequency increases, as well as to the type of circuit and the waveform of the signal current.

A signal conductor should have low resistance at the highest frequency involved, and be capable of carrying the highest peak currents flowing in the circuit with negligible heating. Solid or stranded conductors are suitable for af applications, and a special type of multiple-strand conductor called "Litzendraht" for low- and medium-power rf applications at frequencies up to approximately 3 megacycles per second. At higher frequencies it is advisable to use tubular conductors, which should be silver-plated, if possible, to obtain maximum surface conductivity and to minimize the effects of oxidation.

Leakage (E^2/R) losses are caused primarily by inadequate or improper insulating materials, or by insufficient separation between air-insulated conductors. In the selection of insulating materials for power-tube installations, consideration should be given to the fact that very high peak-signal voltages may be developed in circuits operating at relatively low dc potentials. In addition, the type of insulating material used at any point must be suitable for the temperature and frequency involved.

As a general rule, conductors having enamel, plastic, or fabric coverings should be used only in supply circuits and low-frequency signal circuits operating at low voltages. Supply-circuit conductors should be installed in comparatively cool locations as far from signal conductors and unshielded signal components as possible. Such conductors, when completely insulated, may usually be grouped or cabled together on the chassis or framework of the equipment. When high voltages or very high temperatures are involved, it is generally preferable to use bare conductors which are adequately spaced and supported by insulators of suitable mechanical design.

RF signal conductors, particularly those carrying vhf or uhf currents, should not be insulated, except at points where mechanical support is necessary, because practically all types of surface insulation absorb appreciable energy in the presence of rf fields. These conductors should be isolated from each other, from circuit components, and from the equipment structure.

Losses of signal energy by **radiation** from circuit conductors increase with current and with the length of the conductors, but usually do not become appreciable until conductor length approaches a substantial fraction of a half-wavelength at the operating frequency. Lead length requires careful consideration in vhf and uhf equipment, however, because of the close relationship between practical conductor dimensions and signal wavelengths.

Stray coupling in circuit wiring may produce out-of-phase signal currents in a conductor. These currents cause degeneration losses. Such losses may be minimized by the use of short, direct, circuit connections. These considerations are discussed below under "Circuit Returns."

Cap or wire bulb terminals such as those used on the 807 and 6524 should never be used to support coils, capacitors, or other circuit components because the resulting mechanical stresses may fracture the bulb seals. Connections to bulb terminals should always be made with soft metallic braid or ribbon, or with other types of conductors having good mechanical flexibility and low electrical resistance. Under no circumstances should connections be soldered to cap or wire bulb terminals because the high temperatures developed may soften or crack the bulb seals. The long, flexible, wire terminal leads used on subminiature types such as the 5718, however, may be soldered directly to circuit components, provided speed and care are used to minimize the transmission of heat to the bulb seals.

Circuit Returns

All currents in a power tube (except heater current) originate in and return to the cathode, which is, therefore, a common terminal of all supply and signal circuits associated with the tube. The direct currents drawn by the tube electrodes return to the cathode through the power-supply and bias circuits. Although these circuits also provide return paths to the cathode for signal currents, they usually contain resistive and reactive components which offer considerable impedance to ac signals and thus cause substantial loss of signal energy. When a single power supply is used for more than one stage, its internal impedance may also act as a coupling device between stages and thus introduce undesired degeneration or regeneration. These effects may generally be avoided by the use of separate ac and dc return paths to cathode from each electrode or signal circuit of a tube.

DC circuit returns for a power tube employing fixed bias, grid-resistor bias, or a combination of the two, are made to the cathode terminal of the tube. When cathode-resistor bias is used, either alone or in combination with another type of bias, the dc circuit returns are usually connected to the more negative terminal of the cathode resistor. If the dc voltage drop across the cathode resistor is greater than the bias required, however, the grid-circuit dc return for the tube may be connected to a tap on the cathode resistor which provides the desired bias voltage. When an rf choke coil or a resonant network is connected in series with the cathode of a power tube employing fixed or grid-resistor bias, dc circuit returns are made in the same manner as when cathode-resistor bias is used. In a filamentary-cathode power tube, the heating current creates a voltage drop in the cathode which is equivalent to a bias voltage equal to about one-half the filament voltage. The polarity and value of this drop must be considered in determining the point to be used for dc circuit returns.

When dc filament voltage is applied to a filamentary-cathode tube, all dc circuit returns should be connected to the negative filament terminal of the tube. The use of this point for dc returns provides a small amount of protective bias for the tube because the grid is maintained at a negative potential with respect to the cathode in the event that external bias fails or is accidentally removed.

When ac voltage is applied to a filamentary cathode, dc circuit returns should be made to the mid-point of the filament or filament-supply circuit to minimize hum. A convenient point for these returns is a center tap on the supply winding of the filament transformer, or the junction of two equal resistors connected in series across the filament circuit.

Most heater-cathode tubes have a single cathode terminal which is used for all circuit returns or for connection of a cathode resistor. In some heater-cathode tubes, however, two or more cathode terminals are provided to permit the use of separate ac return leads from the input and output circuits of the tube and thus minimize cathode-lead degeneration. Because these terminals are connected in parallel internally, any one of them may be used as the dc return point of the tube or for connection of a cathode resistor.

When a heater-cathode tube is operated with fixed bias or grid-resistor bias, or with cathode-resistor bias within the maximum heater-cathode voltage rating of the tube, the heater should be connected to the dc return point of the tube. In other cases, the heater should be connected to the tube cathode or to a point having the same dc potential as the cathode. Although either of the heater terminals may generally be used for this connection, it may sometimes be necessary to use a center tap on the heater winding of the supply transformer or a center-tapped resistor across the heater circuit to minimize hum.

The use of separate ac and dc returns in power-tube installations minimizes signal-energy losses in power-supply and bias circuits. It also minimizes degenerative or regenerative effects which may result if common signal-return paths are used for the input and output circuits of a tube or for the circuits of more than one tube. **AC returns** are generally made through capacitors directly to the cathode, or to points having the same ac potential as the cathode, regardless of the location of the dc return point.

In af applications, the grid, plate, and screen-grid circuit returns of the tube may be bypassed individually to the chassis or to a common ground bus (and thus to the cathode), as shown in Fig. 48, by capacitors which have very low impedance at audio frequencies. In this case, the length of the portions of chassis or ground bus used as common ac return paths is not critical because the impedance of such paths at audio frequencies is generally negligible.

At radio frequencies, however, a distance of even a fraction of an inch between points on a chassis or ground bus may represent a substantial impedance and produce undesirable coupling effects.

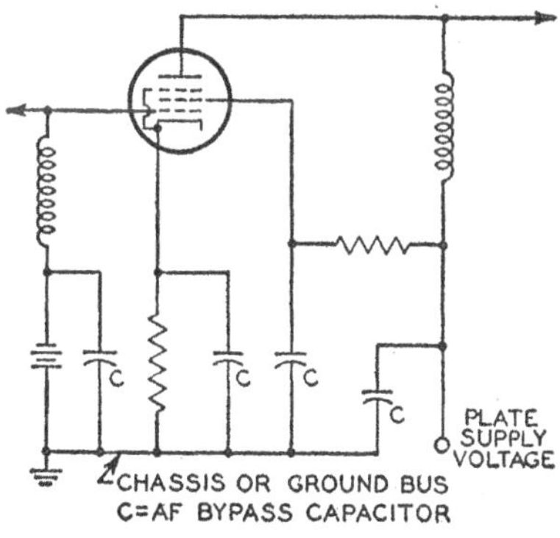

Fig. 48

The ac circuit returns of an rf stage should, therefore, be connected directly to the appropriate cathode terminals of the tube socket or to a single point on the chassis which is at the same ac potential as the cathode. Fig. 49 is a semi-pictorial diagram showing the ac circuit returns required in a high-frequency amplifier stage using a beam power tube. Bypass capacitors are used across each side of the filament center-tap resistor to minimize the rf impedance of the

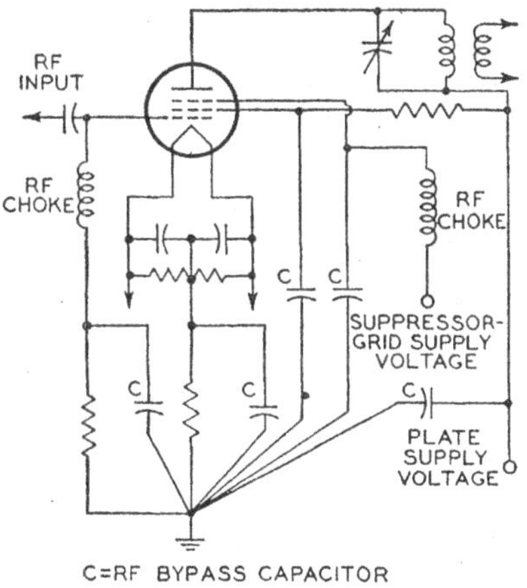

Fig. 49

filament circuit. Capacitors used in rf bypass applications should be specifically designed for use at the required operating frequencies.

Filament or Heater Supply

AC voltage is generally used to heat the cathodes of power tubes because of the convenience and economy with which the relatively low voltages required may be obtained from transformers. The operating voltages applied to thoriated-tungsten or oxide-coated filamentary cathodes should not be permitted to vary more than plus or minus five per cent from the values specified in the tube data. Heater voltages for unipotential cathodes should be maintained within plus or minus ten per cent of rated values unless smaller tolerances are specified in the data for individual tube types. Voltage variations greater than those specified may damage the emitting surface of the cathode, or in other ways cause unsatisfactory tube operation or short life.

When filamentary-cathode power tubes are heated with direct current, any current- or voltage-control devices employed should be placed in the branches of the supply circuit feeding the individual tubes. When alternating current is used, such control devices should be placed in the primary circuits of the filament-supply transformers. When a filamentary cathode is heated by low-frequency alternating current, hum may be introduced into the tube circuit by (1) a periodic variation in the electron emission as the heating current increases and decreases in value; (2) interaction between the magnetic field of the space-charge and that of the filament; and (3) the electrostatic field of the filament. The principal source is usually the electrostatic field of the filament, which induces hum voltages in the signal electrodes of the tube in proportion to the filament voltage and the capacitance between the filament and other electrodes.

Plate Supply

The power-rectifier tubes included in this Manual normally obtain their plate-supply voltage from the secondary windings of high-voltage transformers connected to commercial power lines or to local sources of low-frequency ac voltage. Power-amplifier tubes usually obtain plate voltage from rectifiers provided with suitable filter circuits, although batteries or local dc generators are sometimes used, especially in portable and mobile equipment.

Suppressor-Grid Supply

Voltage for the grid No.3 or suppressor grid of a power pentode may be obtained from any dc source which is substantially free from ripple or other undesirable fluctuations in potential. When an application requires that a suppressor grid draw a varying current, the dc supply should be a battery or other source having good voltage regulation. This requirement is particularly important when a suppressor grid is used as a modulating electrode because the average suppressor-grid current may then vary with the amplitude of the modulating signal.

Screen-Grid Supply

Grid-No.2 or screen-grid voltage for a beam power tube, pentode, or tetrode may be obtained from a separate dc power supply or from the plate supply for the tube. In the latter case, the required voltage may be obtained either from a suitable tap on a voltage divider or through a dropping resistor from the plate-voltage supply point, depending on the type of multigrid tube used and on the application.

A multigrid tube may fail prematurely if its screen-grid current, screen-grid voltage, or total screen-grid input exceeds the maximum value shown in the tube data. Excessive screen-grid current may be drawn if the tube is operated without adequate bias or plate voltage. Because the latter condition is most likely to occur when screen-grid and plate voltages are obtained from separate supplies, such supplies should be designed so that plate voltage is always applied before or simultaneously with screen-grid voltage and removed simultaneously with or after the removal of screen-grid voltage. In addition, any means employed for the reduction of plate voltage should automatically produce a proportional reduction in screen-grid voltage.

The danger of excessive screen-grid voltage is present principally when screen-grid voltage is obtained from the plate supply through a series dropping resistor. In this type of supply circuit, sufficient resistance is connected between the screen grid and the plate supply to assure that the screen-grid voltage and dissipation at the values of screen-grid current, bias, and driving voltage required for full output are within the maximum ratings for the tube. Any condition which reduces the current through the screen-grid dropping resistor to a very low value, therefore, may cause the screen-grid voltage to rise to an excessive value.

Such conditions are most likely to occur in telegraphy transmitters employing "blocked-grid" keying or other methods of keying which cut off or substantially reduce plate and screen-grid currents of multigrid tubes when the key is up. Although Class C Telegraphy ratings for most multigrid tubes permit a rise in screen-grid voltage under key-up conditions, the maximum permissible screen-grid voltage under these conditions is generally substantially less than the plate-supply voltage. Screen-grid voltage for a keyed multigrid amplifier should, therefore, be obtained from a separate supply or a voltage-divider arrangement, rather than by the series-resistor method. In cases where a series-resistor screen-grid supply voltage is used, precautions should be taken to keep the screen-grid voltage within the maximum value specified in the tube data for key-up conditions.

Control-Grid (Bias) Supply

Control-grid voltage or bias for a power tube may be obtained from a separate power supply or a resistor in the grid or cathode circuit. Fixed bias is obtained from an independent battery, dc generator, or rectifier-filter system. Grid-resistor bias is obtained by rectification of a portion of the input signal or driving voltage applied to the tube. Although this type of bias is the most economical, and can provide relatively large bias voltages or voltages which vary with the input signal, it does not provide protection against excessive plate and screen-grid current in the event the driving voltage fails or is removed. Grid-resistor bias, therefore, is usually used in combination with other means to protect the tubes against excessive plate and screen dissipation.

Cathode-resistor bias is obtained from the voltage drop developed across a cathode resistor by the combined dc currents of the tube electrodes. This type of bias provides automatic protection against excessive plate, screen-grid, and control-grid current because any increase in total cathode current produces a corresponding increase in bias voltage. Cathode-resistor bias cannot be used alone if bias voltage equal to or greater than the cutoff voltage is required. Because the effective plate and screen-grid voltages of the tube are reduced by the extent of the voltage drop in the cathode resistor, this type of bias is used principally when relatively small bias voltages are required or as a means of providing a minimum protective bias when the principal operating bias is obtained by the grid-resistor method.

Supply-Voltage Variations

Because a tube may be seriously damaged if its absolute maximum voltage ratings are exceeded, consideration must be given to the variations in electrode voltages which result from line-voltage fluctuations, load variations, and normal manufacturing tolerances in circuit-component values. The operating voltage for each tube electrode should be low enough so that the absolute maximum rated voltages of the tube will not be exceeded under any combination of these variations, or the voltage supplies should have sufficient regulation to permit the use of maximum rated voltages without danger of exceeding the tube ratings.

Protective Devices

Power-tube installations should always be adequately equipped with protective devices to prevent damage to the equipment and/or personal injury. Devices which provide tube and circuit protection include:

(1) fuses or relays which automatically remove power from the equipment, or from a particular circuit, in the event of improper operation;

(2) meters, or facilities for external metering, to permit checking of important circuit operating conditions.

The most common cause of damage to tubes and equipment in power-tube installations is excessive plate or screen-grid current. For adequate protection, therefore, each stage of a power-tube installation should be equipped with fuses or relays which will remove all positive electrode voltages if the plate or screen-grid current reaches a value about 50 per cent above normal. Separate protective devices should be provided for plate and screen-grid circuits of multigrid tubes.

Facilities should be provided for the measurement of plate, screen-grid, and filament (or heater) voltages, and plate, screen-grid and control-grid currents. Control-grid-current measurements are particularly valuable in rf amplifier and frequency-multiplier stages because they facilitate tuning and neutralizing adjustments in addition to providing indications of drive conditions. Because correct filament and heater voltages are essential for maximum tube life, these voltages should always be measured directly at the tube sockets with meters having high accuracy and low power requirements.

For reasons of economy, a single dc milliameter is sometimes placed in the cathode-return lead or the negative high-voltage supply lead of a tube for the measurement of total cathode current. In such cases, the meter should be shunted with a resistor to protect the tube cathode and the meter from high dc potentials with respect to ground in the event of an open circuit in the meter. A shunting resistor having a value of about 100 times the resistance of the meter is generally satisfactory, and introduces an error in meter reading of only about one per cent.

Safety Considerations

Because the rated plate and screen-grid voltages of most power tubes are high enough to be extremely dangerous to the user, care should be taken during maintenance of power-tube equipment to insure that all primary power is disconnected and all exposed circuit parts are effectively grounded. When circuit adjustments are made on "live" equipment, very great care should be taken to avoid contact with any circuit parts which are not at ground potential. Such adjustments should never be made unless another person capable of applying treatment for electric shock is present.

In the design of equipment, personal-safety considerations require the grounding of all operating controls and exposed surfaces, enclosure of all live circuit elements, and the incorporation of "interlock" switches at all points of access to the interior of the equipment. These switches should automatically open the primary circuits of all high-voltage power supplies when access is required.

Rectifier Considerations

Rectifier-type power supplies employing electron tubes are used as sources of plate, screen-grid (grid-No.2), and other dc operating voltages in all types of electronic equipment. They are also used extensively in electroplating, in motor-speed control, and in many other applications requiring economical and conveniently controllable dc power.

The glass envelopes of the rectifier tubes used in such supplies normally show some darkening after continued operation. In addition, mercury-vapor tubes exhibit a blue glow in normal operation. These symptoms are characteristic of such tubes, and should not be considered signs of tube deterioration or failure.

Mercury-Vapor Tubes

A mercury-vapor rectifier tube must be handled with special care to prevent dispersion of the liquid mercury from its normal position at the bottom of the bulb. Spattering of the mercury over other portions of the bulb or on the anode or filament must be avoided because it may lead to internal shorts or arcs when the tube is placed in operation. A mercury-vapor tube should always be transported, stored, and operated in a vertical position with the filament end down, and should never be jarred, shaken, or allowed to rest even momentarily in a horizontal position. The tube should never be rocked or allowed to snap into place in its socket or mounting, and should be protected against excessive equipment vibration.

If spattering occurs, the dispersed mercury must be completely reconcentrated before the tubes are placed in service by means of special preheating and conditioning treatments. In the preheating treatment, the mercury-vapor tube is operated at normal filament voltage, but without anode voltage, for 30 minutes to assure complete vaporization of the mercury content. When filament voltage is removed at the end of this preheating period, most of the vaporized mercury recondenses in a pellet or pool at the bottom of the bulb. The conditioning treatment is then applied to flash out any mercury which may have condensed on the bulb walls or in the vicinity of the anode and filament seals. In this treatment, the tube is operated at normal filament voltage and at about one-sixth normal anode voltage for 5 minutes. The anode voltage is then gradually increased over a period of about 30 minutes to the normal operating value. If an internal flashover occurs at any time during the conditioning treatment, the anode voltage should be reduced until the flashover ceases. It should then be held at this reduced value for a few minutes to assure complete vaporization of the mercury before the treatment is resumed.

Filament Heating Time

Voltage should not be applied to the plates or anodes of vacuum, mercury-vapor, or inert-gas rectifier tubes (except receiving types) until the filaments or cathodes of the tubes have reached normal operating temperature. For gas tubes, this delay is necessary to allow the formation of a plasma (region of electrons and positive ions) which protects the emitting surface against damage from high-velocity positive-ion bombardment. In the case of a mercury-vapor rectifier, the application of anode voltage must also be delayed until the condensed mercury has moved to its normal condensing zone at the bottom of the tube, as discussed above.

Minimum heating times for individual rectifier types are given in the *Tube Types* Section. In each case, the time specified is measured from the instant when the filament voltage reaches its normal operating value and, consequently, may have to be increased if the filament supply has poor regulation.

It should be noted that measurement of the filament voltage of a power-rectifier tube may involve serious personal-safety hazards because the filament is usually a high-voltage terminal of the rectifier circuit. When continuous measurements are required, suitable voltmeters should be permanently incorporated in the equipment. These meters must be insulated to withstand the maximum peak inverse voltage applied to the tubes, and should be recessed in the equipment and protected by glass or plastic viewing panels to prevent any possibility of injury through accidental bodily contact. Portable instruments should not be used for the measurement of rectifier-filament voltages unless adequate personal-safety precautions are taken by the user.

Because a mercury-vapor tube may be severely damaged if the temperature of its filament varies excessively, the filament should be operated from a constant-voltage transformer, or its supply circuit should include under- and over-voltage relays which will open the primary circuit of the rectifier anode supply if the line voltage varies excessively. Relays having small operating delays (less than 10 seconds) may be used in this application to minimize interruptions to operation by normal surges or transient variations in line voltage.

The required delay in application of anode voltage can be obtained conveniently by means of a time-delay relay connected in the primary circuit of the high-voltage transformer, as shown in Fig. 50. This relay should permit adjustment of the delay time to a value sufficient to assure protection for the tubes under the most adverse conditions that can be expected in service.

Mercury Temperature

The life and performance of a mercury-vapor rectifier are critically dependent on the temperature of the condensed mercury. Low ambient temperatures re-

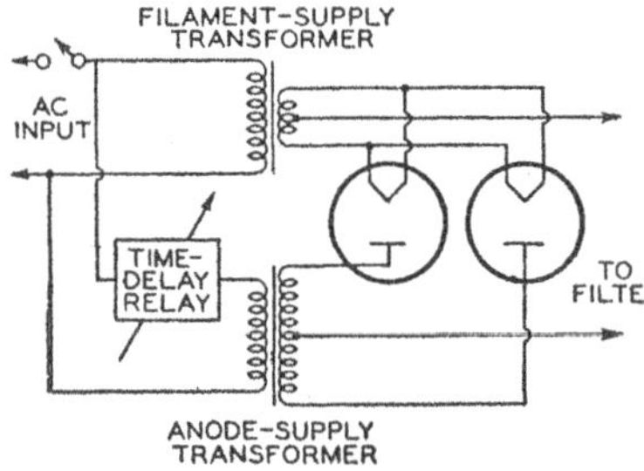

Fig. 50

tard vaporization of the mercury, thus limiting the degree of ionization available at normal filament voltage and raising the anode-cathode potential at which the tube starts to conduct. High ambient temperatures, on the other hand, are conducive to rapid vaporization, but tend to produce over-ionization and thus reduce the peak inverse anode voltage that the tube can withstand without breakdown. Rectifiers using mercury-vapor tubes, therefore, should be equipped with means for measuring condensed-mercury temperatures, and for maintaining these temperatures within limits specified for the tubes employed. Condensed-mercury temperature may be measured with a thermocouple or thermometer attached to the tube by means of a small amount of putty in a region near the bottom of the bulb. The proper measurement zone for each of the mercury-vapor tubes included in this Manual is shown in the *Outlines* Section.

The method used to control condensed-mercury temperature depends on the ambient-temperature conditions under which the tubes operate. If the ambient temperatures are near the minimum values specified in the tube data, some form of heat-conserving enclosure should be provided for the tubes. In extreme cases, it may also be necessary to employ electrical heating, together with suitable means for limiting the maximum temperatures developed. If ambient temperatures are above the maximum values specified in the tube data, forced-air cooling should be employed. The air flow should start when the anode voltage is applied to the tube, and should be directed horizontally onto the bulb about ½ inch above the base at the filament end of the tube. The air flow may be removed simultaneously with the anode voltage. The rise of mercury-vapor temperature above ambient temperature is given as a function of heating time under no-load and/or full-load conditions for mercury-vapor rectifier types in the *Tube Types* Section.

Shielding

Rectifier tubes, particularly mercury-vapor types, should be isolated from transformers and other components which produce strong external magnetic or electrostatic fields. Such fields are generally detrimental to tube life, tend to produce breakdown effects in mercury vapor, and frequently make it difficult to obtain adequate filtering of rectifier output. When tubes cannot be completely isolated from such fields, they should be enclosed in shields of the type described in the *Power-Tube Installation* Section. Mercury-vapor rectifier tubes used to supply transmitters or other types of rf power equipment should also be protected from large rf voltages. Such voltages should be prevented from entering rectifier circuits by rf filters such as that shown in Fig. 51.

Mercury-vapor rectifier tubes occasionally produce multi-frequency oscillations or "hash" which may cause interference in the af stages of associated

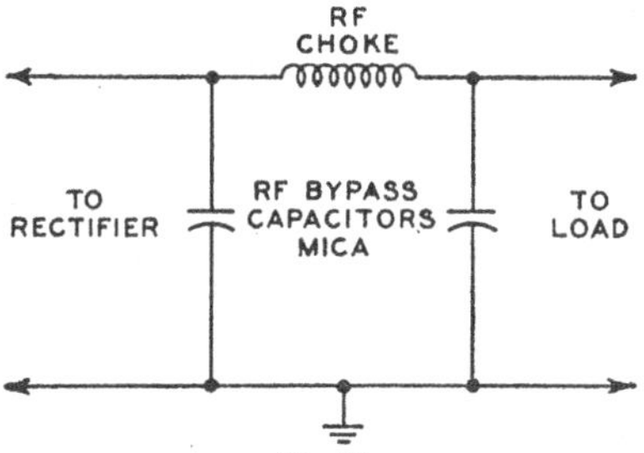

Fig. 51

equipment and in near-by radio receivers. These oscillations are caused by the development of a very steep wave front at the instant conduction begins in each rectifier unit, and may be propagated along internal circuit wiring and external power lines or radiated directly by the tubes. In a receiver, rectifier "hash" can usually be identified as a broadly tunable signal modulated at the rectifier "**ripple**" frequency. (The "ripple" frequency is equal to the power-line frequency times the number of half-wave rectifier units conducting independently.)

In some cases, this type of interference can be minimized by the use of very short leads to the rectifier anodes. It is usually necessary, however, to determine whether the interference is transmitted by radiation or by conduction, and to select the most effective method for its elimination by experiment. Radiation of such interference can usually be minimized by shields of the type used to protect rectifier tubes against external fields. The transfer of such interference to a power line can be minimized by the insertion of a low-pass inductance-capacitance filter in the input circuit of the rectifier, as shown in Fig. 52, or by the use of filament and high-voltage supply

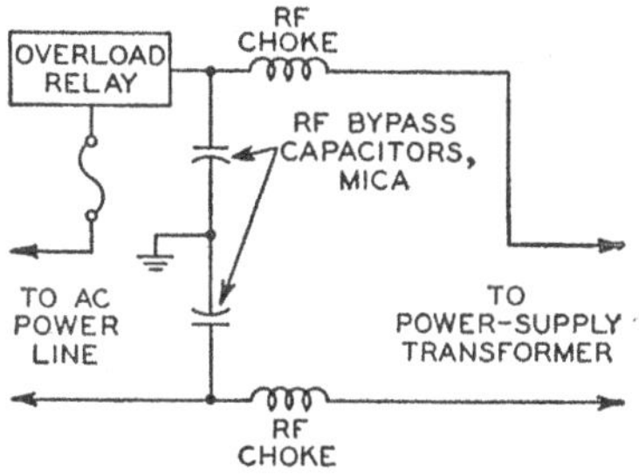

Fig. 52

transformers having electrostatic shields between primary and secondary windings. Low-pass filters of the type shown in Fig. 53 are also useful. The bypass capacitors used in such filters must have a voltage rating at least equal to the peak voltage developed across each half of the transformer secondary (approximately 1.4 times the rms voltage).

Rectifier tubes operated in circuits in which peak inverse voltages are 16000 volts or higher produce X-rays. Because

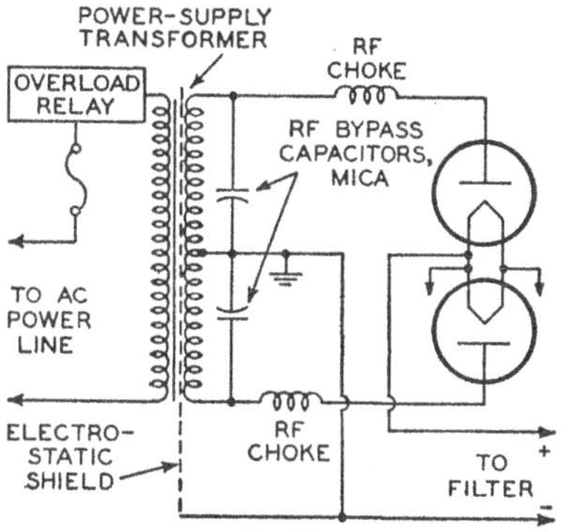

Fig. 53

these rays constitute a serious health hazard, tubes operated in such circuits should be equipped with shielding designed to absorb X-ray radiation.

RCA mercury-vapor and inert-gas rectifier tubes are equipped with internal cathode shields. These shields are connected to a filament or heater terminal designated as the "cathode-shield" or "anode-return" terminal. When two or more gas-rectifier tubes are operated from a common filament or heater supply, the cathode-shield or anode-return terminals of the tubes must be connected to the same side of the supply.

Tube Ratings

Rectifier-tube ratings usually include maximum permissible values for peak inverse anode voltage, peak anode current, average anode current, and fault anode current. Before these ratings are defined and their application to rectifier circuit design is discussed, it is desirable to define certain other terms frequently used in connection with rectifiers.

Forward voltage is voltage applied between the anode and cathode in the direction in which the tube is designed to pass current, *i.e.*, anode positive with respect to cathode. **Inverse voltage** is voltage applied between the anode and cathode in the direction opposite to that in which the tube is designed to pass current, *i.e.*, anode negative with respect to cathode.

Forward current is current flowing through a rectifier as a result of the application of a forward voltage. **Reverse current** is current flowing through a rectifier in the direction opposite to that of normal conduction. The flow of reverse current in a rectifier is an abnormal condition.

Peak inverse anode voltage is the highest instantaneous voltage applied between the anode and cathode during the fraction of any input cycle when the tube is normally not conducting. A maximum peak-inverse-voltage rating indicates the highest value this voltage may attain without danger of arc-back in the tube, electrolysis of glass, and reduced tube life.

Peak anode current is the highest instantaneous value reached by the forward current during the normal conduction interval. A maximum peak-anode-current rating indicates the highest current the tube can safely conduct during this interval. The peak current is determined by the duration of the conduction interval and, therefore, depends on the type of rectifier circuit in which the tube is employed.

Average anode current is the value obtained by integrating the instantaneous anode currents of a rectifier tube over a specified time and averaging the result. A maximum average-anode-current rating indicates the highest average current that should be permitted to flow through the tube in the direction of normal conduction. This current may be measured by means of a dc meter inserted in the anode circuit of the tube. When the rectifier load is constant, the average anode current may be read directly on the meter. When the rectifier load is varying, the meter readings should be averaged over the period specified in the tube data (usually 15 to 30 seconds).

Fault anode current is the highest current flowing through a rectifier tube in the forward direction under abnormal or fault conditions, *e.g.*, during a load short circuit or an arc-back in an associated tube. A maximum fault-current rating indicates the highest current that should be permitted to flow through the tube in the direction of normal conduction over a period not exceeding 0.1 second under fault conditions. Rectifier circuits should be designed to limit fault currents to values within the maximum ratings because even a single fault current of the maximum value will materially shorten or terminate the life of the tube.

Rectifier tubes of the same type can be connected in parallel to provide increased output current. When mercury-vapor or inert-gas types are operated in parallel, it is necessary to employ a resistor or a small inductance in the anode circuit of each tube to assure equal division of the total load current. Stabilizing resistors for high-voltage circuits should produce an average voltage drop of not less than 50 volts. Stabilizing inductors should have a value of approximately one-sixth henry each for a supply frequency of 50 to 60 cycles per second. Stabilizing inductors are generally preferable to resistors because they minimize power losses and help to limit the peak anode currents in the tubes. Center-tapped inductors (interphase reactors) can be used as stabilizing elements for pairs of parallel tubes. These inductors assure simultaneous starting as well as equal division of current. Vacuum rectifier tubes do not generally require the use of stabilizing devices when operated in parallel.

Corresponding filament terminals of mercury-vapor or inert-gas rectifiers operated in parallel must be connected together. Failure to observe this precaution will seriously unbalance the voltage drops in the paralleled tubes and may make it necessary to use undesirably high stabilizing impedances.

Circuits

The most suitable type of rectifier circuit for a particular application depends on the dc voltage and current requirements, the amount of rectifier "ripple" that can be tolerated in the output, and the type of ac power available.

The half-wave single-phase circuit shown in Fig. 54 delivers only one pulse of current for each cycle of the ac input

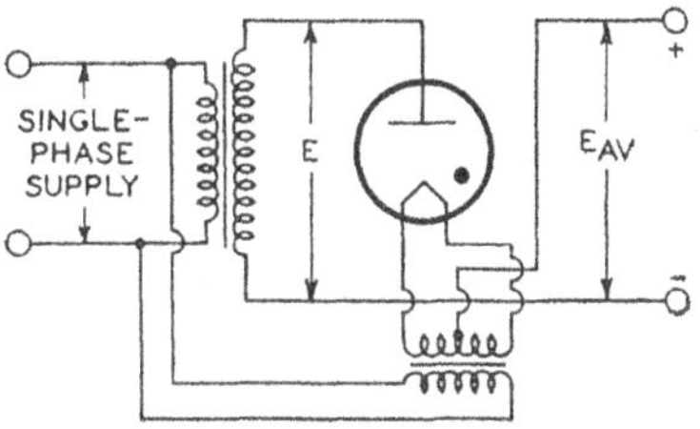

Fig. 54

voltage. Because its output contains a very high percentage of ripple, this type of circuit is used principally in low-voltage, high-current applications (*e.g.*, in power supplies for ac/dc receivers) and in low-current, high-voltage applications (*e.g.*, in ultor-voltage supplies for kinescopes and other types of cathode-ray tubes).

A full-wave single-phase circuit using two half-wave rectifier tubes is shown in Fig. 55, and a series single-phase circuit in Fig. 56. Although the bridge circuit requires four half-wave rectifier tubes and three filament transformers (or three independent filament windings), it can deliver twice as much output voltage as the two-tube circuit for the same anode-transformer voltage, and does not require a center-tapped high-voltage winding.

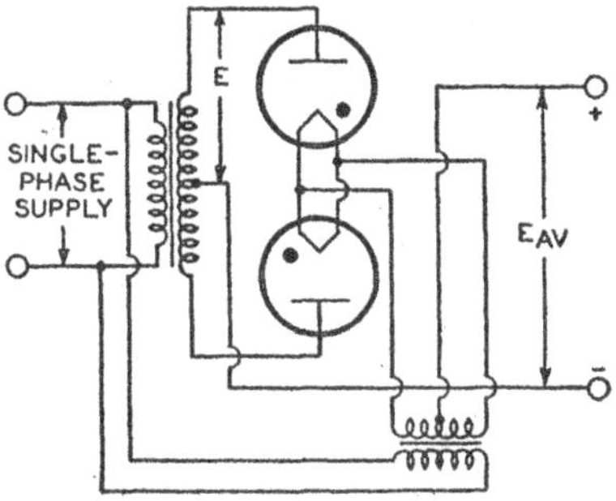

Fig. 55

Fig. 57 shows a half-wave three-phase circuit using three rectifier tubes. This circuit delivers three current pulses per cycle and its output, therefore,

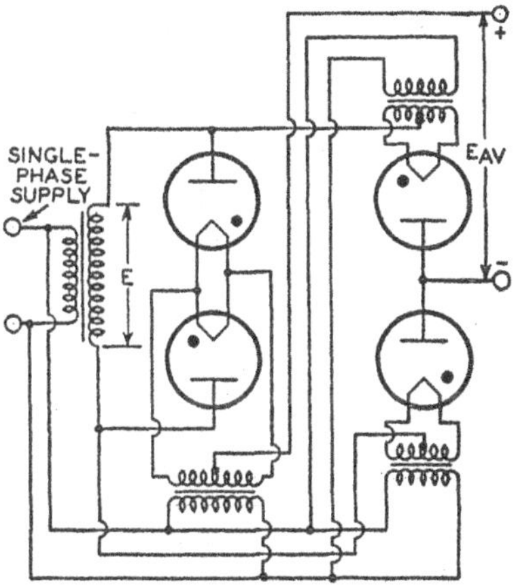

Fig. 56

contains a smaller percentage of ripple than that of a full-wave single-phase circuit.The parallel three-phase circuit employing six half-wave rectifier tubes

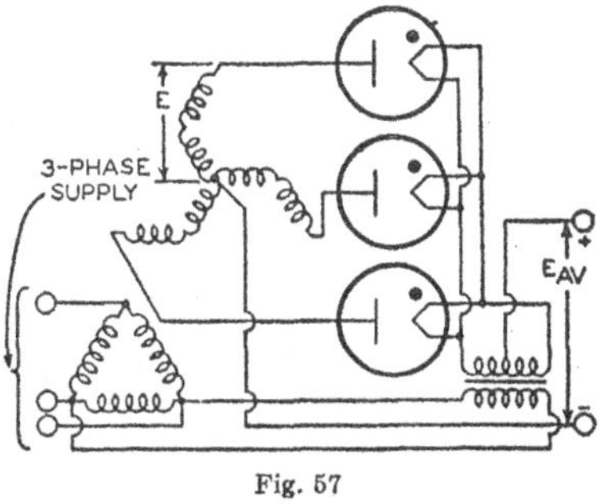

Fig. 57

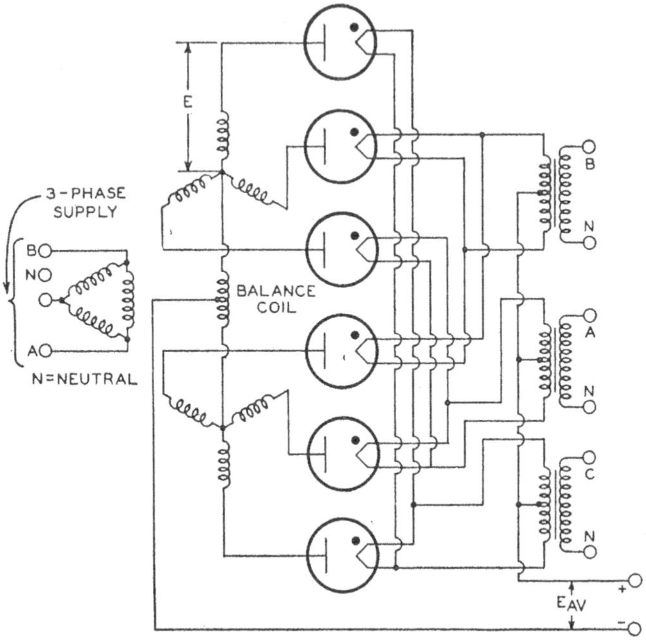

Fig. 58

shown in Fig. 58 delivers six current pulses per cycle. This circuit delivers twice as much output current as the circuit shown in Fig. 57 for the same average anode current per tube. The balance coil used in this circuit assures equal division of the load current and proper phasing in (or simultaneous starting of) the parallel branches.

In the series three-phase circuit shown in Fig. 59, two half-wave rectifier tubes are connected in series across each leg of the high-voltage transformer. This circuit delivers twice as much output voltage as the half-wave three-phase circuit shown in Fig. 57 for the same transformer voltage and peak inverse anode voltage per tube. Figs. 60 and 61 show

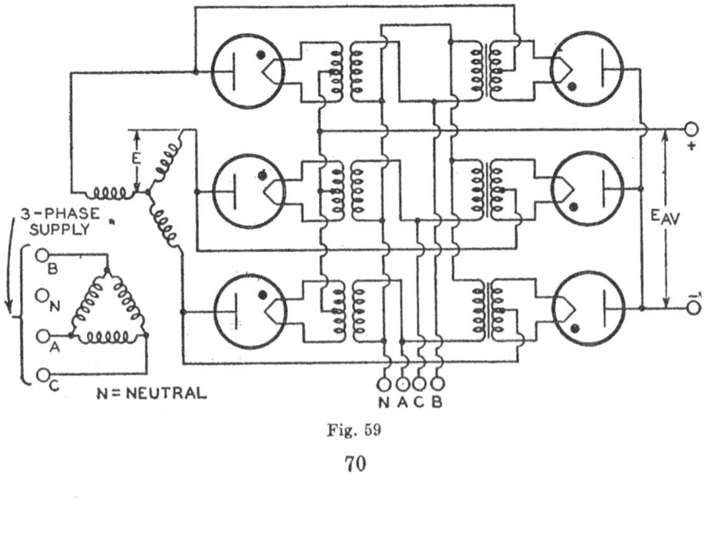

Fig. 59

half-wave four-phase and six-phase circuits, respectively.

Quadrature Operation

The filament current of a rectifier tube is composed of two components: the normal heating current supplied by the filament transformer, and the anode current, the greater part of which flows through the most negative portion of the filament. When the filament-supply voltage and anode voltage of a rectifier are in phase (the normal relationship when both voltages are obtained from the same ac supply line), the two components of the filament current reach peak value simultaneously during each conduction interval, and cause a localized increase in filament temperature which may seriously shorten the life of the tube.

In single-phase rectifier circuits, which have a conduction interval per tube of 180 degrees, the ratio of peak anode current to peak filament-supply current is relatively small and the effects of "in-phase" operation are usually negligible. In polyphase rectifier circuits having conduction intervals per tube of 120 degrees or less, however, the ratio of peak anode current to peak filament-supply current is relatively large, and

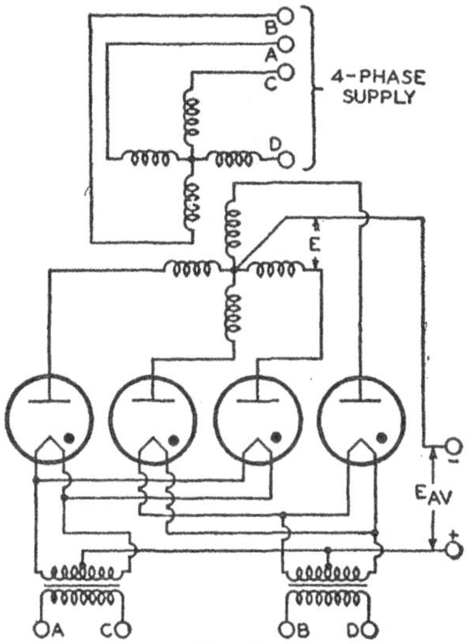

Fig. 60

the use of in-phase filament and anode voltages may result in extremely short tube life.

This difficulty can be minimized by the use of "Quadrature Operation." In this method of operation, the peak value of the total filament current is minimized

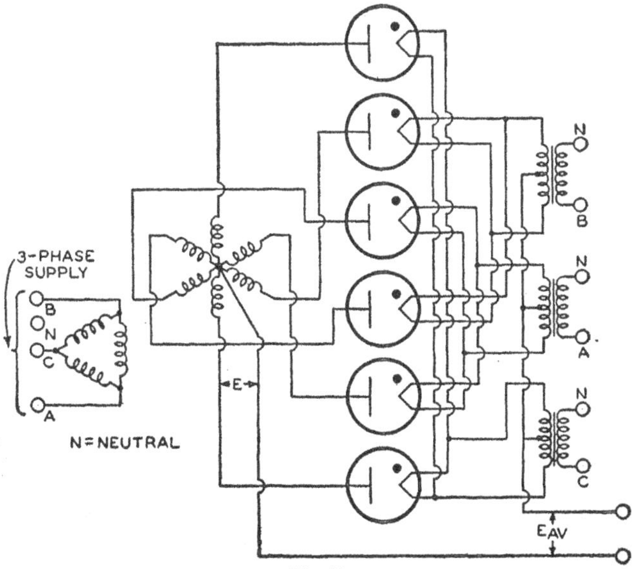

Fig. 61

by supplying the filament of each rectifier tube with voltage out of phase with its anode voltage. Although the ideal phase relationship between filament-supply voltage and anode voltage is 90 degrees (true "Quadrature"), substantial benefits are also realized at phase angles of 60 or 120 degrees, which are readily obtainable in three-phase and six-phase rectifier circuits.

Table IV gives the voltage, frequency, current, and power ratios for the basic rectifier circuits shown in Figs. 54 through 61. These ratios apply for sinusoidal ac input voltages. Current and power ratios given for inductive loads apply only when a filter choke is used between the output of the rectifier and any capacitor in the filter circuit. This table does not take into consideration voltage drops which occur in the power transformer, the rectifier tubes, or the filter components under load conditions. When a particular tube type has been selected for use in a specific rectifier circuit, the ratios given in Table IV can be used in conjunction with the tube data to determine the parameters and characteristics of the circuit.

Example of the Use of Table IV

Problem. Select the most suitable type of rectifier tube for use in a full-wave single-phase circuit which must de-

TABLE IV

RATIO	Fig. 54	Fig. 55	Fig. 56	Fig. 57	Fig. 58*	Fig. 59	Fig. 60	Fig. 61
Voltage Ratios								
E/E_{av}	2.22	1.11	1.11	0.854	0.854	0.427	0.785	0.74
E_{bmi}/E	1.41	2.83	1.41	2.45	2.45	2.45	2.83	2.83
E_{bmi}/E_{av}	3.14	3.14	1.57	2.09	2.09	1.05	2.22	2.09
E_m/E_{av}	3.14	1.57	1.57	1.21	1.05	1.05	1.11	1.05
E_r/E_{av}	1.11	0.472	0.472	0.177	0.04	0.04	0.094	0.04
Frequency Ratio								
f_r/f	1	2	2	3	6	6	4	6
Current Ratios								
I_b/I_{av}	1	0.5	0.5	0.33	0.167	0.33	0.25	0.167
Resistive Load								
I_p/I_{av}	1.57	0.785	0.785	0.587	0.294	0.587	0.503	0.408
I_{pm}/I_{av}	3.14	1.57	1.57	1.21	0.52	1.05	1.11	1.05
I_{pm}/I_b	3.14	3.14	3.14	3.63	3.14	3.14	4.5	6.3
Inductive Load ■								
I_p/I_{av}	–	0.707	0.707	0.577	0.289	0.577	0.500	0.408
I_{pm}/I_{av}	–	1	1	1	0.5	1	1	1
Power Ratios								
Resistive Load								
P_{as}/P_{dc}	3.49	1.74	1.24	–	–	–	–	–
P_{ap}/P_{dc}	2.69	1.23	1.24	–	–	–	–	–
P_{al}/P_{dc}	2.69	1.23	1.24	–	–	–	–	–
Inductive Load ■								
P_{as}/P_{dc}	–	1.57	1.11	1.71	1.48	1.05	1.57	1.81
P_{ap}/P_{dc}	–	1.11	1.11	1.21	1.05	1.05	1.11	1.29
P_{al}/P_{dc}	–	1.11	1.11	1.21	1.05	1.05	1.11	1.05

* Bleeder current of 2-per-cent full-load current will provide exciting current for balance coil and thus avoid poor regulation at light loading.

■ The use of a large filter-input choke is assumed.

E=transformer secondary voltage (rms)
E_{av}=average dc output voltage
E_{bmi}=peak inverse anode voltage
E_m=peak dc output voltage
E_r=major ripple voltage (rms)
I_{av}=average dc output current
I_b=average anode current
I_p=anode current (rms)
I_{pm}=peak anode current
f=supply frequency
f_r=major ripple frequency
P_{al}=line volt-amperes
P_{ap}=transformer primary volt-amperes
P_{as}=transformer secondary volt-amperes
P_{dc}=dc power ($E_{av} \times I_{av}$)

NOTE: Conditions assumed include sine-wave supply, zero voltage drop in tubes, no losses in transformer and circuit, no back emf in the load circuit, and no phase-back.

liver a dc voltage (E_{av}) of 2500 volts at an average dc current (I_{av}) of 500 milliamperes to the input of a filter. Also determine the rms voltage (E) that must be delivered by each half of the high-voltage transformer secondary winding.

Procedure. (1) Determine the maximum peak inverse anode voltage which each rectifier tube must withstand. From Table IV, the ratio of peak inverse voltage (E_{bmi}) to dc output voltage in single-phase full-wave circuits is 3.14.

$E_{bmi} = 3.14 \times 2500 = 7850$ volts.

(2) Determine the average anode current (I_b) in each tube. From Table IV, I_b in a full-wave single-phase circuit is one-half the total dc output current.

$I_b = 0.5 \times 500 = 250$ milliamperes.

(3) Select a tube having suitable voltage and current ratings from the Rectifier-Tube Selection Guide in the *Charts* Section. The 866-A, which has a maximum peak-inverse anode-voltage rating of 10000 volts and a maximum average-anode-current rating of 250 milliamperes, meets the requirements. (Although the 872-A, which has a maximum peak-inverse anode-voltage rating of 10000 volts and a maximum average-anode-current rating of 1.25 amperes, would also be satisfactory, the 866-A is the more economical type for this application.)

(4) Determine the rms voltage (E) which must be developed by each half of the high-voltage transformer secondary for the rectifier to deliver 2500 volts dc to the filter at the specified load current of 500 milliamperes under full-load conditions.

$E = 1.11 \times (2500 + 15) = 2790$ volts (1)

The second term within the parentheses represents the voltage drop in the 866-A. For exact calculation of E, the full-load voltage drop in one half of the high-voltage secondary winding must also be added to the values within the parentheses.

Regulation

The voltage drops in filter-choke windings or current-limiting resistors which follow the rectifier, as well as those in the rectifier tubes and transformer windings, become a very important consideration when a rectifier filter is required to supply a varying load. Except for the drop in a gas-tube rectifier, which is substantially constant at all anode-current values up to the maximum rating for the tube, these drops vary with load current and cause a corresponding variation in output voltage. This variation is known as the **voltage regulation** of the supply, and is usually expressed as the per-cent change in output voltage for load-current variations between zero and the maximum value. For example, a power supply which has a no-load output of 1000 volts and a full-load output of 900 volts has a voltage regulation of 10 per cent. The regulation of well-designed rectifier-type power supplies is usually 10 per cent or less.

For good voltage regulation, the voltage drops in all sections of the supply should be held to a minimum. Voltage drops can be minimized by the use of transformers and chokes having generous overload ratings and low-resistance windings, mercury-vapor or inert-gas rectifier tubes or vacuum types having close anode-cathode spacing, and choke-input filters employing "swinging" chokes of the proper value. In addition, a "bleeder" resistor drawing about 10 per cent of the total output current should be permanently connected across the output of the supply. Although this resistor reduces the maximum useful output current slightly, it prevents the output voltage from rising excessively when the external load is reduced, and thus improves regulation and provides a substantial measure of protection for the filter capacitors. It also discharges the filter capacitors when the equipment is switched off and thus minimizes shock hazards.

Good regulation is desirable even when substantially constant output voltage under varying load conditions is not a primary requirement. Because good regulation minimizes variations in the voltage across the output terminals of a power supply, its effect is similar to that obtained when a very large bypass capacitance is connected across the output of the supply, *i.e.*, the amount of ac ripple in the output is substantially reduced. The internal impedance of the supply is also reduced, so that there is less danger of undesirable coupling and feedback in associated equipment when the supply is used for two or more stages.

Filters

The filter employed to minimize ripple in the output of a rectifier may be either a choke-input or a capacitor-input type. Careful consideration must be given to the selection and design of the filter if the maximum ratings of the tubes are not to be exceeded.

One of the most important considerations in the choice and design of a filter is its effect on the peak current in the rectifier circuit, and particularly on the current surge which occurs when the rectifier circuit is turned on. The sudden application of anode voltage to a rectifier causes a sudden flow or surge of current. The maximum value of this current is determined by the instantaneous amplitude of the ac input voltage and the surge impedance of the rectifier circuit. If the rectifier output is shunted by a large capacitor, the surge impedance is low and, therefore, the surge current may reach dangerously high values. On the other hand, if a relatively large choke is connected between the rectifier and the first filter capacitor, the surge impedance is high, and the surge current usually does not exceed the normal peak current through the tubes.

Choke-input filters limit surge and normal peak currents and, therefore, make it possible to obtain maximum continuous dc output current from rectifier tubes under the operating conditions most favorable for long tube life. They also provide the best regulation and are especially recommended for use with rectifiers employing mercury-vapor and inert-gas tubes or vacuum tubes having closely spaced electrodes. An additional advantage of choke-input filters is that their performance can be predicted accurately by calculation.

Capacitor-input filters provide the highest dc output voltages obtainable from given transformers and rectifier-tube combinations. They cause high current surges when the circuit is turned on, however, and have poor voltage regulation. In addition, the dc load current obtainable from a given rectifier-tube-and-transformer combination is less when a capacitor-input filter is used than when a choke-input filter is used.

When a capacitor-input filter is used, a current-limiting resistor should be connected between the rectifier tubes and the filter to limit current surges. The total resistance, R_t, required to limit the surge current to a safe value, including the effective resistance of the power-transformer secondary (or one half of the secondary of a full-wave transformer) is a function of the dc output voltage (E_{av}) and the rated peak anode current (I_{pm}) of the tube.

$$R_t = \frac{K \times E_{av}}{I_{pm}}$$

The factor K is equal to 3.14 for the circuit shown in Fig. 54, 1.57 for the circuits shown in Figs. 55 and 56, 1.21 for the circuit of Fig. 57, 1.11 for Fig. 60, and 1.05 for Figs. 59 and 61. The balance coil used in the circuit shown in Fig. 58 limits the peak anode current so that a limiting resistor is not needed. The current-limiting resistor may be short-circuited after the rectifier-filter system has been switched on to avoid a reduction in useful dc output voltage. The resistor must be employed, however, each time the circuit is switched on. Capacitor-input filters may be used in rectifier circuits employing mercury-vapor or inert-gas rectifier tubes only when a current-limiting resistor is used as described above.

Design of Choke-Input Filters

The filter-design charts shown in Figs. 62 and 63 permit quick determination of inductance and capacitance values for choke-input filters for use with full-wave single-phase rectifier circuits operating from 60-cycle supplies. For other supply frequencies, the inductance and capacitance values indicated by these charts should be multiplied by the ratio 60/f, where f is the supply frequency used.

The chart shown in Fig. 62 is used to determine component values for single-section choke-input filters or for the first section of a multisection choke-input filter. Single-section and double-section choke input filters are shown in Fig. 64. The R_L curves in Fig. 62 are used to determine the minimum value of choke inductance required. The equivalent load resistance (R_L) in ohms is equal to the dc output voltage (E_{av}) of the rectifier in volts divided by the load current (I_b)

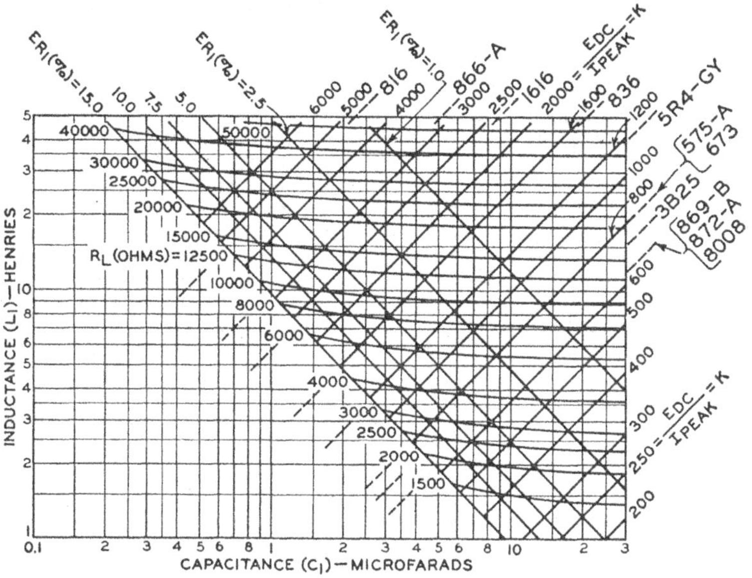

Fig. 62

in amperes. A dc output voltage equal to 90 per cent of the rms voltage (E) per rectifier-tube anode is used in this calculation (from Table IV, $E/E_{av} = 1.11$). This value does not include the voltage drops in the power transformer, filter choke, or rectifier tubes. The load current used must assure operation of each rectifier tube within its maximum average-anode-current rating. Inductance and capacitance values must always lie in the region of the chart above the applicable R_L curve.

The K curves in Fig. 62 indicate combinations of minimum filter inductance (L_1) and maximum filter capacitance (C_1) which will keep the peak anode currents (I_{pm}) of the rectifier tubes within their maximum ratings at a given rms anode voltage. The factor K is equal to the dc voltage from the rectifier tubes at the input to the filter (in volts) divided by the maximum peak-anode-current rating of the rectifier tubes (per anode, in amperes). The K curves shown in Fig. 62 represent the following relation:

$$L_1 = C_1 \times (K/1000)$$

Filter component values must always lie in the region of the chart to the left of the proper K line.

When a particular rectifier tube is used at its maximum peak-inverse-anode-voltage rating and maximum peak-anode-current rating simultaneously, the applicable K line may be determined directly by placing a ruler across the appropriate pair of dashed lines shown in Fig. 62. When a tube is used at voltages below its maximum peak-inverse anode-voltage rating, a lower value of K determined from the above equation must be used.

The R_L and K curves, therefore, indicate limiting values of inductance and capacitance which will assure that average and peak anode-current ratings of the rectifier tubes will not be exceeded. Filter-component values can now be chosen within the wedge-shaped portion of the chart outlined by the appropriate R_L and K curves on or above the E_{R1} line for the maximum percentage of ripple which can be tolerated in the output of the filter section.

In power supplies for cw transmitters, a ripple of not more than 5 per cent is usually satisfactory. Power supplies for variable-frequency oscillators and phone transmitters generally should have ripple of 0.25 per cent or less. Power-supply ripple in high-gain speech amplifiers and receivers should not exceed

0.1 per cent to prevent hum modulation of output signals.

The most economical method of obtaining ripple voltages below 1 per cent

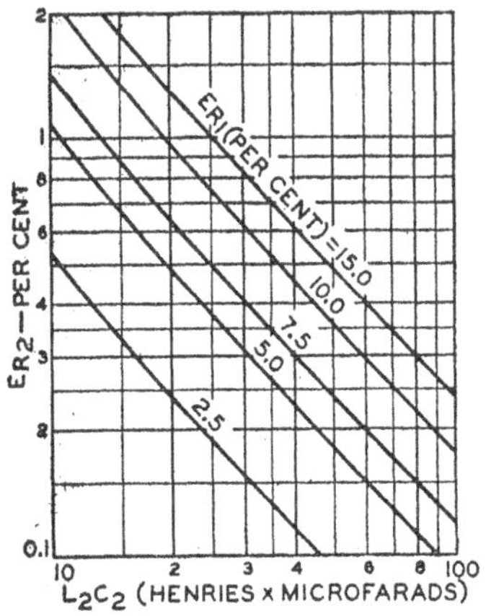

Fig. 63

is by the use of double-section filters of the type shown in Fig. 64(b). Values of L_2 and C_2 for the second section of such filters are determined from the chart shown in Fig. 63. After the value of E_{R1} for the first section is determined, the values of L_2 and C_2 (as a product) for any desired ripple percentage E_{R2} at the output of the second filter section may be determined from the appropriate E_{R1} curve in Fig. 63. Although any values of inductance and capacitance having the indicated product $L_2 \times C_2$ will provide

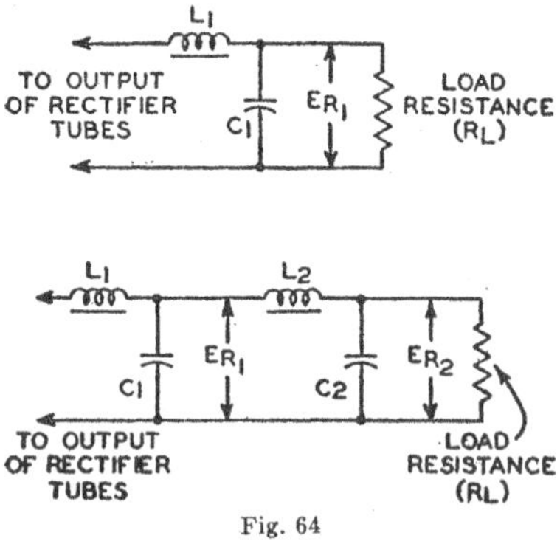

Fig. 64

the desired filtering, serious instability may result if the combination selected is resonant at or near the ripple frequency. The inductance of L_2, therefore, should always be greater than

$$\frac{3 \times (C_1 + C_2)}{2 \times (C_1 \times C_2)}$$

For applications in which the load resistance (R_L) varies over a wide range, some means should be used to limit the resulting variation in output voltage. A bleeder resistor may be inserted across the filter output to restrict the range over which the effective load varies or an input choke having an inductance determined by the maximum load resistance attained may be used. The most economical method for minimizing output-voltage variations, however, is by the use of a "swinging" input choke.

The inductance of a well-designed swinging choke varies inversely with load current. The required minimum and maximum inductance for the choke can be determined from Fig. 62 at the intersections of the appropriate K curve with the curves for maximum and minimum R_L. It is generally most economical to select low values of swinging-choke inductance and obtain the required smoothing by the use of additional filter sections employing non-swinging ("smoothing") chokes.

Examples of Filter Design

Single-Section Filter

Problem: A full-wave rectifier operating from a 60-cycle source and employing two 872-A mercury-vapor tubes has a dc output voltage of 3200 volts. Design a single-section choke-input filter which will (*a*) limit output ripple to 5 per cent at a load current equal to the combined maximum dc load-current ratings of the tubes ($2 \times 1.25 = 2.5$ amperes); (*b*) keep the peak anode current of each tube within its maximum peak-anode-current rating (5 amperes).

Procedure: $R_L = 3200/2.5 = 1280$ ohms. The value $K = 3200/5 = 640$. The curve for $K = 640$ in Fig. 62 would lie between the curves for $K = 600$ and $K = 800$ and, consequently, would be above the position where the curve for $R_L = 1270$ would be shown. Therefore, any combination of inductance and capacitance along the curve $E_{R1} = 5$ per cent to the left of $K = 640$ will satisfy the requirements. A 5-henry choke and a 5-microfarad capacitor would be a suitable combination.

Two-Section Filter

Problem: A 60-cycle full-wave rectifier employing two 866-A mercury-vapor tubes delivers 2500 volts dc at full load to the input terminals of the filter. Design a two-section filter which will (*a*) limit the output ripple to 0.5 per cent at a load current equal to the combined maximum dc load-current ratings of the tubes ($2 \times 0.25 = 0.5$ ampere); (*b*) keep the peak anode current of each tube within its maximum peak-anode-current rating (1.0 ampere). Because the voltage regulation must be good from no load to full load, the input choke shall be of the "swinging" type.

Procedure: At maximum load, $R_L = 2500/0.5 = 5000$ ohms. $K = (2500 \times 1.11)/1.0 = 2775$. Because the curve in Fig. 62 for $R_L = 5000$ ohms would be completely below the curve for $K = 2775$, the maximum-load value of R_L (minimum R_L) need not be considered in the selection of constants for the first filter section. If an E_{R_1} of 10 per cent at the output of the first filter section is assumed to be satisfactory, the minimum swinging-choke inductance and the corresponding value for the first-section filter capacitor are selected along the curve $E_{R_1} = 10$ per cent to the left of the curve for $K = 2775$. Suitable values would be $L_1 = 13.5$ henries and $C_1 = 1$ microfarad. The maximum inductance of the swinging choke should be as high as practical. If a maximum value of 25 henries is chosen, the minimum-load value of R_L (maximum R_L) at which the regulating action of the choke will be effective is indicated by the point at which the 1-microfarad line intersects the line for 25 henries. This point corresponds to an R_L of 26000 ohms. Therefore, a bleeder having a resistance of not more than 26000 ohms should be used to prevent the dc output voltage from rising excessively when the load is removed. The bleeder draws a current of 2500/26000, or 0.096 ampere, and is required to dissipate 2500×0.096, or 240 watts. Because the maximum average current which can be supplied by two 866-A's in a full-wave circuit is 0.5 ampere, the useful load current available from the rectifier filter combination is $0.500 - 0.096 = 0.404$ ampere, or 404 milliamperes.

The second filter section (L_2C_2) must reduce the ripple from the value of 10 per cent at the output of the first filter section to a value of 0.5 per cent. From Fig. 63, the value of the product L_2C_2 at the intersection of the curve for $E_{R_1} = 10$ per cent with the line for $E_{R_2} = 0.5$ per cent is 37. If C_2 is chosen to be 2 microfarads, then L_2 should have an inductance of 18.5 henries. The value chosen for L_2 should be checked to determine whether resonance effects will be present, *i.e.*, L_2 should be equal to, or greater than, $3\times(1+2)/[2\times(1\times2)]=9/4 = 2.25$. Because the value of 18.5 henries selected for L_2 is considerably greater than 2.25, the filter design is satisfactory.

Interpretation of Tube Data

The tube data given in the *Tube Types* Section include maximum ratings, typical operation values, characteristics, and characteristics curves.

A maximum rating, as applied to a tube, is a limit on a particular operating parameter (such as voltage, current, temperature, or frequency) or on a combination of parameters. Operation above these maximum ratings may not only impair the performance of a tube but also shorten its life considerably.

RCA power tubes may carry as many as three different kinds of ratings, based on operating conditions encountered in different types of service. The three general types of service may be defined as follows:

Continuous Commercial Service (CCS) covers applications involving continuous tube operation in which maximum dependability and long tube life are the primary considerations.

Intermittent Commercial and Amateur Service (ICAS) covers applications in which high tube output is a more important consideration than long tube life. The term "Intermittent Commercial" in this title applies to types of service in which the operating or "on" periods do not exceed 5 minutes each, and are followed by "off" or stand-by periods of the same or greater duration. The term "Amateur Service" covers other applications where operation is of an infrequent or highly intermittent nature, as well as the use of tubes in "amateur" transmitters. ICAS ratings generally are considerably higher than CCS ratings. Although the ability of a tube to produce greater output power is usually accompanied by a reduction in tube life, the equipment designer may decide that a small tube operated at its ICAS ratings meets his requirements better than a larger tube operated within CCS ratings.

Intermittent Mobile Service (IMS) covers applications in which very high power output for short periods is required from equipment of the smallest practical size and weight. Tube ratings for IMS service are based on the premise that transmitter "on" periods do not exceed 15 seconds each, and are followed by "off" periods of at least 60 seconds duration. In equipment tests, however, maximum "on" periods of not more than 5 minutes each followed by "off" periods of at least 5 minutes are permissible, provided the total "on" time of such test periods does not exceed 10 hours during the life of the tube. Although tubes operated under IMS ratings may have a life of only about 100 hours, the use of these ratings is economically justified where high power must be obtained intermittently from very small tubes.

Each maximum rating of a tube must be considered with respect to all other ratings given for that tube, so that the use of any one maximum rating will not cause any other maximum rating to be exceeded. For example, if the product of the maximum plate-voltage and maximum plate-current ratings exceeds the maximum permissible dc plate input, then either the plate voltage or the plate current, or both, must be reduced. As an illustration, the maximum CCS ratings for Class C Telegraphy operation of type 812-A are: plate volts, 1250 *max;* plate milliamperes, 175 *max;* plate input, 175 watts *max*. It is apparent that when the maximum plate voltage of 1250 volts is used, the dc plate current must be reduced to 140 milliamperes or less if operation is to be within the 175-watt maximum plate-input rating. On the other hand, if the maximum plate current of 175 milliamperes is to be used, it will be necessary to reduce the plate voltage to 1000 volts or less to avoid exceeding the 175-watt maximum input rating.

The tube ratings given in this Manual are "**Absolute Maximum**" ratings, unless otherwise indicated. The equipment designer must select operating values which are sufficiently below these absolute-maximum ratings so that no rating will ever be exceeded under any usual condition of supply-voltage variation, load variation, or manufacturing variation in the equipment itself.

A few of the low-power tubes listed in this Manual are rated under the "**Design-Center**" system. This system, which is used principally for tubes intended for home-instrument applications, is designed to provide satisfactory average performance in the greatest number of equipments on the premise that they will not be adjusted to local power-supply conditions at time of installation. Equipment for use on ac or dc power lines should be designed so that the design-center maximum values are not exceeded at a line-voltage-center value of 117 volts. In equipment designed for use with storage-battery-with-charger supply or similar supplies, plate voltages, screen-grid supply voltages, dissipations, and rectifier output currents should never exceed 90 per cent of the design-center maximum ratings for a terminal potential at the battery source of 2.2 volts per cell. Equipment for use with "B" batteries should be designed so that under no condition of battery voltage will the plate voltages, screen-grid supply voltages, or dissipations ever exceed the maximum rated values by more than 10 per cent.

Values shown in tube data under "Typical Operation" should not be interpreted as ratings. These values represent operating conditions within the maximum ratings of a tube that are suitable for a particular application, and do not imply that the tube cannot be operated satisfactorily under other conditions in the same application. The choice of the most suitable tube operating conditions for any particular application should be based on a careful consideration of all pertinent factors.

The values for grid-bias voltages, other electrode voltages, and electrode supply voltages are given with reference to a specified datum point as follows: For tube types having filaments heated with dc, the negative filament terminal is taken as the datum point to which other electrode voltages are referred. For types having filaments heated with ac, the filament mid-point (*i.e.*, the center tap on the filament-transformer secondary, or the mid-point on a resistor shunting the filament) is taken as the datum point. For types having indirectly heated unipotential cathodes, the cathode is taken as the datum point.

Electrode voltage and current ratings are in general self-explanatory, but a brief explanation of other ratings will aid in the understanding and interpretation of tube data.

Plate Input is the total power supplied to the plate. It is the product of the dc plate voltage (E_b) and the direct current flowing in the plate circuit (I_b).

Plate Dissipation is the power lost in the form of heat as a result of electron bombardment of the plate. It is the difference between the power supplied to the plate of the tube (plate input) and the power delivered by the tube to the load circuit.

Power Output is the output obtainable from the tube itself and is equal to plate-input power minus plate dissipation. The useful power actually delivered to the tube load, however, depends on the circuit efficiency, the operating frequency, and other variable factors.

Grid-No. 2 (Screen-Grid) Input is the dc power supplied to the screen grid of a multigrid tube, and is the product of the screen-grid voltage and screen-grid current. This power is dissipated in the form of heat by the screen grid as a result of electron bombardment.

Grid (or Grid-No. 1) Driving Power is the actual signal-power input to the control grid plus the power lost in the bias supply. It is given by the formula $W_d = 0.9 E_g I_c$, where W_d is the grid driving power in watts, E_g is the peak signal voltage applied to the grid in volts, and I_c is the average grid current in amperes. This value does not include signal-power losses that occur in the tube, grid-tank circuit, socket, or wiring, or tube losses caused by electron transit-time effects (except where the value given in the tube data is for a specific operating frequency).

Peak Heater-Cathode Voltage ratings are given only for tubes that have separate cathode and heater terminals. These ratings indicate the highest instantaneous voltage that may be applied between a heater and cathode without breakdown of the insulation between these electrodes.

Charts

RCA transmitting tubes are classified in this Section according to the types of service for which they are designed. The maximum frequency for full input is given in Charts I and II. Most tube types, however, can be operated above this frequency provided the plate voltage and plate input are reduced. Chart I shows the relationship between operating frequency and the maximum permissible percentage of maximum rated plate voltage and plate input.

1. Power Tubes for Class C Telegraphy Service#

Type No.	Maximum Plate Ratings (per tube) *Absolute Values Except as Noted* — CCS: Dissipation Watts	CCS: DC Volts	CCS: Input Watts	ICAS: Dissipation Watts	ICAS: DC Volts	ICAS: Input Watts	Per Cent of Maximum Plate Volts and Input for Indicated Frequencies: %	Mc	Type No.
TRIODES:									
958-A	0.6■	185■	0.95■	–	–	–	–	–	958-A
3A5	1	135	2	–	–	–	100	40	3A5
6F4	2■	150■	–	–	–	–	–	–	6F4
6026	3	150	3.3	–	–	–	100	400	6026
5876	6.25	360	9	–	–	–	100	1700	5876
5893	7	320	11	8	400	16	100	1000	5893
6263	8	300	13	13	400	22	100	500	6263
6264	8	330	13	13	400	22	100	500	6264
5556	10	350	14	–	–	–	100	6	5556
							75	15	
							50	30	
809	25	750	75	30	1000	100	100	60	809
							88	70	
							50	120	
8025-A	40▲	1000▲	75▲	30	1000	50	100	500	8025-A
							70•	600	
811-A, 812-A	45	1250	175	65	1500	260	100	30	811-A, 812-A
							89	60	
							70	80	
							55	100	
826	45	1000	95	55	1000	130	100	250	826
							80	300	
808	50	1500	200	75	2000	300	100	30	808
							75	60	
							50	130	
834	50	1250	125	–	–	–	100	100	834
							80	170	
							53	350	
826▲	60	1000	125	75	1250	175	100	250	826▲
							80	300	
8005	75	1250	240	85	1500	300	100	60	8005
							75	80	
							60	100	
2C39-A	100	1000	–	–	–	–	100	2500	2C39-A
810, 8000	125	2000	500	175	2500	750	100	30	810, 8000
							70	60	
							50	100	
805	125	1500	315	–	–	–	100	30	805
							82	45	
							55	80	
5713	250	1500	450	–	–	–	100	220	5713
6161▲	250	1600	400	–	–	–	100	900	6161▲
							80	1200	
							71	1400	
							62.5	1650	
							62.5	2000	
833-A	300	3000	1250	350	3300	1500	100	30	833-A
							90	50	
							72	75	
833-A▲	400	4000	1800	450	4000	2000	100	20	833-A▲
							83	50	
							65	75	
5786	600	3000	1500	–	–	–	100	160	5786
6383▲	600	1500	600	–	–	–	100	2000	6383▲

Ratings apply also for Class C FM Telephony Service.
■ Design-Center Value.
▲ With forced-air cooling.
• Refers to plate volts only.
♦ Refers to plate input only.
■ Push-pull type.

1. Power Tubes for Class C Telegraphy Service (cont.)#

Type No.	Maximum Plate Ratings (per tube) *Absolute Value Except as Noted* CCS Dissipation Watts	CCS DC Volts	CCS Input Watts	ICAS Dissipation Watts	ICAS DC Volts	ICAS Input Watts	Per Cent of Maximum Plate Volts and Input for Indicated Frequencies %	Mc	Type No.
PENTODES:									
5618	–	–	–	5	300	– –	100 90♦	100 165	5618
802	10	500	25	13	600	7.5	100 77 55	30 55 100	802
1613	10	350	17.5	–	–	33	100 90 85	45 60 90	1613
BEAM POWER TUBES:									
2E24 2E26	10	500	30	13.5	600	–	100 83 75 68	125 150 160 175	2E24 2E26
837	12	500	32	–	–	40	100 76 62	20 40 60	837
5763 6417	12	300	15	13.5	350	17	100 80♦	50 175	5763 6417
832-A□	15	750	36	20	750	50	100 89	200 250	832-A□
6146 6159 6883	20	600	67 5	25	750	90	100 79♦ 67• 66♦ 53•	60 120 175	6146 6159 6883
6524□ 6850	20	500	70	25	600	85	100 78♦ 79• 51♦ 76•	100 220 470	6524□ 6850
815□	20	400	60	25	500	75	100 80 70	125 175 200	815□
807 1625	25	600	60	30	750	75	100 80 55	60 80 125	807 1625
829-B□	30	750	90	40	750	120	100 89	200 250	829-B□
829-B□▲	40	750	120	45	750	150	100 89	200 250	829-B□▲
5894□	40	600	120	–	–	–	100 96 90 83	250 300 400 500	5894□
814	50	1250	180	65	1500	225	100 80 64	30 50 75	814
4-65A	65	3000	–	–	–	–	100	50	4-65A
828	70	1250	200	80	1500	270	100 80 65	30 50 75	828
4E27/ 8001	75	4000	300	–	–	–	100 75 50	75 120 150	4E27/ 8001
813	100	2000	360	125	2250	500	100 87 75 50	30 45 60 120	813
4E27A/ 5-125B	125	4000	–	–	–	–	100	75	4E27A/ 5-125B
4-125A/ 4D21	125	3000	–	–	–	–	100 80• 64• 56•	120 150 200 250	4-125A/ 4D21
4X150A▲ 4X150D▲	150	1250	–	–	–	–	100	500	4X150A▲ 4X150D▲
4-250A/ 5D22	250	4000	–	–	–	–	100 85• 74•	110 125 150	4-250A/ 5D22
4X500A▲	500	4000	–	–	–	–	100	120	4X500A▲
827-R▲	800	3500	1500	–	–	–	100	110	827-R▲

■ ▲ • ♦ □ See preceding page.

II. Power Tubes for Plate-Modulated Class C Telephony Service

Type No.	Maximum Plate Ratings (per tube) *Absolute Values* CCS Dissipation Watts	CCS DC Volts	CCS Input Watts	ICAS Dissipation Watts	ICAS DC Volts	ICAS Input Watts	Maximum Frequency For Full Input⊕ Mc	Type No.
TRIODES:								
5893	5	260	8.5	5.5	320	10.5	1000	5893
6263	5.5	275	9	9	330	15	500	6263
8025-A	27▲	800▲	50▲	20	800	33	500	8025-A
811-A 812-A	30	1000	115	45	1250	175	30	811-A 812-A
826	30	800	60	45	1000	95	250	826
826▲	40	800	75	60	1000	125	250	826▲
8005	50	1000	160	75	1250	240	60	8005
2C39-A	70	600	–	–	–	–	2500	2C39-A
810 8000	85	1600	335	125	2000	500	30	810 8000
6161▲	167	1300	270	–	–	–	900	6161▲
833-A	200	2500	835	250	3000	1000	30	833-A
833-A▲	270	3000	1250	350	4000	1800	20	833-A▲
5786▲	400	2500	1000	–	–	–	160	5786▲
6383▲	400	1200	400	–	–	–	2000	6383▲
BEAM POWER TUBES:								
2E24 2E26	6.7	400	20	9	500	27	125	2E24 2E26
5763 6417	8	250	10	12	300	15	50	5763 6417
832-A□	10	600	22	15	600	36	200	832-A□
6146 6159 6883	13.3	480	45	16.7	600	67.5	60	6146 6159 6883
6524□ 6850□	13.5	400	45	16.7	500	55	100	6524□ 6850□
815□	13.5	325	40	20	400	60	125	815□
807 1625	16.5	475	40	25	600	60	60	807 1625
829-B□	21	600	67.5	28	600	90	200	829-B□
829-B□▲	28	600	90	40	600	120	200	829-B□▲
5894□	27	450	72	–	–	–	250	5894□
4-65A	45	2500	–	–	–	–	50	4-65A
828	47	1000	135	70	1250	200	30	828
813	67	1600	240	100	2000	400	30	813
4-125A/ 4D21	85	2500	–	–	–	–	120	4-125A/ 4D21
4X150A 4X150D	100	1000	–	–	–	–	500	4X150A 4X150D
4-250A/ 5D22	165	3200	–	–	–	–	110	4-250A/ 5D22

⊕ Reduction in ratings at higher frequencies are given in Chart I, **Power Tubes For Class C Telegraphy Service**.

▲ With forced-air cooling.
□ Push-pull type.

III. Power Tubes for AF Power Amplifier and Modulator Service

Maximum Plate Ratings (per tube)
Absolute Values

Type No.	CCS Dissipation Watts	CCS DC Volts	CCS Input Watts	ICAS Dissipation Watts	ICAS DC Volts	ICAS Input Watts	Class of Service	Type No.
TRIODES:								
5556	7.5	350	–	–	–	–	A	5556
811-A 812-A	45	1250	165	65	1500	235	B	811-A 812-A
8005	75	1250	225	85	1500	250	B	8005
845	100	1250	150	–	–	–	AB$_1$	845
810 8000	125	2500	425	175	2750	510	B	810 8000
805	125	1500	315	–	–	–	B	805
833-A▲	300 400	3500 4000	1125 1600	350 450	3300 4000	1300 1800	B	833-A
5786▲	600	4000	1500	–	–	–	B	5786
6383▲	600	1500	600	–	–	–	A	6383
PENTODES:								
5618	–	–	–	5	300	–	A$_1$	5618
802	15	500	15	18	600	18	A	802
BEAM POWER TUBES:								
2E26	10	400	30	12.5	500	37.5	AB$_2$	2E26
2E24	10	400	30	13.5	500	37.5	AB$_2$	2E24
6146 6159 6883	20	600	62.5	25	750	90	AB$_2$	6146 6159 6883
6524□ 6850□	20	500	70	25	600	85	AB$_2$	6524□ 6850□
815□	20	400	60	25	500	75	AB$_2$	815□
1614	21	375	40	25	550	60	AB$_1$	1614
807 1625	25	600	60	30	750	90	AB$_2$	807 1625
829-B□	30	750	100	–	–	–	AB$_1$	829-B□
5894□	40	600	120	–	–	–	B	5894□
4-65A□	65	3000	–	–	–	–	AB$_2$	4-65A□
828	70	1750	225	80	2000	270	AB$_1$	828
813	100	2250	360	125	2500	450	AB$_1$	813
4-125A/ 4D21	125	3000	–	–	–	–	AB$_2$	4-125A/ 4D21
4X150A 4X150D	150	1250	–	–	–	–	AB$_2$	4X150A 4X150D
4-250A/ 5D22	250	400	–	–	–	–	AB$_1$ AB$_2$	4-250A/ 5D22

▲ With forced-air cooling. □ Push-pull type.

IV. Power Tubes for Special Applications

Type No.	Description	Applications	Features
3C33	Twin Power Triode	Control Amplifier	
3E29	Twin Beam Power Tube	Rectangular-Wave Pulse Modulator	For use with duty factors between 0.0001 and 1.0 at a maximum averaging time of 1200 microseconds.
4C33	Power Triode	Class C Plate-Pulsed Oscillator	Compact, forced-air-cooled radiator type used with full input up to 625 Mc.
5618	Power Pentode	Frequency Multiplier	Seven-pin miniature type used as doubler or tripler up to 80 Mc.
5763	Beam Power Tube	Frequency Multiplier	Nine-pin miniature type used as doubler or tripler up to 175 Mc.
5794	Fixed-Tuned Oscillator Triode	Radiosonde Service	Pencil type having integral resonators for use at 1680 Mc.
5893	Medium-Mu Triode	Plate-Pulsed Oscillator and Frequency Doubler	Pencil type used as oscillator up to 3300 Mc and as doubler up to 1000 Mc.
5946	Power Triode	Plate-Pulsed Oscillator and Amplifier	Compact, forced-air-cooled radiator type used with full input up to 1300 Mc and with reduced input up to 2000 Mc.
6026	Oscillator Triode	Radiosonde Service	Subminiature type for use at 400 Mc.
6161	Power Triode	Frequency Multiplier	Compact, forced-air-cooled radiator type used with full input up to 900 Mc.
6264	Medium-Mu Triode	Frequency Multiplier	Pencil type used as tripler up to 510 Mc at altitudes up to 60,000 feet.
6293	Beam Power Tube	Rectangular-Wave Pulse Modulator	For use with duty factors up to 1.0 at a maximum averaging time of 10,000 microseconds.
6383	Power Triode	Frequency Multiplier	Compact, liquid-and-forced-air-cooled type used as doubler up to 900 Mc.
6417	Beam Power Tube	Frequency Multiplier	Identical with type 5763 except for 12.6-volt, 0.375-ampere heater.
6524	Twin Beam Power Tube	Frequency Tripler	Used with full input up to 100 Mc and with reduced input up to 470 Mc.
6562	Fixed-Tuned Oscillator Triode	Radiosonde Service	Pencil type having integral resonators and external cathode tab for use at 1680 Mc.
6850	Twin Beam Power Tube	Frequency Tripler	Identical with type 6524 except for 12.6-volt, 0.625-ampere heater.

V. Rectifier Tubes

Unless otherwise specified, maximum ratings are absolute values

Maximum Anode or Plate Amperes: Average	Maximum Anode or Plate Amperes: Peak	Maximum Peak Inverse Anode or Plate Volts	Temperature Range♦ °C	Filament (F) or Heater (H): Volts	Filament (F) or Heater (H): Amperes	Type No.
Half-Wave Mercury-Vapor Types:						
0.125	0.5	7500	20 to 60	2.5 F	2.0	816
0.25	1.0	10000 5000▲	25 to 60 25 to 70	2.5 F	5.0	866-A
0.5	2.0	2500	25 to 70			
1.25	5.0	10000 5000	20 to 60 20 to 70	5.0 F	7.5	872-A 8008
1.50 1.75 2.50■	6.0 7.0 10.0■	15000 10000 15000■ 10000■	20 to 50 20 to 60 20 to 50 20 to 60	5.0 F	10.0	575-A 673
2.50	15.0	5000 2000	30 to 60 30 to 80	5.0 H	4.5	5558
4.0 6.4	16.0 40.0	10000 3000	25 to 50 40 to 80	5.0 H	10.0	5561
Half-Wave Gas Types:						
0.25 0.50●	1.0 2.0●	10000 5000●	−75 to +90	2.5 F	5.0	3B28
0.50●	2.0●	4500●	−75 to +90	2.5 F	5.0	3B25
Half-Wave Vacuum Types:						
0.130	0.800	6000	–	2.5 F	5.0	1616
0.25	1.0	5000	–	2.5 H	5.0	836
Full-Wave Vacuum Types:						
0.150△ 0.175△ 0.250	0.650	2800 2400 2100	–	5.0 F	2.0	5R4-GY♠

♦ Operating condensed-mercury temperature range for mercury-vapor types; ambient-temperature range for gas types.

▲ For frequency of power supply of 1000 cps maximum.

■ Quadrature operation.

● For frequency of power supply of 500 cps maximum.

△ With capacitor input to filter.

♠ Maximum ratings for this type are on design-center basis.

VI. Receiving Tubes for Class C Telegraphy Service

Ratings apply only for use as rf power amplifier and oscillator in amateur service

	Maximum ICAS Ratings, *Absolute Values*										
Power Output (Approx.)	DC Plate Volts	DC Grid-No.2 Volts	DC Grid-No.1 Volts▲	DC Plate Ma	DC Grid-No.2 Ma	DC Grid-No.1 Ma	Grid-No.2 Input Watts	Plate Dissipation Watts	Maximum Frequency for Full Input Mc	Mu-Factor, Grid No.2 to Grid No.1	Type No.
4.0	375	250	−100	15	4.0	3.0	1.0	3.5	60	9.5	6AK6
5.5	350	–	−100	25	–	8.0	–	5.0	60	18■	6C4
7.5	375	250	−75	30	9.0	5.0	1.5	9.0	30	22	6AG7
11.0	350	250	−100	47	7.0	5.0	2.0	8.0	60	10	6AQ5
11.0	350	250	−100	47	7.0	5.0	2.0	8.0	30	9	6V6
14.0	400	275	−100	50	11.0	5.0	3.0	12.5	30	7	6F6
28.0	400	300	−125	100	12.0	5.0	3.5	21.0	30	8	6L6

▲ Grid-No.1-circuit resistance must not exceed 0.1 megohm.
■ For triode, this value is the amplification factor.

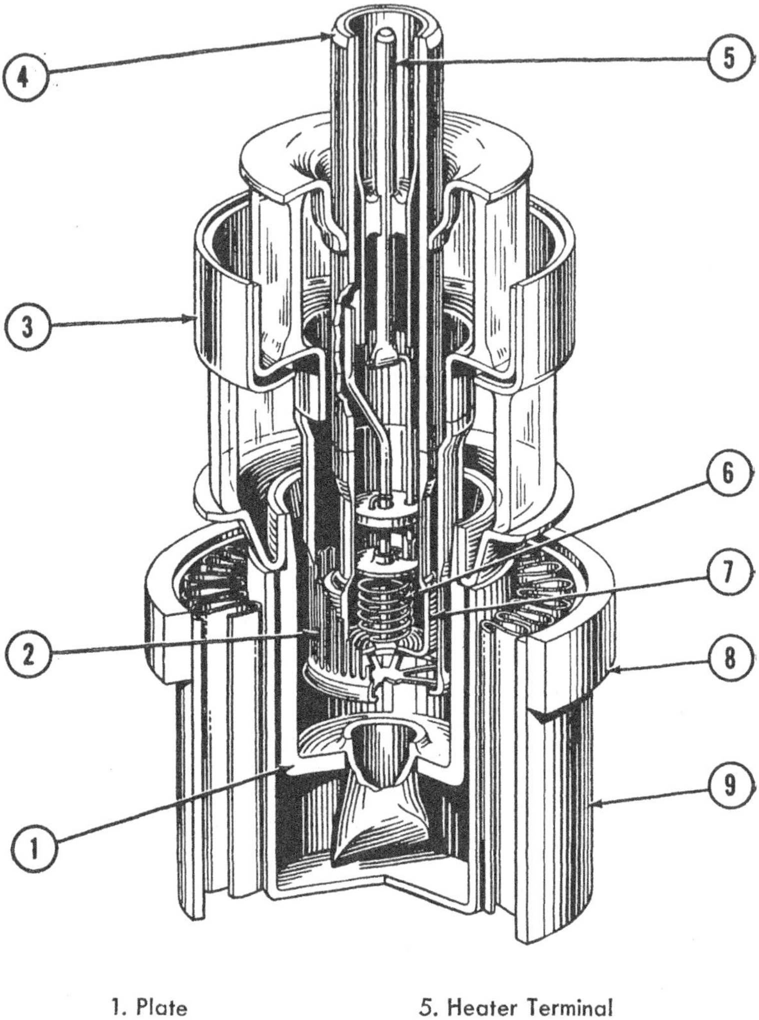

1. Plate
2. Grid
3. Grid Terminal
4. Cathode and Heater Terminal
5. Heater Terminal
6. Heater
7. Cathode
8. Plate Terminal
9. Air-Cooled Radiator

Structure of RCA-6161 UHF Power Triode

RCA Tube Types

This section contains technical descriptions of RCA tubes used in transmitting, industrial, and amateur equipment. It includes data on current types, as well as information on those RCA discontinued types in which there may still be some interest as to characteristics.

In choosing tube types for the design of new electronic equipment, the designer is referred to the inside back cover for information regarding the availability of the latest **RCA Preferred Types List** and for a listing of **RCA Tube Types Not Recommended for New Equipment Design.**

Tube types are listed in this section according to the numerical-alphabetical-numerical sequence of their type designations. For **Legend for Base and Envelope Connection Diagrams,** see inside back cover.

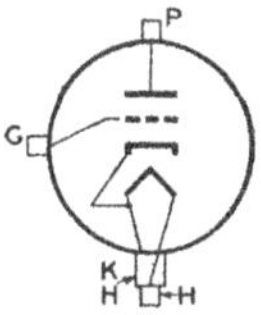

UHF POWER TRIODE

2C39-A

Forced-air-cooled type used as rf power amplifier, oscillator, and frequency multiplier. May be used at full input up to 2500 Mc and at higher frequencies in cathode-drive circuits of the coaxial-cylinder type. Class C Telegraphy maximum CCS plate dissipation, 100 watts.

Heater Voltage (ac/dc)°	6.3 ± 10%	volts
Heater Current	1.0	ampere
Transconductance*	22000	μmhos
Amplification Factor	100	
Direct Interelectrode Capacitances:		
Grid to plate	1.95	μμf
Grid to cathode and heater	6.5	μμf
Plate to cathode and heater	0.035 *max*	μμf

° Because the cathode is subjected to considerable back bombardment as the frequency is increased with resultant increase in temperature, the heater voltage should be reduced depending on operating conditions and frequency to prevent overheating of the cathode and resultant short life.

* Plate volts, 600; plate milliamperes, 70.

PLATE-MODULATED RF POWER AMPLIFIER—Class C Telephony

Carrier conditions per tube for use with a maximum modulation factor of 1.0

Maximum CCS Ratings:

DC Plate Voltage	600• *max*	volts
Grid Voltage:		
DC	−150 *max*	volts
Peak Negative RF	400 *max*	volts
Peak Positive RF	30 *max*	volts
DC Grid Current	50 *max*	ma
DC Cathode Current	100 *max*	ma
Grid Input	2 *max*	watts
Plate Dissipation	70 *max*	watts

•For less than 100-per-cent modulation, it is permissible to use a higher dc plate voltage provided the sum of the peak positive modulation voltage and the dc plate voltage does not exceed 1200 volts.

RF POWER AMPLIFIER AND OSCILLATOR—Class C Telegraphy#

Maximum CCS Ratings:

DC Plate Voltage	1000 *max*	volts
Grid Voltage:		
DC	−150 *max*	volts
Peak Negative RF	400 *max*	volts
Peak Positive RF	30 *max*	volts

DC GRID CURRENT	50 *max*	ma
DC CATHODE CURRENT	125 *max*	ma
GRID INPUT	2 *max*	watts
PLATE DISSIPATION	100 *max*	watts

Typical Operation as Amplifier in Cathode-Drive Circuit at 500 Mc:

DC Plate Voltage	800	volts
DC Grid Voltage	−45	volts
DC Plate Current	80	ma
DC Grid Current (Approx.)	35	ma
Driver Power Output (Approx.)	6	watts
Useful Power Output (Approx.)	27	watts

Typical Operation as Oscillator at 2500 Mc:

DC Plate Voltage	900	volts
DC Grid Voltage (Approx.)	−22	volts
DC Plate Current	90	ma
DC Grid Current (Approx.)	27	ma
Useful Power Output (Minimum)	12	watts

Key-down conditions per tube without amplitude modulation. Amplitude modulation essentially negative may be used if the positive peak of the audio-frequency envelope does not exceed 115 per cent of the carrier conditions.

OPERATING CONSIDERATIONS

Type 2C39-A, with its ring-type seals of graduated diameters, is useful either in cavity or parallel-line circuits of compact fixed and mobile equipment. Requires special mounting which should support the tube by the plate-terminal flange only. May be mounted in any position. Flexible connectors of the spring-contact type are required for all terminal connections. OUTLINE 69, *Outlines* Section.

Cooling of the 2C39-A is accomplished by passing a stream of clean air through the radiator and by directing streams of air onto the cathode and heater seals, the grid seal, and the plate seal. Adequate air must be provided to prevent the temperature of the seals and radiator from exceeding 175°C.

LIGHTHOUSE TRIODE

2C40

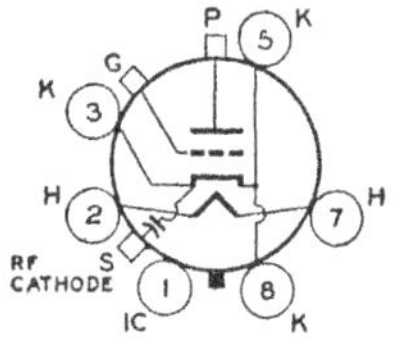

Disk-seal type used as rf amplifier at frequencies up to 1200 Mc and as cw oscillator at frequencies up to 3370 Mc. Requires Octal socket and may be mounted in any position. OUTLINE 7, *Outlines* Section. Heater volts (ac/dc), 6.3; amperes, 0.75. Direct interelectrode capacitances: grid to plate, 1.3 μμf; grid to cathode, shell, and heater, 2.1 μμf; plate to cathode, shell, and heater, (with shield having diameter of 2⅜ inches in plane of grid-disk terminal), 0.03 *max* μμf; cathode to shell, 70 μμf. Maximum CCS ratings as RF POWER AMPLIFIER AND OSCILLATOR, CLASS C TELEGRAPHY: dc plate volts, 500 *max;* dc plate milliamperes, 25 *max;* plate dissipation, 6.5 *max* watts; peak heater-cathode volts, ± 90 *max.* Characteristics as CLASS A_1 AMPLIFIER: plate-supply volts, 250; cathode resistor, 200 ohms; plate milliamperes, 17.5; transconductance, 5000 μmhos; amplification factor, 36; plate resistance (approx.), 7200 ohms. The 2C40 is used principally for renewal purposes.

LIGHTHOUSE TRIODE

2C43

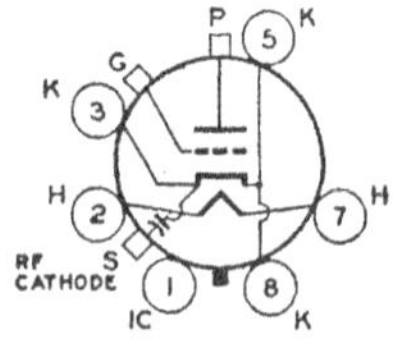

Disk-seal type used as rf amplifier and cw oscillator at frequencies up to 1500 Mc. OUTLINE 10, *Outlines* Section. Requires Octal socket and may be mounted in any position. Heater volts (ac/dc), 6.3; amperes, 0.9. Direct interelectrode capacitances: grid to plate, 1.7 μμf; grid to cathode, shell, and heater, 2.8 μμf; plate to cathode, shell, and heater (with shield having diameter of 2⅜ inches in plane of grid-disk terminal), 0.05 *max* μμf; cathode to shell, 70 μμf. Maximum CCS ratings as RF POWER AMPLIFIER AND OSCILLATOR, CLASS C TELEGRAPHY: dc plate volts, 500 *max;* dc plate milliamperes, 40 *max;* plate dissipation, 12 *max* watts; peak heater-cathode volts, ± 90 *max*. Characteristics as CLASS A_1 AMPLIFIER: plate-supply volts, 250; cathode resistor, 100 ohms; plate milliamperes, 20; transconductance, 8000 μmhos; amplification factor, 48; plate resistance (approx.) 5600 ohms. The 2C43 is used principally for renewal purposes.

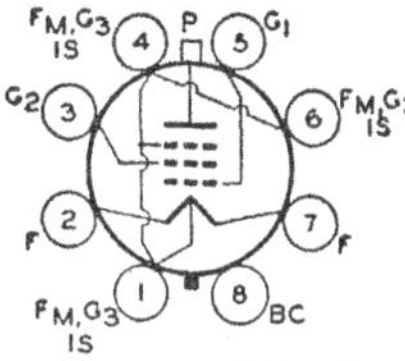

BEAM POWER TUBE

2E24

Glass-octal type having quick-heating coated filament used as af power amplifier and modulator and as rf power amplifier and oscillator in mobile- and emergency-communications equipment. May be used with full input up to 125 Mc and with reduced input up to 175 Mc. Class C Telegraphy maximum plate dissipation, CCS 10 watts, ICAS 13.5 watts.

Filament Voltage (ac/dc)	6.3 ± 10%	volts
Filament Current	0.65	ampere
Filament Heating Time	less than 2	seconds
Transconductance*	3200	μmhos
Mu-Factor, Grid No.2 to Grid No.1**	7.5	
Direct Interelectrode Capacitances:		
Grid No.1 to plate	0.11 *max*	μμf
Grid No.1 to filament mid-tap, grid No.3, internal shield, grid No.2, and base sleeve	8.5	μμf
Plate to filament mid-tap, grid No.3, internal shield, grid No.2, and base sleeve	6.5	μμf
Bulb Temperature (At hottest point)	210 *max*	°C

* Plate volts, 500; grid-No.2 volts, 200; plate milliamperes, 16.
** Plate and grid-No.2 volts, 200; plate milliamperes, 16.

AF POWER AMPLIFIER AND MODULATOR—Class AB_2

Maximum Ratings:	*CCS*	*ICAS*	
DC Plate Voltage	400 *max*	500 *max*	volts
DC Grid-No.2 (screen-grid) Voltage	200 *max*	200 *max*	volts
Maximum-Signal DC Plate Current■	75 *max*	75 *max*	ma
Maximum-Signal Plate Input■	30 *max*	37.5 *max*	watts
Maximum-Signal Grid-No.2 Input■	2.5 *max*	2.5 *max*	watts
Plate Dissipation■	10 *max*	13.5 *max*	watts

Typical Operation (*Values are for 2 tubes*):			
DC Plate Voltage	400	500	volts
DC Grid-No.2 Voltage	125	125	volts
DC Grid-No.1 (Control-Grid) Voltage†	–15	–15	volts
Peak AF Grid-No.1-to-Grid-No.1 Voltage	82	82	volts
Zero-Signal DC Plate Current	18	20	ma
Maximum-Signal DC Plate Current	150	150	ma
Zero-Signal DC Grid-No.2 Current	0.6	0.6	ma
Maximum-Signal DC Grid-No.2 Current	26	28	ma
Effective Load Resistance (Plate to plate)	7000	9000	ohms
Maximum-Signal Driving Power (Approx.)	0.43	0.46	watt
Maximum-Signal Power Output (Approx.)	42	54	watts

Maximum Circuit Values (CCS or ICAS conditions):		
Grid-No.1-Circuit Resistance	30000‡ *max*	ohms

■ Averaged over any audio-frequency cycle of sine-wave form.
† For ac filament supply.
‡ For operation at less than maximum ratings, this value may be as high as 100000 ohms.

PLATE-MODULATED RF POWER AMPLIFIER—Class C Telephony

Carrier conditions per tube for use with a maximum modulation factor of 1.0

Maximum Ratings:	*CCS*	*ICAS*	
DC Plate Voltage	400 *max*	500 *max*	volts

DC GRID-NO.2 VOLTAGE	200 *max*	200 *max*	volts
DC GRID-NO.1 VOLTAGE	−175 *max*	−175 *max*	volts
DC PLATE CURRENT	60 *max*	70 *max*	ma
DC GRID-NO.1 CURRENT	3.5 *max*	3.5 *max*	ma
PLATE INPUT	20 *max*	27 *max*	watts
GRID-NO.2 INPUT	1.7 *max*	2.3 *max*	watts
PLATE DISSIPATION	6.7 *max*	9 *max*	watts

Typical Operation:

DC Plate Voltage	400	500	volts
DC Grid-No.2 Voltage⊙	180	180	volts
From a series resistor of	27500	40000	ohms
DC Grid-No.1 Voltage†♂	−45	−45	volts
From a grid-No.1 resistor of	18000	18000	ohms
Peak RF Grid-No.1 Voltage	61	62	volts
DC Plate Current	50	54	ma
DC Grid-No.2 Current	8	8	ma
DC Grid-No.1 Current (Approx.)	2.5	2.5	ma
Driving Power (Approx.)	0.15	0.16	watt
Power Output (Approx.)	13.5	18	watts

Maximum Circuit Values (CCS or ICAS conditions):

Grid-No.1-Circuit Resistance	30000‡ *max*	ohms

⊙ Obtained preferably from separate source modulated along with plate supply, or from the modulated plate supply through series resistor of value shown.
† For ac filament supply.
♂ Obtained preferably from grid-No.1 resistor or from a combination of grid-No.1 resistor with either fixed supply or cathode resistor.
‡ For operation at less than maximum ratings, this value may be as high as 100000 ohms.

RF POWER AMPLIFIER AND OSCILLATOR—Class C Telegraphy# and RF POWER AMPLIFIER—Class C FM Telephony

Maximum Ratings:

	CCS	*ICAS*	
DC PLATE VOLTAGE	500 *max*	600 *max*	volts
DC GRID-NO.2 VOLTAGE	200 *max*	200 *max*	volts
DC GRID-NO.1 VOLTAGE	−175 *max*	−175 *max*	volts
DC PLATE CURRENT	75 *max*	85 *max*	ma
DC GRID-NO.1 CURRENT	3.5 *max*	3.5 *max*	ma
PLATE INPUT	30 *max*	40 *max*	watts
GRID-NO.2 INPUT	2.5 *max*	2.5 *max*	watts
PLATE DISSIPATION	10 *max*	13.5 *max*	watts

Typical CCS Operation:

	125 Mc		*160 Mc*	
DC Plate Voltage	400	500	350	volts
DC Grid-No.2 Voltage⊕	200	190	170	volts
From a series resistor of	20000	29000	18000	ohms
DC Grid-No.1 Voltage†•	−45	−45	−50	volts
From a grid-No.1 resistor of	15000	15000	16500	ohms
Peak RF Grid-No.1 Voltage	62	65	70	volts
DC Plate Current	75	60	85	ma
DC Grid-No.2 Current	10	10.5	10	ma
DC Grid-No.1 Current	3	3	3	ma
Driving Power (Approx.)	0.19	0.2	2	watts
Power Output (Approx.)	20	20	16.5	watts

Typical ICAS Operation:

	125 Mc	
DC Plate Voltage	600	volts
DC Grid-No.2 Voltage⊕	195	volts
From a series resistor of	40500	ohms
DC Grid-No.1 Voltage† •	−50	volts
From a grid-No.1 resistor of	16700	ohms
Peak RF Grid-No.1 Voltage	71	volts
DC Plate Current	66	ma
DC Grid-No.2 Current	10	ma
DC Grid-No.1 Current	3	ma
Driving Power (Approx.)	0.21	watt
Power Output (Approx.)	27	watts

Maximum Circuit Values (CCS or ICAS conditions):

Grid-No.1-Circuit Resistance . . . . . 30000‡ *max* ohms

Key-down conditions per tube without amplitude modulation. Amplitude modulation essentially negative may be used if the positive peak of the audio-frequency envelope does not exceed 115 per cent of the carrier conditions.

⊕ Obtained preferably from separate source, or from the plate-supply voltage with a voltage divider, or through a series resistor of value shown. Grid-No.2 voltage must not exceed 600 volts under key-up conditions.

† For ac filament supply.

• Obtained from fixed supply, by grid-No.1 resistor, by cathode resistor, or by combination methods.

‡ For operation at less than maximum ratings, this value may be as high as 100000 ohms.

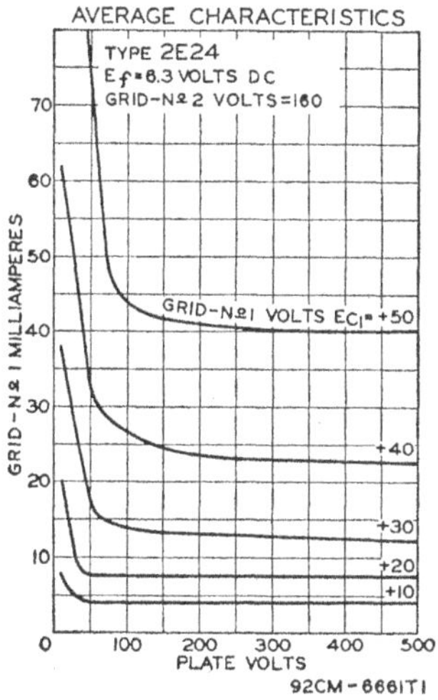

OPERATING CONSIDERATIONS

Type 2E24 requires Octal socket and may be mounted in vertical position with base up or down, or in horizontal position with pins 3 and 7 in vertical plane. Effective rf grounding and simplified shielding of input from output are facilitated by the base sleeve with separate base-pin connection and the single base-pin connection for filament mid-tap, grid No.3, and internal shield. OUTLINE 15, *Outlines* Section.

For operation at 150 Mc, plate voltage and plate input should be reduced to 83 per cent of maximum ratings; at 160 Mc, to 75 per cent; at 175 Mc, to 68 per cent. Plate shows no color when the tube is operated at maximum CCS or ICAS ratings.

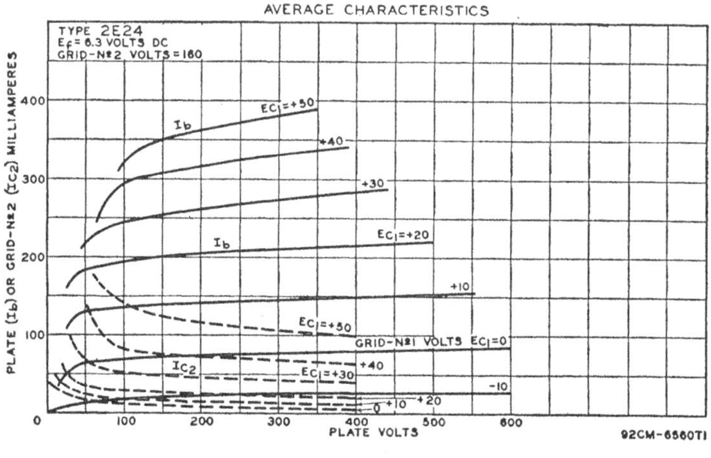

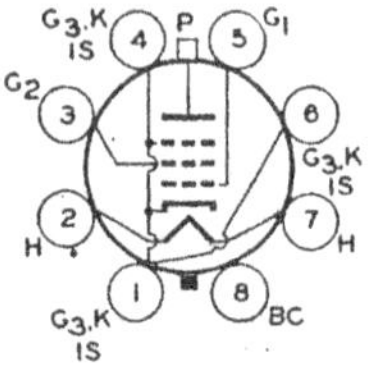

2E26

BEAM POWER TUBE

Glass-octal heater-cathode type used as af power amplifier and modulator and as rf power amplifier and oscillator. May be used with full input up to 125 Mc and with reduced input up to 175 Mc. Class C Telegraphy maximum plate dissipation, CCS 10 watts, ICAS 13.5 watts.

Heater Voltage (ac/dc)	6.3 ±10%	volts
Heater Current	0.8	ampere
Transconductance*	3500	μmhos
Mu-Factor, Grid No.2 to Grid No.1**	6.5	
Direct Interelectrode Capacitances:■		
Grid No.1 to plate	0.20 *max*	μμf
Grid No.1 to cathode, grid-No.3, internal shield, grid-No.2, and heater	13	μμf
Plate to cathode, grid-No.3, internal shield, grid-No.2, and heater	7	μμf
Bulb Temperature (At hottest point)	210 *max*	°C

* Plate volts, 500; grid-No.2 volts, 200; plate milliamperes, 20.
** Plate and grid-No.2 volts, 200; plate milliamperes, 20.
■ Base sleeve connected to ground.

AF POWER AMPLIFIER AND MODULATOR—Class AB_2

Maximum Ratings:	*CCS*	*ICAS*	
DC Plate Voltage	400 *max*	500 *max*	volts
DC Grid-No.2 (screen-grid) Voltage	200 *max*	200 *max*	volts
Maximum-Signal DC Plate Current■	75 *max*	75 *max*	ma
Maximum-Signal Plate Input■	30 *max*	37.5 *max*	watts
Maximum-Signal Grid-No.2 Input■	2.5 *max*	2.5 *max*	watts
Plate Dissipation■	10 *max*	12.5 *max*	watts
Peak Heater-Cathode Voltage:			
Heater negative with respect to cathode	100 *max*	100 *max*	volts
Heater positive with respect to cathode	100 *max*	100 *max*	volts

Typical Operation (*Values are for 2 tubes*):			
DC Plate Voltage	400	500	volts
DC Grid-No.2 Voltage▲†	125	125	volts
DC Grid-No.1 (Control-Grid) Voltage	−15	−15	volts
Peak AF Grid-No.1-to-Grid No.1 Voltage	60	60	volts
Zero-Signal DC Plate Current	20	22	ma
Maximum-Signal DC Plate Current	150	150	ma
Maximum-Signal DC Grid-No.2 Current	32	32	ma
Effective Load Resistance (Plate to plate)	6200	8000	ohms
Maximum-Signal Driving Power (Approx.)	0.36	0.36	watt
Maximum-Signal Power Output (Approx.)	42	54	watts

Maximum Circuit Values (CCS or ICAS conditions):		
Grid-No.1-Circuit Resistance:		
For fixed-bias operation	30000‡ *max*	ohms
For cathode-bias operation	Not recommended	

■ Averaged over any audio-frequency cycle of sine-wave form.
▲ Preferably obtained from a separate source or from the plate-supply voltage with a voltage divider.
† In applications requiring the use of grid-No.2 voltages above 135 volts, provisions should be made for adjustment of grid-No.1 bias for each tube separately. The necessity for this adjustment at lower grid-No.2 voltages depends on the distortion requirements and on whether the plate-dissipation rating is exceeded at zero-signal plate current.
‡ For operation at less than maximum ratings, this value may be as high as 100000 ohms.

PLATE-MODULATED RF POWER AMPLIFIER—Class C Telephony

Carrier conditions per tube for use with a maximum modulation factor of 1.0

Maximum Ratings:	*CCS*	*ICAS*	
DC Plate Voltage	400 *max*	500 *max*	volts
DC Grid-No.2 Voltage	200 *max*	200 *max*	volts
DC Grid-No.1 Voltage	−175 *max*	−175 *max*	volts
DC Plate Current	60 *max*	70 *max*	ma
DC Grid-No.1 Current	3.5 *max*	3.5 *max*	ma
Plate Input	20 *max*	27 *max*	watts

GRID-NO.2 INPUT	1.7 *max*	2.3 *max*	watts
PLATE DISSIPATION	6.7 *max*	9 *max*	watts
PEAK HEATER-CATHODE VOLTAGE:			
Heater negative with respect to cathode	100 *max*	100 *max*	volts
Heater positive with respect to cathode	100 *max*	100 *max*	volts

Typical Operation:

DC Plate Voltage	400	500	volts
DC Grid-No.2 Voltage⊙	160	180	volts
From series resistor of	32000	35500	ohms
DC Grid-No.1 Voltage♂	–50	–50	volts
From grid-No.1 resistor of	20000	20000	ohms
Peak RF Grid-No.1 Voltage	60	60	volts
DC Plate Current	50	54	ma
DC Grid-No 2 Current	7.5	9	ma
DC Grid-No.1 Current (Approx.)	2.5	2.5	ma
Driving Power (Approx.)	0.15	0.15	watt
Power Output (Approx.)	13.5	18	watts

Maximum Circuit Values (CCS or ICAS conditions):

Grid-No.1-Circuit Resistance . . . . . 30000‡ *max* ohms

⊙ Obtained preferably from separate source modulated along with plate supply, or from the modulated plate supply through series resistor of value shown.
♂ Obtained from the grid-No.1 resistor or from a combination of grid-No.1 resistor with either fixed supply or cathode resistor.
‡ For operation at less than maximum ratings, this value may be as high as 100000 ohms.

RF POWER AMPLIFIER AND OSCILLATOR—Class C Telegraphy#
and
RF POWER AMPLIFIER—Class C FM Telephony

Maximum Ratings:

	CCS	*ICAS*	
DC PLATE VOLTAGE	500 *max*	600 *max*	volts
DC GRID-NO.2 VOLTAGE	200 *max*	200 *max*	volts
DC GRID-NO.1 VOLTAGE	–175 *max*	–175 *max*	volts
DC PLATE CURRENT	75 *max*	85 *max*	ma
DC GRID-NO.1 CURRENT	3.5 *max*	3.5 *max*	ma
PLATE INPUT	30 *max*	40 *max*	watts
GRID-NO.2 INPUT	2.5 *max*	2.5 *max*	watts
PLATE DISSIPATION	10 *max*	13.5 *max*	watts
PEAK HEATER-CATHODE VOLTAGE:			
Heater negative with respect to cathode	100 *max*	100 *max*	volts
Heater positive with respect to cathode	100 *max*	100 *max*	volts

Typical CCS Operation:

	125 Mc		*160 Mc*	
DC Plate Voltage	400	500	300	volts
DC Grid-No.2 Voltage⊕	190	185	170	volts
From series resistor of	19000	28500	21500	ohms
DC Grid-No.1 Voltages♠	–30	–40	–75	volts
From grid-No.1 resistor of	10000	13500	30000	ohms
Peak RF Grid-No.1 Voltage	41	50	85	volts
DC Plate Current	75	60	75	ma
DC Grid-No.2 Current	11	11	6	ma
DC Grid-No.1 Current (Approx.)	3	3	2.5	ma
Driving Power (Approx.)	0.12	0.15	1.5	watts
Power Output (Approx.)	20	20	13	watts

Typical ICAS Operation:

	125 Mc	*160 Mc*	
DC Plate Voltage	600	350	volts
DC Grid-No.2 Voltage⊕	185	200	volts
From series resistor of	41500	21500	ohms
DC Grid-No 1 Voltage♠	–45	–90	volts
From grid-No.1 resistor of	15000	30000	ohms
Peak RF Grid-No.1 Voltage	57	105	volts
DC Plate Current	66	85	ma
DC Grid-No.2 Current	10	7	ma
DC Grid-No.1 Current (Approx.)	3	3	ma
Driving Power (Approx.)	0.17	2	watts
Power Output (Approx.)	27	16.5	watts

Maximum Circuit Values (CCS or ICAS conditions):

Grid-No.1-Circuit Resistance..300000 *max*‡ ohms

Key-down conditions per tube without amplitude modulation. Amplitude modulation essentially negative may be used if the positive peak of the audio-frequency envelope does not exceed 115 per cent of the carrier conditions.

⊕ Obtained preferably from separate source, or from the plate-supply voltage with a voltage divider, or through a series resistor of value shown. Grid-No.2 voltage must not exceed 600 volts under key-up conditions.

♠ Obtained from fixed supply, by grid-No.1 resistor, by cathode resistor, or by combination methods.

‡ For operation at less than maximum ratings, this value may be as high as 100000 ohms.

OPERATING CONSIDERATIONS

Type 2E26 requires Octal socket and may be mounted in any position. Effective rf grounding and simplified shielding are facilitated by the base sleeve with separate base-pin connection and the single base-pin connection for cathode, grid No.3, and internal shield. OUTLINE 15, *Outlines* Section.

For operation at 150 Mc, plate voltage and plate input should be reduced to 83 per cent of maximum ratings; at 160 Mc, to 75 per cent; at 175 Mc, to 68 per cent. Plate shows no color when the tube is operated at maximum CCS or ICAS ratings.

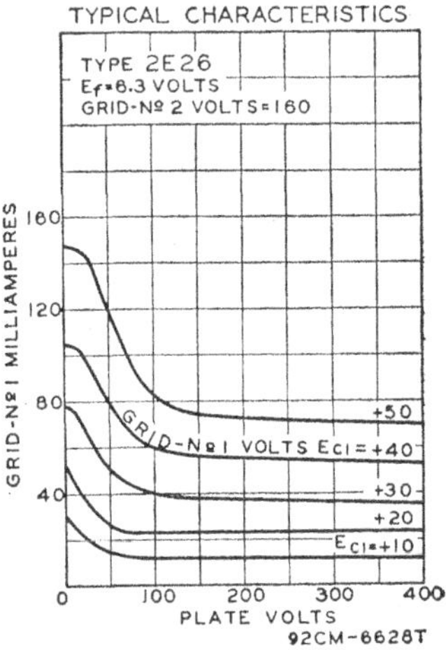

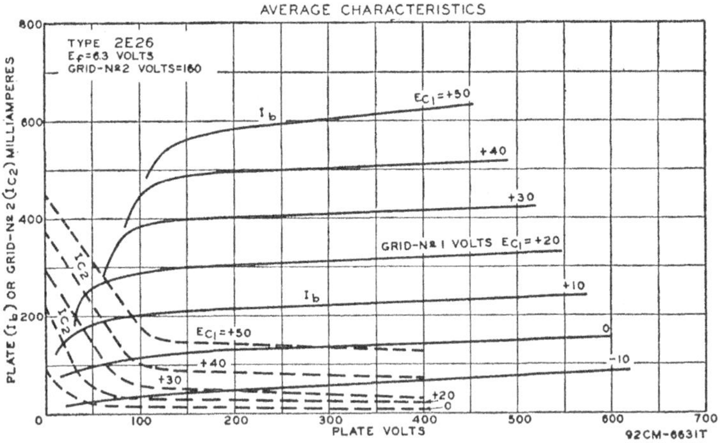

POWER PENTODE

3A4

Seven-pin miniature type having coated filament used as rf power amplifier in light-weight, compact, portable, low-power, battery-operated equipment. May be used at full input up to 10 Mc. Class C maximum CCS plate dissipation, 2 watts.

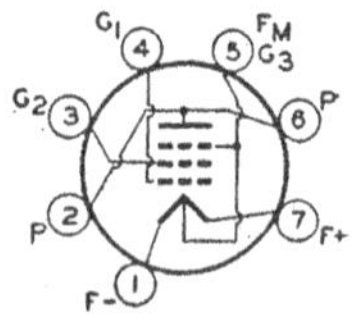

	Series	Parallel	
Filament Arrangement			
Filament Voltage (dc)	2.8	1.4	volts
Filament Current	0.1	0.2	ampere
Transconductance*	2250		μmhos
Plate Resistance (Approx.)*	80000		ohms
Direct Interelectrode Capacitances:			
Grid No.1 to plate	0.34 *max*		μμf
Grid No.1 to filament mid-tap, grid No.3, and grid No.2	4.8		μμf
Plate to filament mid-tap, grid No.3, and grid No.2	4.2		μμf

* Plate volts, 150; grid-No.2 volts, 90; grid-No.1 volts, -8.4.

RF POWER AMPLIFIER—Class C

Maximum CCS Ratings, *Design-Center Values:*

DC Plate Voltage	150 *max*	volts
DC Grid-No.2 (screen-grid) Voltage	135 *max*	volts
DC Grid-No.1 (control-grid) Voltage	-30 *max*	volts
DC Plate Current	20 *max*	ma
DC Grid-No.1 Current	0.25 *max*	ma
Total DC Cathode Current■	25 *max*	ma
Plate Input	3 *max*	watts
Grid-No.2 Input	0.9 *max*	watt
Plate Dissipation	2 *max*	watts

Typical Operation at 10 Mc *(with Parallel Filament Arrangement):*

DC Plate Voltage	150	volts
DC Grid-No.2 Voltage	135	volts
Grid-No.1 Resistor	0.2	megohm
DC Plate Current	18.3	ma
DC Grid-No.2 Current	6.5	ma
DC Grid-No.1 Current	0.13	ma
Power Output (Approx.)	1.2	watts

■ For each 1.4-volt filament section.

OPERATING CONSIDERATIONS

Type 3A4 requires miniature seven-contact socket and may be mounted in any position. Outline 6, *Outlines* Section.

The filament power supply may be obtained from dry-cell batteries, from storage batteries, or from a power line. With dry-cell battery supply, the filament may be connected either directly across a battery rated at a terminal potential of 1.5 volts, or in series with the filaments of similar tubes across a power supply consisting of dry cells in series. In any case, the voltage across each 1.4-volt section of filament should not exceed 1.6 volts.

With power-line or storage-battery supply, the filament may be operated in series with the filaments of other tubes of the same filament-current rating. For such operation, design adjustments should be made so that, with tubes of rated characteristics operating with all electrode voltages applied and on a normal line voltage of 117 volts or on a normal storage-battery voltage of 2.0 volts per cell (without a charger) or 2.2 volts per cell (with a charger), the voltage drop across each 1.4-volt section of filament will be maintained within a range of 1.25 to 1.4 volts with a center of 1.3 volts.

For series operation of the sections, a shunting resistor must be connected across the section between pins 1 and 5 to bypass any cathode current in this section which is in excess of the rated maximum per section. When other tubes in a series-filament arrangement contribute to the filament current of the 3A4, an additional shunting resistor may be required across the entire filament (pins 1 and 7).

For series-filament arrangement, filament voltage is applied between pins 1 and 7. For parallel-filament arrangement, filament voltage is applied between pin 5 and pins 1 and 7 connected together. In series-filament arrangement, the grid-No.1 voltage is referred to pin 1. In parallel-filament arrangement, the grid-No.1 voltage is referred to pin 5.

Plate of the 3A4 shows no color when the tube is operated at maximum CCS ratings.

3A5

MEDIUM-MU TWIN TRIODE

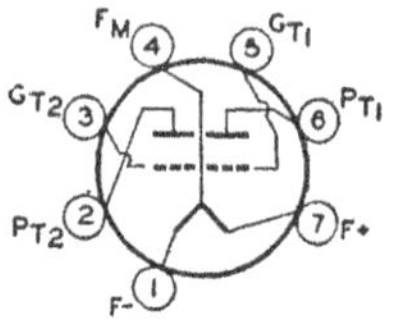

Seven-pin miniature type having coated filament used as rf power amplifier and oscillator in light-weight, compact, portable, low-power, battery-operated equipment. May be used at full input up to 40 Mc. Class C Telegraphy maximum CCS plate dissipation (each unit), 1 watt. Requires miniature seven-contact socket and may be mounted in any position. OUTLINE 6, *Outlines* Section. For filament considerations, refer to type 3A4, noting that for type 3A5 pin 4 is the filament mid-tap. Plates of the 3A5 show no color when the tube is operated at CCS ratings.

	Series		Parallel	
FILAMENT ARRANGEMENT	*Series*		*Parallel*	
FILAMENT VOLTAGE (DC)	2.8		1.4	volts
FILAMENT CURRENT	0.11		0.22	ampere
TRANSCONDUCTANCE*		1800		μmhos
AMPLIFICATION FACTOR*		15		
PLATE RESISTANCE (Approx.)*		8300		ohms
DIRECT INTERELECTRODE CAPACITANCES (Each unit):				
Grid to plate		3.2		μμf
Grid to filament mid-tap		0.9		μμf
Plate to filament mid-tap		1.0		μμf
Plate to plate		0.32		μμf

* Plate volts, 90; grid volts, −2.5; plate milliamperes, 3.7.

RF POWER AMPLIFIER AND OSCILLATOR—Class C Telegraphy#
and
RF POWER AMPLIFIER—Class C FM Telephony

Maximum CCS Ratings, *Design-Center Values for each unit:*

DC PLATE VOLTAGE	135 *max*	volts
DC GRID VOLTAGE	−30 *max*	volts
DC PLATE CURRENT	15 *max*	ma
DC GRID CURRENT	2.5 *max*	ma
PLATE INPUT	2 *max*	watts
PLATE DISSIPATION	1 *max*	watt

Typical Push-Pull Operation (*Values are for both units*):

DC Plate Voltage	135	volts
DC Grid Voltage•	−20	volts
From grid resistor of	4000	ohms
From cathode resistor of	570	ohms
Peak RF Grid-to-Grid Voltage	90	volts
DC Plate Current	30	ma
DC Grid Current (Approx.)	5	ma
Driving Power (Approx.)	0.2	watt
Power Output (Approx.)	2	watts

\# Key-down conditions per tube without amplitude modulation. Amplitude modulation essentially negative may be used if the positive peak of the audio-frequency envelope does not exceed 115 per cent of the carrier conditions.

• Obtained by fixed supply, by grid resistor, by cathode resistor, or by combination methods.

3B25

HALF-WAVE GAS RECTIFIER

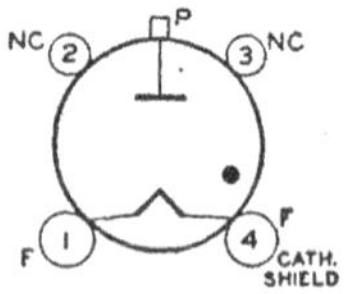

Xenon-filled rectifier of the coated-filament type. May be used in equipment subject to wide range of ambient temperature (−75° to +90°C). Maximum peak inverse anode volts, 4500; maximum average anode amperes, 0.5. Requires Small four-contact socket and may be mounted in any position. OUTLINE 36, *Outlines* Section.

Filament Voltage (ac)°	2.5	volts
Filament Current	5.0	amperes
Peak Tube Voltage Drop (Approx.)	10	volts

° Filament voltage must be applied at least 30 seconds before application of anode voltage.

HALF-WAVE RECTIFIER

Maximum Ratings:

Peak Inverse Anode Voltage	4500 *max*	volts
Anode Current:		
Peak	2.0 *max*	amperes
Average∅	0.5 *max*	ampere
Fault, for duration of 0.1 second maximum	20 *max*	amperes
Frequency of Power Supply	500 *max*	cps
Ambient-Temperature Range	−75 to +90	°C

∅ Averaged over any period of 30 seconds maximum.

Operating Values:

Circuit (For circuit figures, refer to *Rectifier Considerations* Section)	*Fig.*	*Max. Trans. Sec. Volts (RMS)* E	*Approx. DC Output Volts To Filter* Eav	*Max. DC Output Amperes* Iav	*Max. DC Output KW To Filter* Pdc
In-Phase Operation					
Half-Wave Single-Phase	54	3100	1400	0.5	0.7
Full-Wave Single-Phase	55	1500	1400	1.0	1.4
Series Single-Phase	56	3100	2850	1.0	2.8
Half-Wave Three-Phase	57	1800	2150	1.5	3.2
Quadrature Operation					
Parallel Three-Phase	58	1800	2150	3.0	6.4
Series Three-Phase	59	1800	4300	1.5	6.4
Half-Wave Four-Phase	60	1500	2000	1.8* 2.0∎	3.6* 4.0∎
Half-Wave Six-Phase	61	1500	2150	1.9* 2.0∎	4.1* 4.3∎

* Resistive Load ∎ Inductive Load

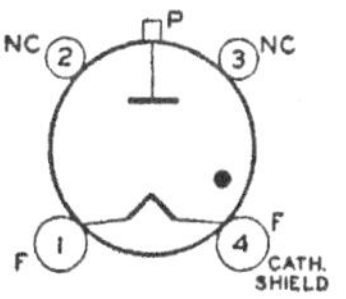

HALF-WAVE GAS RECTIFIER

3B28

Xenon-filled rectifier of the coated-filament type. May be used in equipment subject to wide range of ambient temperature (−75° to +90°C). Rating I: maximum peak inverse anode volts, 10,000; maximum average anode amperes, 0.25. Rating II: maximum peak inverse anode volts, 5000; maximum average anode amperes, 0.5. Requires Small four-contact socket and may be mounted in any position. Outline 33, *Outlines* Section.

Filament Voltage (ac)°	2.5	volts
Filament Current	5.0	amperes
Peak Tube Voltage Drop (Approx.)	10	volts

° Filament voltage must be applied at least 10 seconds before the application of anode voltage.

HALF-WAVE RECTIFIER

Maximum Ratings:

Peak Inverse Anode Voltage	5000 *max*	10000 *max*	volts
Anode Current:			
Peak	2 *max*	1 *max*	amperes
Average∅	0.5 *max*	0.25 *max*	ampere
Fault, for duration of 0.1 second maximum	20 *max*	20 *max*	amperes
Frequency of Power Supply	500 *max*	60 *max*	cps
Ambient-Temperature Range	−75 to +90	−75 to +90	°C

∅ Averaged over any period of 30 seconds maximum.

Operating Values:

Circuit (For circuit figures, refer to *Rectifier Considerations* Section)	*Fig.*	*Max. Trans. Sec. Volts (RMS)* *E*	*Approx. DC Output Volts To Filter* *Eav*	*Max. DC Output Amperes* *Iav*	*Max. DC Output KW To Filter* *Pdc*
		In-Phase Operation			
		7000●	3200	0.25	0.8
Half-Wave Single-Phase..	54	3500▲	1600	0.5	0.8
		3500●	3200	0.5	1.6
Full-Wave Single-Phase...	55	1700▲	1600	1.0	1.6
		7000●	6400	0.5	3.2
Series Single-Phase......	56	3500▲	3200	1.0	3.2
		4000●	4800	0.75	3.6
Half-Wave Three-Phase...	57	2000▲	2400	1.5	3.6
		Quadrature Operation			
		4000●	4800	1.5	7.2
Parallel Three-Phase.....	58	2000▲	2400	3.0	7.2
		4000●	9600	0.75	7.2
Series Three-Phase.......	59	2000▲	4800	1.5	7.2
		3500●	4500	0.9* 1.0■	4.0* 4.5■
Half-Wave Four-Phase....	60	1700▲	2250	1.8* 2.0■	4.0* 4.5■
		3500●	4800	0.95* 1.0■	4.5* 4.8■
Half-Wave Six-Phase.....	61	1700▲	2400	1.9* 2.0■	4.5* 4.8■

● For maximum peak inverse anode voltage of 10000 volts.
▲ For maximum peak inverse anode voltage of 5000 volts.
* Resistive load.
■ Inductive load.

3C33

TWIN POWER TRIODE

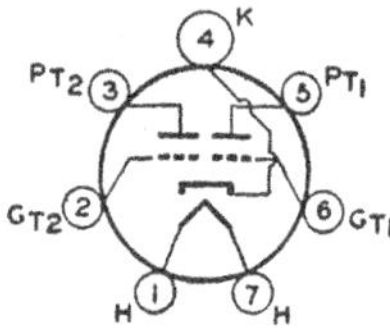

Heater-cathode type containing two high-perveance units used as industrial control amplifier and voltage regulator. Control Amplifier maximum CCS plate dissipation (each unit), 15 watts. Requires Septar seven-contact socket and may be mounted in vertical position with base up or down, or in horizontal position with pins 1 and 4 in vertical plane. OUTLINE 16, *Outlines* Section. Plates show no color when the tube is operated at maximum CCS ratings.

HEATER VOLTAGE (AC/DC)	12.6 ± 10%	volts
HEATER CURRENT	1.125	amperes
AMPLIFICATION FACTOR (Each unit)*	11	
DIRECT INTERELECTRODE CAPACITANCES (Each unit):		
Grid to plate	5.4	μμf
Grid to cathode and heater	7.8	μμf
Plate to cathode and heater	4.2	μμf

* Grid volts, –200; plate milliamperes, 90.

CONTROL AMPLIFIER SERVICE

Values are for each unit

Maximum CCS Ratings:

PEAK PLATE VOLTAGE	±2000 *max*	volts
DC GRID VOLTAGE	–200 *max*	volts
PEAK CATHODE CURRENT	500 *max*	ma
AVERAGE PLATE CURRENT	120 *max*	ma
AVERAGE GRID CURRENT	7.5 *max*	ma
PLATE DISSIPATION	15 *max*	watts
PEAK HEATER-CATHODE VOLTAGE:		
Heater negative with respect to cathode	100 *max*	volts
Heater positive with respect to cathode	100 *max*	volts
BULB TEMPERATURE (At hottest point)	250 *max*	°C

Maximum Circuit Values:

Grid-Circuit Resistance:		
When grid potential is always negative	0.5 *max*	megohm
When grid potential swings positive	0.03 *max*	megohm

TWIN BEAM POWER TUBE

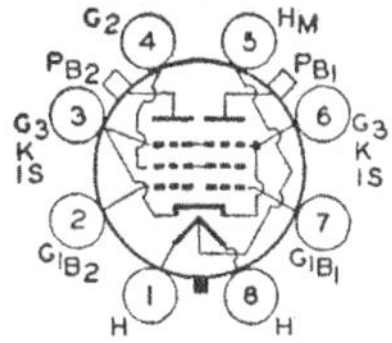

3E22

Glass-octal heater-cathode type used as push-pull rf power amplifier and oscillator in intermittent mobile-service applications. May be used with full input up to 15 Mc. OUTLINE 25, *Outlines* Section. Heater volts (ac/dc), 12.6 ± 10% (series), 6.3 ± 10% (parallel); amperes, 0.8 (series), 1.6 (parallel). Direct interelectrode capacitances (each unit): grid No.1 to plate, 0.22 *max* μμf; grid No.1 to cathode, grid No.3, internal shield, grid No.2, and heater, 14 μμf; plate to cathode, grid No.3, internal shield, grid No.2, and heater, 8.5 μμf. Maximum IMS ratings as PUSH-PULL RF POWER AMPLIFIER AND OSCILLATOR, CLASS C TELEGRAPHY (per tube): dc plate volts, 600 *max;* dc grid-No.2 volts, 225 *max;* dc grid-No.1 volts, -175 *max;* dc plate milliamperes, 175 *max;* dc grid-No.1 milliamperes, 11 *max;* plate input, 100 *max* watts; grid-No.2 input, 6 *max* watts; plate dissipation, 35 *max* watts; peak heater-cathode volts, ± 100 *max*. Plates show no color when the tube is operated at maximum IMS ratings during the normal cycle of 15 seconds on, 1 minute off. The 3E22 is used principally for renewal purposes.

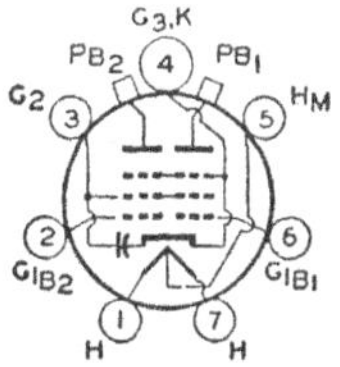

TWIN BEAM POWER TUBE

3E29

Heater-cathode type containing two high-perveance units used as rectangular-wave pulse modulator. Modulator Service maximum CCS plate dissipation (per tube), 15 watts. Requires Septar seven-contact socket and may be mounted in vertical position with base up or down, or in horizontal position with pins 2 and 6 in vertical plane. OUTLINE 22, *Outlines* Section. Plates show no color when the tube is operated at maximum CCS ratings.

HEATER ARRANGEMENT	*Series*	*Parallel*	
HEATER VOLTAGE (AC/DC)	12.6°	6.3°	volts
HEATER CURRENT	1.125	2.25	amperes
TRANSCONDUCTANCE (Each unit, approx.)*	8500		μmhos
MU-FACTOR, Grid No.2 to Grid No.1 (Each unit)**	9		
DIRECT INTERELECTRODE CAPACITANCES (Each unit):			
Grid No.1 to plate (with external shield)	0.12 *max*		μμf
Grid No.1 to cathode, grid No.3, grid No.2, and heater mid-tap	14.0		μμf
Plate to cathode, grid No.3, grid No.2, and heater mid-tap	7.0		μμf

° Should not deviate more than +10% or —5% from value shown.
* Plate volts, 250; grid-No.2 volts, 175; plate milliamperes, 60.
** Plate and grid-No.2 volts, 225; plate milliamperes, 60.

MODULATOR—Rectangular-Wave Modulation

Values are for both units in parallel

Maximum CCS Ratings:

For Duty Factor■ between 0.0001 and 1.0 and Maximum Averaging Time of 1200 Microseconds in Any Interval

DC PLATE-SUPPLY VOLTAGE▲	5000 *max*	volts
INSTANTANEOUS PLATE VOLTAGE	5750 *max*	volts
DC GRID-NO.2 (SCREEN-GRID) SUPPLY VOLTAGE▲	850 *max*	volts
DC GRID-NO.1 (CONTROL-GRID) SUPPLY VOLTAGE▲	−225 *max*	volts
INSTANTANEOUS GRID-NO.1 VOLTAGE	−600 *max*	volts
PEAK POSITIVE GRID-NO.1 VOLTAGE	250 *max*	volts
PEAK PLATE CURRENT	[illegible] *max*	amperes
PEAK GRID-NO.2 CURRENT	3.5 *max*	amperes
PEAK GRID-NO.1 CURRENT	4 *max*	amperes
PLATE INPUT	85 *max*	watts
GRID-NO.2 INPUT	3 *max*	watts
GRID-NO.1 INPUT	1 *max*	watt
PLATE DISSIPATION	15 *max*	watts
PEAK HEATER-CATHODE VOLTAGE:		
Heater negative with respect to cathode	100 *max*	volts
Heater positive with respect to cathode	100 *max*	volts

▲ Duty factor is defined as the "on" time in microseconds divided by 1200 microseconds. Pulse duration is defined as the time interval between the two points on the pulse at which the instantaneous value is 70 per cent of the peak value. The peak value is defined as the maximum value of a smooth curve through the average of the fluctuations over the top portion of the pulse.

▲ For tube protection, it is essential that sufficient dc resistance be used in the plate-supply circuit, the grid-No.2-supply circuit, and the grid-No.1-supply circuit so that the short-circuit current is limited to 0.5 ampere in each circuit.

● For a duty factor between 0.0001 and 0.001, the rated peak plate current is 10 amperes maximum. For higher duty factors, the peak plate current must be reduced. The rated peak plate current for a duty factor of 1.0 is 0.3 ampere approx.

4-65A

BEAM POWER TUBE

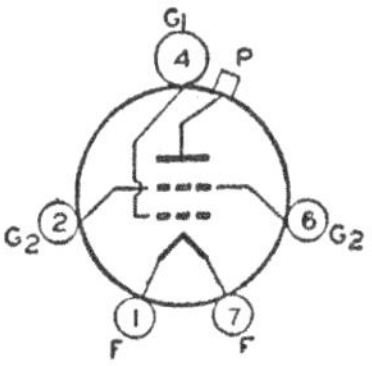

Small, thoriated-tungsten-filament type used as af power amplifier and modulator and as rf power amplifier and oscillator. May be used with full input up to 50 Mc and with reduced input up to 250 Mc. Class C Telegraphy maximum CCS plate dissipation, 65 watts. Requires Septar seven-contact socket and may be mounted in vertical position only, base up or down. OUTLINE 23, *Outlines* Section. Plate shows an orange-red color when the tube is operated at maximum CCS ratings.

FILAMENT VOLTAGE (AC/DC)	6.0	volts
FILAMENT CURRENT	3.5	amperes
TRANSCONDUCTANCE*	4000	μmhos
MU-FACTOR, Grid No.2 to Grid No.1	5	
DIRECT INTERELECTRODE CAPACITANCES:		
Grid No.1 to plate	0.12 *max*	μμf
Grid No.1 to filament and grid No.2	8	μμf
Plate to filament and grid No.2	2.1	μμf

* Plate volts, 500; grid-No.2 volts, 250; plate milliamperes, 125.

AF POWER AMPLIFIER AND MODULATOR—Class AB_2

Maximum CCS Ratings:

DC PLATE VOLTAGE	3000 *max*	volts
DC GRID-NO.2 (SCREEN-GRID) VOLTAGE	600 *max*	volts
MAXIMUM-SIGNAL DC PLATE CURRENT**	150 *max*	ma
MAXIMUM-SIGNAL DC GRID-NO.2 INPUT**	10 *max*	watts
PLATE DISSIPATION**	65 *max*	watts

** Averaged over any audio-frequency cycle of sine-wave form.

PLATE-MODULATED RF POWER AMPLIFIER—Class C Telephony

Carrier conditions per tube for use with a maximum modulation factor of 1.0

Maximum CCS Ratings:

DC PLATE VOLTAGE	2500 *max*	volts
DC GRID-NO.2 (SCREEN-GRID) VOLTAGE	400 *max*	volts
DC GRID-NO.1 (CONTROL-GRID) VOLTAGE	–500 *max*	volts
DC PLATE CURRENT	120 *max*	ma
GRID-NO.2 INPUT	10 *max*	watts
GRID-NO.1 INPUT	5 *max*	watts
PLATE DISSIPATION	45 *max*	watts

Typical Operation:

DC Plate Voltage	600	1000	1500	2000	2500	volts
DC Grid-No.2 Voltage⊙	250	250	250	250	250	volts
DC Grid-No.1 Voltage®	–120	–125	–125	–130	–135	volts
Peak AF Grid-No.2 Voltage⊙	250	250	250	250	250	volts
Peak RF Grid-No.1 Voltage	215	220	220	225	215	volts
DC Plate Current	120	120	120	120	110	ma
DC Grid-No.2 Current (Approx.)	40	40	40	40	25	ma
DC Grid-No.1 Current (Approx.)	15	16	16	16	12	ma
Driving Power (Approx.)	3.2	3.5	3.5	3.6	2.6	watts
Power Output	45	90	140	195	230	watts

⊙ Obtained from unmodulated plate supply through a series resistor, by the use of an af reactor in the positive grid-No.2 supply lead, or from a separate winding on the modulation transformer. With the series-resistor or reactor method, the af variations in grid-No.2 current resulting from variations in plate voltage as the plate is modulated automatically produce the grid-No.2 modulation voltage.

® Obtained from grid-No.1 resistor or from suitable combination of grid-No.1 resistor and fixed supply.

RF POWER AMPLIFIER AND OSCILLATOR—Class C Telegraphy#
and
RF POWER AMPLIFIER—Class C FM Telephony

Maximum CCS Ratings:

DC Plate Voltage	3000 *max*	volts
DC Grid-No.2 Voltage	400 *max*	volts
DC Grid-No.1 Voltage	-500 *max*	volts
DC Plate Current	150 *max*	ma
Grid-No.2 Input	10 *max*	watts
Grid-No.1 Input	5 *max*	watts
Plate Dissipation	65 *max*	watts

Typical Operation:

DC Plate Voltage	600	1000	1500	2000	3000	volts
DC Grid-No.2 Voltage	250	250	250	250	250	volts
DC Grid-No.1 Voltage	−75	−80	−85	−90	−100	volts
Peak RF Grid-No.1 Voltage	170	175	180	190	170	volts
DC Plate Current	150	150	150	140	115	ma
DC Grid-No.2 Current (Approx.)	40	40	40	40	22	ma
DC Grid-No.1 Current (Approx.)	18	17	18	11	10	ma
Driving Power (Approx.)	3.1	3.0	3.2	2.1	1.7	watts
Power Output	45	95	165	215	280	watts

Key-down conditions per tube without amplitude modulation. Amplitude modulation essentially negative may be used if the positive peak of the audio-frequency envelope does not exceed 115 per cent of the carrier conditions.

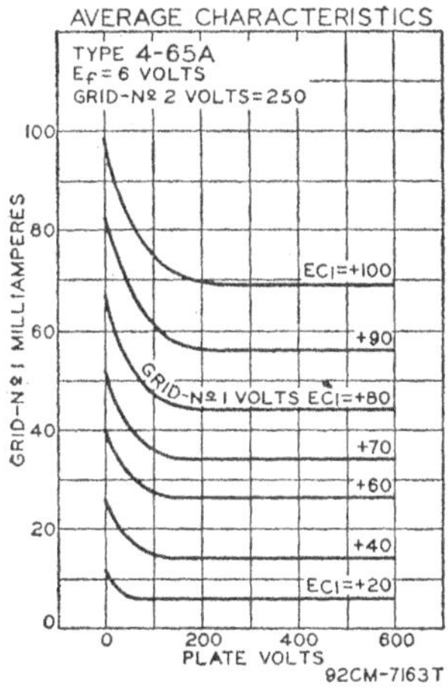

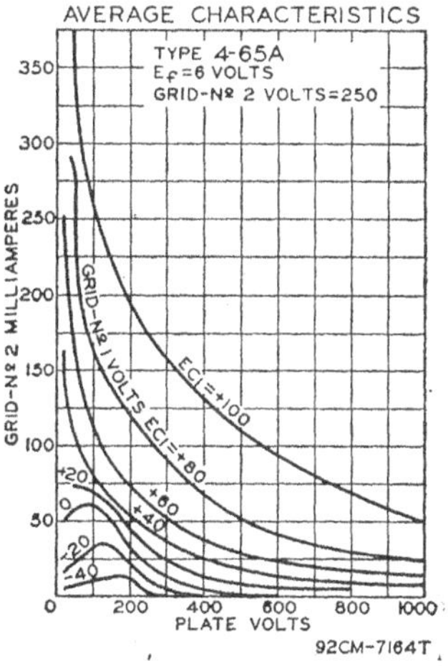

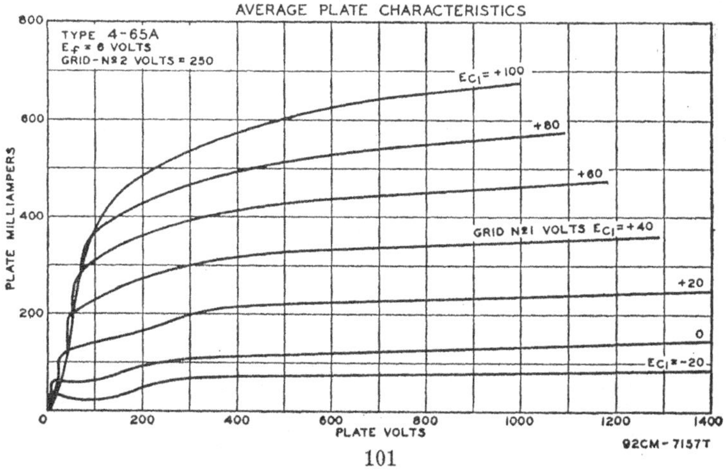

4-125A/ 4D21

BEAM POWER TUBE

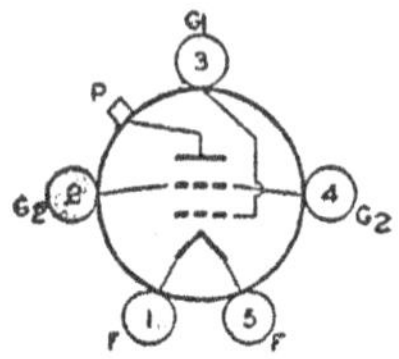

Forced-air-cooled, thoriated-tungsten-filament type used as af power amplifier and modulator and as rf power amplifier and oscillator. May be used with full input up to 120 Mc and with reduced input up to 250 Mc. Class C Telegraphy maximum CCS plate dissipation, 125 watts.

Filament Voltage (ac/dc)	5.0	volts
Filament Current	6.5	amperes
Transconductance*	2450	μmhos
Mu-Factor, Grid No.2 to Grid No.1	5.9	
Direct Interelectrode Capacitances:		
Grid-No.1 to plate (Base shell connected to ground)	0.05	μμf
Grid No.1 to filament, grid No.2, and base shell	10.8	μμf
Plate to filament, grid No.2, and base shell	3.1	μμf

* Plate volts, 2500; grid-No.2 volts, 400; plate milliamperes, 50.

AF POWER AMPLIFIER AND MODULATOR—Class AB2

Maximum CCS Ratings:

DC Plate Voltage	3000 *max*	volts
DC Grid-No.2 (screen-grid) Voltage	400 *max*	volts
Maximum-Signal DC Plate Current■	225 *max*	ma
Grid-No.2 Input■	20 *max*	watts
Plate Dissipation■	125 *max*	watts

Typical Operation *(Values are for 2 tubes):*

DC Plate Voltage	1500	2000	2500	volts
DC Grid-No.2 Voltage	350	350	350	volts
DC Grid-No.1 (Control-Grid) Voltage	-41	-45	-43	volts
Peak AF Grid-No.1-to-Grid-No.1 Voltage	282	210	178	volts
Zero-Signal DC Plate Current	87	72	93	ma
Maximum-Signal DC Plate Current	400	300	260	ma
Zero-Signal DC Grid-No.2 Current	0	0	0	ma
Maximum-Signal DC Grid-No.2 Current	34	5	6	ma
Effective Load Resistance (Plate to plate)	7200	13600	22200	ohms
Maximum-Signal Average Driving Power (Approx.)	2.5	1.4	1	watts
Maximum-Signal Peak Driving Power (Approx.)	5.2	3.1	2.4	watts
Total Harmonic Distortion	2.5	1	2.2	per cent
Maximum-Signal Power Output (Approx.)	350	350	400	watts

Maximum Circuit Values:

Grid-No.1-Circuit Resistance	0.25 *max*	megohm

■ Averaged over any audio-frequency cycle of sine-wave form.

PLATE-MODULATED RF POWER AMPLIFIER—Class C Telephony

Carrier conditions per tube for use with a maximum modulation factor of 1.0

Maximum CCS Ratings:

DC Plate Voltage	2500 *max*	volts
DC Grid-No.2 Voltage	400 *max*	volts
DC Grid-No.1 Voltage	-500 *max*	volts
DC Plate Current	200 *max*	ma
Grid-No.2 Input	20 *max*	watts
Grid-No.1 Input	5 *max*	watts
Plate Dissipation	85 *max*	watts

Typical Operation:

DC Plate Voltage	2000	2500	volts
DC Grid-No.2 Voltage⊙	350	350	volts
DC Grid-No.1 Voltage♂	-220	-210	volts
Peak AF Grid-No.2 Voltage	210	210	volts
Peak RF Grid-No.1 Voltage (Approx.)	375	360	volts
DC Plate Current	150	152	ma
DC Grid-No.2 Current	33	30	ma
DC Grid-No.1 Current	10	9	ma
Driving Power (Approx.)	3.8	3.3	watts
Power Output (Approx.)	225	300	watts

⊙ Obtained preferably from separate source modulated along with plate supply, or from the modulated plate supply through a series resistor.
♂ Obtained preferably from grid-No.1 resistor or from a combination of grid-No.1 resistor with either fixed supply or cathode resistor.

RF POWER AMPLIFIER AND OSCILLATOR—Class C Telegraphy#
and
RF POWER AMPLIFIER—Class C FM Telephony

Maximum CCS Ratings:

DC PLATE VOLTAGE	3000 *max*	volts
DC GRID-No.2 VOLTAGE	400 *max*	volts
DC GRID-No.1 VOLTAGE	-500 *max*	volts
DC PLATE CURRENT	225 *max*	ma
GRID-No.2 INPUT	20 *max*	watts
GRID-No.1 INPUT	5 *max*	watts
PLATE DISSIPATION	125 *max*	watts

Typical Operation:

DC Plate Voltage	2000	2500	3000	volts
DC Grid-No.2 Voltage	350	350	350	volts
DC Grid-No.1 Voltage	-100	-150	-150	volts
Peak RF Grid-No.1 Voltage (Approx.)	230	320	280	volts
DC Plate Current	200	200	167	ma
DC Grid-No.2 Current	50	40	30	ma
DC Grid-No.1 Current	12	12	9	ma
Driving Power (Approx.)	2.8	3.8	2.5	watts
Power Output (Approx.)	275	375	375	watts

Key-down conditions per tube without amplitude modulation. Amplitude modulation essentially negative may be used if the positive peak of the audio-frequency envelope does not exceed 115 per cent of the carrier conditions.

OPERATING CONSIDERATIONS

Type 4-125A requires Special Metal-Shell Giant five-contact socket such as E. F. Johnson Co. socket No. 122-275, or equivalent, and may be mounted in vertical position only, base up or down. OUTLINE 30, *Outlines* Section.

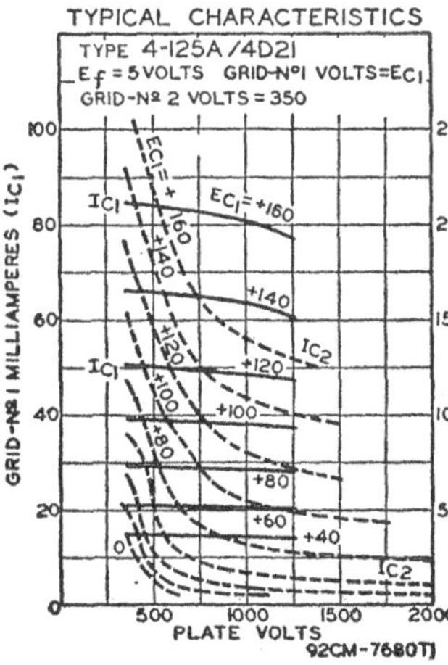

For operation at 150 Mc, plate voltage should be reduced to 80 per cent of maximum rating; at 200 Mc, to 64 per cent; at 250 Mc, to 56 per cent. Plate shows an orange-red color when the tube is operated at maximum CCS ratings.

Adequate cooling must be provided for the seals and envelope of the 4-125A. In CCS applications, the temperature of the plate seal, as measured on the top of the plate cap, should not exceed 170°C. Use of a heat-radiating connector such as Eimac HR-6, or equivalent, on the plate cap is required when the ambient temperature exceeds 30°C. At frequencies above 30 Mc, special attention should be given to adequate cooling of the bulb and plate seal. A small fan directed toward the upper part of the bulb will generally provide sufficient cooling.

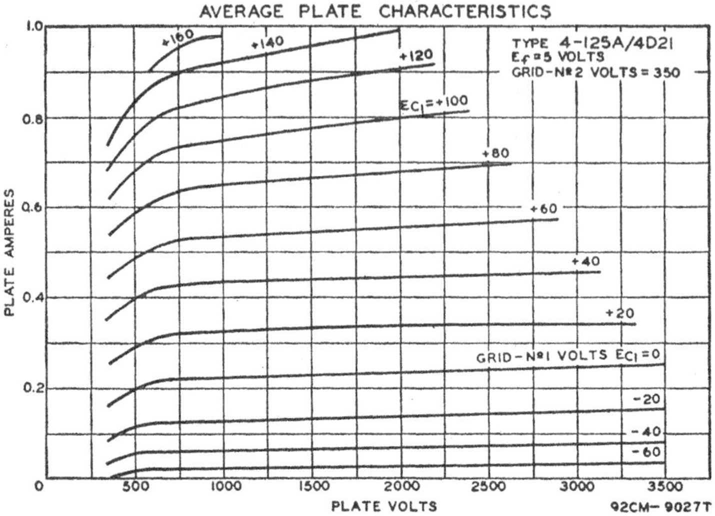

4-250A/ 5D22

BEAM POWER TUBE

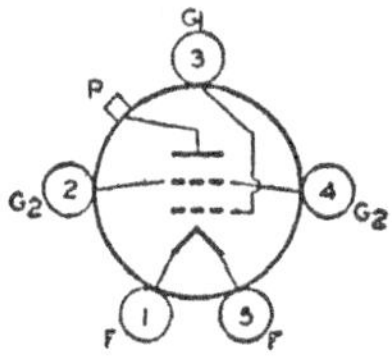

Forced-air-cooled thoriated-tungsten-filament type used as af power amplifier and modulator and as rf power amplifier and oscillator. May be used with full input up to 110 Mc and with reduced input up to 150 Mc. Class C Telegraphy maximum CCS plate dissipation, 250 watts.

Filament Voltage (ac/dc)	5.0	volts
Filament Current	14.5	amperes
Transconductance*	4000	μmhos
Mu-Factor, Grid No.2 to Grid No.1	5.1	
Direct Interelectrode Capacitances:		
Grid No.1 to plate (Base shell connected to ground)	0.14 *max*	μμf
Grid No.1 to filament, grid No.2, and base shell	12.7	μμf
Plate to filament, grid No.2, and base shell	4.5	μμf

* Plate volts, 2500; grid-No.2 volts, 500; plate milliamperes, 100.

AF POWER AMPLIFIER AND MODULATOR—Class AB1

Maximum Ratings:

DC Plate Voltage	4000 *max*	volts
DC Grid-No.2 (screen-grid) Voltage	600 *max*	volts
Maximum-Signal DC Plate Current•	350 *max*	ma
Grid-No.2 Input•	35 *max*	watts
Grid-No.1 (control-grid) Input•	10 *max*	watts
Plate Dissipation•	250 *max*	watts

Typical Operation *(Values are for 2 tubes):*

DC Plate Voltage	1500	2000	2500	3000	volts
DC Grid-No.2 Voltage▲	600	600	600	600	volts
DC Grid-No.1 Voltage⊕	−95	−104	−110	−116	volts
Peak AF Grid-No.1-to-Grid-No.1 Voltage	128	176	180	186	volts
Zero-Signal DC Plate Current	120	110	120	120	ma
Maximum-Signal DC Plate Current	400	405	430	417	ma
Zero-Signal DC Grid-No.2 Current	−0.4	−0.3	−0.3	−0.2	ma
Maximum-Signal DC Grid-No.2 Current	23	22	13	10.5	ma
Effective Load Resistance (Plate to plate)	6250	9170	11400	15000	ohms
Maximum-Signal Driving Power	0	0	0	0	watts
Total Harmonic Distortion	4	2.5	2	2.5	per cent
Maximum-Signal Power Output (Approx.)	310	460	625	750	watts

• Averaged over any audio-frequency cycle of sine-wave form.
▲ Obtained from a source having good regulation.
⊕ Total effective grid-No.1-circuit resistance should not exceed 0.25 megohm.

AF POWER AMPLIFIER AND MODULATOR—Class AB_2

Maximum Ratings:

DC PLATE VOLTAGE	4000 *max*	volts
DC GRID-NO.2 VOLTAGE	600 *max*	volts
MAXIMUM-SIGNAL DC PLATE CURRENT●	350 *max*	ma
GRID-NO.2 INPUT●	35 *max*	watts
GRID-NO.1 INPUT●	10 *max*	watts
PLATE DISSIPATION●	250 *max*	watts

Typical Operation (*Values are for 2 tubes*):

DC Plate Voltage	1500	2000	2500	3000	volts
DC Grid-No.2 Voltage▲	300	300	300	300	volts
DC Grid-No.1 Voltage♂	–48	–48	–51	–53	volts
Peak AF Grid-No.1-to-Grid-No.1 Voltage	192	198	200	198	volts
Zero-Signal DC Plate Current	100	120	120	125	ma
Maximum-Signal DC Plate Current	485	510	500	473	ma
Zero-Signal DC Grid-No.2 Current	0	0	0	0	ma
Maximum-Signal DC Grid-No.2 Current	34	26	23	33	ma
Effective Load Resistance (Plate to plate)	5400	8000	10900	16000	ohms
Maximum-Signal Average Driving Power (Approx.)	2.1	2.3	2.2	1.9	watts
Maximum-Signal Peak Driving Power (Approx.)	4.7	5.5	4.8	4.6	watts
Total Harmonic Distortion	3	4	4	4.5	per cent
Maximum-Signal Power Output (Approx.)	428	650	840	1040	watts

● Averaged over any audio-frequency cycle of sine-wave form.
▲ Obtained from a source having good regulation.
♂ Obtained from fixed supply having dc resistance not exceeding 250 ohms.

PLATE-MODULATED RF POWER AMPLIFIER—Class C Telephony

Carrier conditions per tube for use with a maximum modulation factor of 1.0

Maximum CCS Ratings:

DC PLATE VOLTAGE	3200 *max*	volts
DC GRID-NO.2 VOLTAGE	600 *max*	volts
DC GRID-NO.1 VOLTAGE	–500 *max*	volts
DC PLATE CURRENT	275 *max*	ma
GRID-NO.2 INPUT	35 *max*	watts
GRID-NO.1 INPUT	103 *max*	watts
PLATE DISSIPATION	165 *max*	watts

Typical Operation:

DC Plate Voltage	2500	3000	volts
DC Grid-No.2 Voltage⊙	400	400	volts
DC Grid-No.1 Voltage♂	–200	–310	volts
Peak AF Grid-No.2 Voltage	350	350	volts
Peak RF Grid-No.1 Voltage (Approx.)	255	365	volts
DC Plate Current	200	225	ma
DC Grid-No.2 Current	30	30	ma
DC Grid-No.1 Current (Approx.)	9	9	ma
Driving Power (Approx.)	2.2	3.2	watts
Power Output (Approx.)	375	510	watts

⊙ Obtained preferably from separate source modulated along with plate supply, or from the modulated plate supply through a series resistor.
♂ Obtained preferably from grid-No.1 resistor or from a combination of grid-No.1 resistor with either fixed supply or cathode resistor.

RF POWER AMPLIFIER AND OSCILLATOR—Class C Telegraphy#
and
RF POWER AMPLIFIER—Class C FM Telephony

Maximum CCS Ratings:

DC PLATE VOLTAGE	4000 *max*	volts
DC GRID-NO.2 VOLTAGE	600 *max*	volts
DC GRID-NO.1 VOLTAGE	–500 *max*	volts
DC PLATE CURRENT	350 *max*	ma
GRID-NO.2 INPUT	35 *max*	watts
GRID-NO.1 INPUT	10 *max*	watts
PLATE DISSIPATION	250 *max*	watts

Typical Operation:

DC Plate Voltage	2500	3000	4000	volts
DC Grid-No.2 Voltage	500	500	500	volts
DC Grid-No.1 Voltage	–150	–180	–225	volts
Peak RF Grid-No.1 Voltage (Approx.)	220	265	303	volts

DC Plate Current	300	345	312	ma
DC Grid-No.2 Current	60	60	45	ma
DC Grid-No.1 Current (Approx.)	9	10	9	ma
Driving Power (Approx.)♦	1.7	2.6	2.5	watts
Power Output (Approx.)	575	800	1000	watts

Key-down conditions per tube without amplitude modulation. Amplitude modulation essentially negative may be used if the positive peak of the audio-frequency envelope does not exceed 115 per cent of the carrier conditions.

♦ Increased driving power is required at frequencies above 40 Mc.

OPERATING CONSIDERATIONS

Type 4-250A requires Special Metal-Shell Giant five-contact socket and may be mounted in vertical position only, base up or down. OUTLINE 37, *Outlines* Section.

For operation at 125 Mc, plate voltage should be reduced to 85 per cent of maximum rating; at 150 Mc, to 74 per cent. Plate shows an orange-red color when the tube is operated at maximum CCS ratings.

Cooling requirements for seals and envelope are the same as those for the 4-125A/4D21.

BEAM POWER TUBE

4-1000A

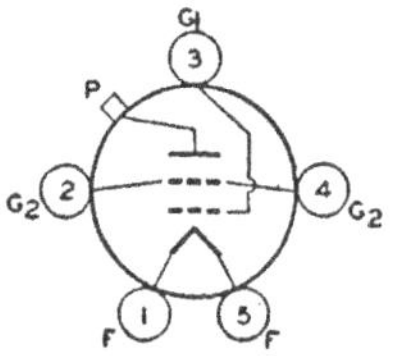

Forced-air-cooled thoriated-tungsten-filament type used as af power amplifier and modulator and as rf power amplifier and oscillator at frequencies up to 110 Mc. OUTLINE 58, *Outlines* Section. Filament volts (ac/dc), 7.5; amperes, 21. Direct interelectrode capacitances: grid No.1 to plate (with base shell connected to ground), 0.24 μμf; grid No.1 to filament, grid No.2, and base shell, 27.2 μμf; plate to filament, grid-No.2, and base shell, 7.6 μμf. Maximum CCS ratings as RF POWER AMPLIFIER AND OSCILLATOR (up to 110 Mc): dc plate volts, 6000 *max;* dc grid-No.2 (screen-grid) volts, 1000 *max;* dc grid-No.1 (control-grid) volts, -500 *max;* dc plate milliamperes, 700 *max;* grid-No.2 input, 75 *max* watts; grid-No.1 input, 25 *max* watts plate dissipation, 1000 *max* watts. Characteristics as CLASS A_1 AMPLIFIER; plate volts, 2500; grid-No.2 volts, 500; plate milliamperes, 300; transconductance, 10,000 μmhos; mu-factor, grid No.2 to grid No.1, 7. Plate shows an orange-red color when tube is operated at maximum CCS ratings. The 4-1000A is used principally for renewal purposes.

POWER TRIODE

4C33

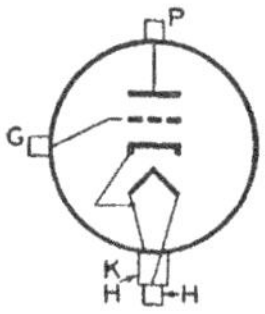

Forced-air-cooled heater-cathode type used as Class C plate-pulsed oscillator. May be used with full input up to 625 Mc. Class C maximum CCS plate dissipation, 250 watts. Requires special mounting designed for use in circuits of the coaxial-cavity type and may be mounted in vertical position only, base up or down. OUTLINE 74, *Outlines* Section, except that grid-flange thickness is 0.040 ± 0.005 inch and outside diameter of air-cooled radiator is 2 ± 0.005 inch.

HEATER VOLTAGE (AC/DC)°	5.0	volts
HEATER CURRENT	9.1	amperes
HEATER STARTING CURRENT	16 *max*	amperes
AMPLIFICATION FACTOR	25	
DIRECT INTERELECTRODE CAPACITANCES:		
Grid to plate	13	μμf
Grid to cathode and heater	34	μμf
Plate to cathode and heater	0.7	μμf

° Heater voltage must be applied for a minimum time of 2 minutes before application of plate voltage.

PLATE-PULSED OSCILLATOR—Class C

Maximum CCS Ratings:

PEAK PLATE-PULSE SUPPLY VOLTAGE	13000 *max*	volts
PEAK GRID VOLTAGE	-2000 *max*	volts
PEAK PLATE CURRENT FROM PULSE SUPPLY	30 *max*	amperes
PEAK RECTIFIED GRID CURRENT	4 *max*	amperes

DC PLATE CURRENT	30 *max*	ma
DC GRID CURRENT	4 *max*	ma
PEAK PLATE INPUT	390000 *max*	watts
PLATE DISSIPATION	250 *max*	watts
PULSE LENGTH	5 *max*	μsec

BEAM POWER TUBE

4D21

See type 4-125A/4D21.

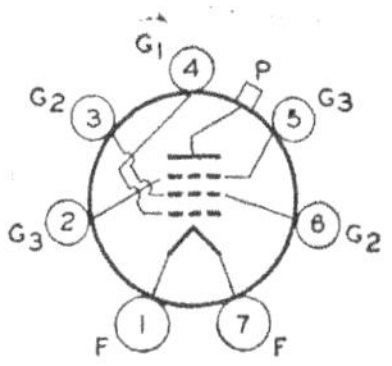

BEAM POWER TUBE

4E27/ 8001

Thoriated-tungsten-filament type used as af power amplifier and modulator and as rf power amplifier and oscillator. May be used with full input up to 75 Mc. For operation at 120 Mc, plate voltage and plate input should be reduced to 75 per cent of maximum ratings; at 150 Mc, to 50 per cent. Class C Telegraphy maximum CCS plate dissipation, 75 watts. Requires Giant seven-contact socket and may be mounted in vertical position only, base up or down. OUTLINE 34, *Outlines* Section. Plate shows an orange-red color when the tube is operated at maximum CCS ratings.

FILAMENT VOLTAGE (AC/DC)	5.0	volts
FILAMENT CURRENT	7.5	amperes
TRANSCONDUCTANCE (For plate current of 75 milliamperes)	2800	μmhos
DIRECT INTERELECTRODE CAPACITANCES:		
Grid to plate (Base shell connected to ground)	0.06	μμf
Grid No.1 to filament, grid No.3, grid No.2, internal shield, and base shell	12	μμf
Plate to filament, grid No.3, grid No.2, internal shield, and base shell	6.5	μμf

RF POWER AMPLIFIER AND OSCILLATOR—Class C Telegraphy#
and
RF POWER AMPLIFIER—Class C FM Telephony

Maximum CCS Ratings:

DC PLATE VOLTAGE	4000 *max*	volts
DC GRID-NO.2 (SCREEN-GRID) VOLTAGE	750 *max*	volts
DC GRID-NO.1 (CONTROL-GRID) VOLTAGE	−500 *max*	volts
DC PLATE CURRENT	150 *max*	ma
DC GRID-NO.2 CURRENT	30 *max*	ma
DC GRID-NO.1 CURRENT	25 *max*	ma
PLATE INPUT	300 *max*	watts
GRID-NO.2 INPUT	25 *max*	watts
PLATE DISSIPATION	75 *max*	watts

Key-down conditions per tube without amplitude modulation. Amplitude modulation essentially negative may be used if the positive peak of the audio-frequency envelope does not exceed 115 per cent of the carrier conditions.

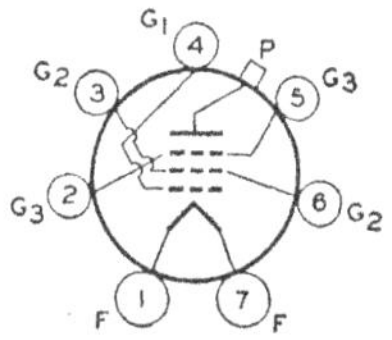

BEAM POWER TUBE

4E27A/ 5-125B

Thoriated-tungsten-filament type used as af power amplifier and modulator and as rf power amplifier and oscillator. May be used at full input up to 75 Mc. Class C Telegraphy maximum CCS plate dissipation, 125 watts. Requires Giant seven-contact socket such as E. F. Johnson Co. socket No. 122-237, or equivalent, and may be mounted in vertical position only, base up or down. OUTLINE 35, *Outlines* Section. Plate shows a cherry-red color when the tube is operated at maximum CCS ratings.

FILAMENT VOLTAGE (AC/DC)	5.0	volts
FILAMENT CURRENT	7.5	amperes
TRANSCONDUCTANCE*	2150	μmhos
MU-FACTOR, Grid No.2 to Grid No.1	5	
DIRECT INTERELECTRODE CAPACITANCES:		
Grid No.1 to plate (Base shell connected to ground)	0.1 *max*	μμf
Grid No.1 to filament, grid No.3, grid No.2, and base shell	10.5	μμf
Plate to filament, grid No.3, grid No.2, and base shell	4.7	μμf

* Plate volts, 2500; grid-No.2 volts, 500; grid-No.3 volts, 0; plate milliamperes, 50.

RF POWER AMPLIFIER AND OSCILLATOR—Class C Telegraphy#
and
RF POWER AMPLIFIER—Class C FM Telephony

Maximum CCS Ratings:

DC Plate Voltage	4000 *max*	volts
DC Grid-No.2 (screen-grid) Voltage	750 *max*	volts
DC Grid-No.1 (control-grid) Voltage	-500 *max*	volts
DC Plate Current	200 *max*	ma
Grid-No.3 (suppressor-grid) Input	20 *max*	watts
Grid-No.2 Input	20 *max*	watts
Grid-No.1 Input	5 *max*	watts
Plate Dissipation	125 *max*	watts

Key-down conditions per tube without amplitude modulation. Amplitude modulation essentially negative may be used if the positive peak of the audio-frequency envelope does not exceed 115 per cent of the carrier conditions.

4X150A
4X150D

BEAM POWER TUBE

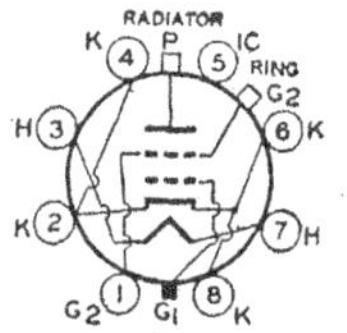

Forced-air-cooled heater-cathode types having integral plate radiators used as af power amplifiers and modulators and as rf power amplifiers and oscillators. May be used with full input up to 500 Mc. Class C Telegraphy maximum CCS plate dissipation, 150 watts.

	4X150A	*4X150D*	
Heater Voltage (ac/dc)	6.0 ± 10%	26.5 ± 10%	volts
Heater Current	2.6	0.58	amperes
Heating Time (Minimum)		30	seconds
Mu-Factor, Grid No.2 to Grid No.1**		5	
Direct Interelectrode Capacitances:			
Grid No.1 to plate		0.06 *max*	μμf
Grid No.1 to cathode, grid No.2, and heater		15.7	μμf
Plate to cathode, grid No.2, and heater		4.3	μμf

** Grid-No.2 volts, 300; grid-No.2 milliamperes, 50.

AF POWER AMPLIFIER AND MODULATOR—Class AB_2

Maximum CCS Ratings:

DC Plate Voltage	1250 *max*	volts
DC Grid-No.2 (screen-grid) Voltage	400 *max*	volts
Maximum-Signal DC Plate Current♦	250 *max*	ma
Grid-No.2 Input♦	12 *max*	watts
Grid-No.1 (control-grid) Input	2 *max*	watts
Plate Dissipation♦	150 *max*	watts
Peak Heater-Cathode Voltage:		
Heater negative with respect to cathode	150 *max*	volts
Heater positive with respect to cathode	150 *max*	volts

Typical Operation (*Values are for 2 tubes*):

DC Plate Voltage	600	800	1000	1250	volts
DC Grid-No.2 Voltage	300	300	300	300	volts
DC Grid-No.1 Voltage	-41	-43	-43	-44	volts
Peak AF Grid-No.1-to-Grid-No.1 Voltage	94	96	98	100	volts
Zero-Signal DC Plate Current	185	160	165	180	ma
Maximum-Signal DC Plate Current	485	490	495	475	ma
Zero-Signal DC Grid-No.2 Current	0	0	0	0	ma
Maximum-Signal DC Grid-No.2 Current	80	75	70	65	ma
Effective Load Resistance (Plate to plate)	2600	3500	4600	5600	ohms
Maximum-Signal Driving Power (Approx.)	0.15	0.15	0.15	0.15	watt
Maximum-Signal Power Output (Approx.)	170	240	315	425	watts

♦ Averaged over any audio-frequency cycle of sine-wave form.

PLATE-MODULATED RF POWER AMPLIFIER—Class C Telephony

Carrier conditions per tube for use with a maximum modulation factor of 1.0

Maximum CCS Ratings:

DC Plate Voltage	1000 *max*	volts

DC GRID-NO.2 VOLTAGE	300 *max*	volts
DC GRID-NO.1 VOLTAGE	–250 *max*	volts
DC PLATE CURRENT	200 *max*	ma
GRID-NO.2 INPUT	12 *max*	watts
GRID-NO.1 INPUT	2 *max*	watts
PLATE DISSIPATION	100 *max*	watts
PEAK HEATER-CATHODE VOLTAGE:		
Heater negative with respect to cathode	150 *max*	volts
Heater positive with respect to cathode	150 *max*	volts

Typical Operation at 165 Mc:

DC Plate Voltage	400	600	800	1000	volts
DC Grid-No. 2 Voltage (Modulated approximately 55 per cent) ⊙	250	250	250	250	volts
DC Grid-No.1 Voltage	–90	–95	–100	–105	volts
Peak AF Grid-No.2 Voltage (For 100-per-cent modulation)	140	150	160	170	volts
Peak RF Grid-No.1 Voltage	110	120	120	125	volts
DC Plate Current	200	200	200	200	ma
DC Grid-No.2 Current	40	35	25	20	ma
DC Grid-No.1 Current (Approx.)	7	8	10	15	ma
Driving Power (Approx.)	1	1	1.5	2	watts
Power Output (Approx.)	55	80	100	140	watts

Maximum Circuit Values:

Grid-No.1-Circuit Resistance	25000 *max*	ohms

⊙ DC grid-No.2 voltage must be modulated approximately 55 per cent in phase with the plate modulation in order to obtain 100-per-cent modulation of the 4X150A or 4X150D. The use of a series grid-No.2 resistor or reactor may not give satisfactory performance and is therefore not recommended.

RF POWER AMPLIFIER AND OSCILLATOR—Class C Telegraphy# and RF POWER AMPLIFIER—Class C FM Telephony

Maximum CCS Ratings:

DC PLATE VOLTAGE	1250 *max*	volts
DC GRID-NO.2 VOLTAGE	300 *max*	volts
DC GRID-NO.1 VOLTAGE	–250 *max*	volts
DC PLATE CURRENT	250 *max*	ma
GRID-NO.2 INPUT	12 *max*	watts
GRID-NO.1 INPUT	2 *max*	watts
PLATE DISSIPATION	150 *max*	watts
PEAK HEATER-CATHODE VOLTAGE:		
Heater negative with respect to cathode	150 *max*	volts
Heater positive with respect to cathode	150 *max*	volts

Typical Operation at 165 Mc:

DC Plate Voltage	600	750	1000	1250	volts
DC Grid-No.2 Voltage	250	250	250	250	volts
DC Grid-No.1 Voltage	–75	–80	–80	–90	volts
Peak RF Grid-No.1 Voltage	91	96	95	106	volts
DC Plate Current	200	200	200	200	ma
DC Grid-No.2 Current	37	37	31	20	ma
DC Grid-No.1 Current (Approx.)	11	11	10	11	ma
Driving Power (Approx.)	1	1.1	1	1.2	watts
Power Output (Approx.)	85	110	150	195	watts

Typical Operation at 500 Mc with Coaxial Cavity:

DC Plate Voltage	600	800	1000	1250	volts
DC Grid-No.2 Voltage	250	250	250	280	volts
DC Grid-No.1 Voltage	–110	–110	–110	–115	volts
DC Plate Current	170	200	200	200	ma
DC Grid-No.2 Current	6	7	7	5	ma
DC Grid-No.1 Current (Approx.)	6	10	10	10	ma
Driver Power Output (Approx.)	15	20	25	30	watts
Useful Power Output (Approx.)	50	95	120	140	watts

Maximum Circuit Values:

Grid-No.1-Circuit Resistance	25000 *max*	ohms

Key-down conditions without amplitude modulation. Amplitude modulation essentially negative may be used if the positive peak of the audio-frequency envelope does not exceed 115 per cent of the carrier conditions.

OPERATING CONSIDERATIONS

Types 4X150A and 4X150D require Eimac 4X150A Air-System eight-contact socket, or equivalent, and may be mounted in any position. OUTLINE 70, *Outlines* Section.

Terminal arrangement facilitates use of these tubes in circuits of the coaxial-cavity type. Grid-No.2 contact ring provides effective isolation of output from input at higher frequencies.

Adequate forced-air cooling must be provided to limit the temperature of the radiator, as measured on metal surface between radiator core and glass envelope, and that of the envelope and base seals to 150°C. The air flow must be applied before or simultaneously with electrode voltages and may be removed simultaneously with them. A minimum air flow of 7.5 cubic feet per minute is required through socket and radiator when tube is operated at maximum CCS ratings.

Because the cathode is subjected to considerable back bombardment as the frequency is increased with resultant increase in temperature, the heater voltage should be reduced depending on operating conditions and frequency to prevent overheating of the cathode and resultant short life.

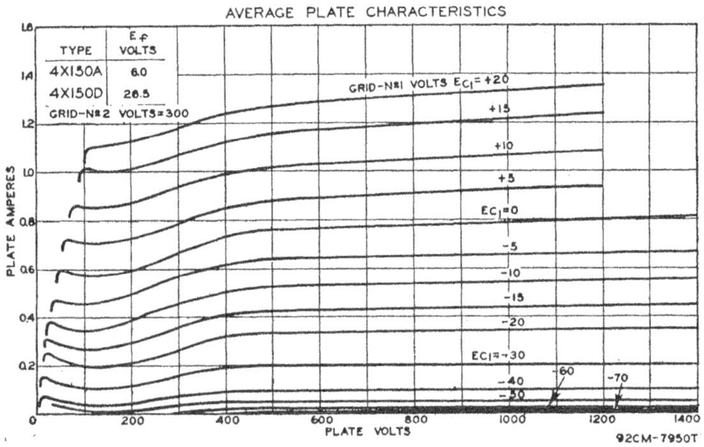

BEAM POWER TUBE

4X500A

Forced-air-cooled type having integral plate radiator and thoriated-tungsten filament used as rf power amplifier and oscillator. May be used with full input up to 120 Mc. Class C Telegraphy maximum CCS plate dissipation, 500 watts.

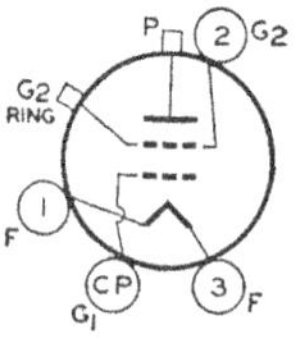

FILAMENT VOLTAGE (AC/DC)	5.0	volts
FILAMENT CURRENT	13.5	amperes
TRANSCONDUCTANCE*	5200	µmhos
MU-FACTOR, Grid No.2 to Grid No.1	6.2	
DIRECT INTERELECTRODE CAPACITANCES:		
Grid No.1 to plate	0.05	µµf
Grid No.1 to filament and grid No.2	12.8	µµf
Plate to filament and grid No.2	5.6	µµf

* Plate volts, 2500; grid-No.2 volts, 500; plate milliamperes, 200.

RF POWER AMPLIFIER AND OSCILLATOR—Class C Telegraphy#
and
RF POWER AMPLIFIER—Class C FM Telephony

Maximum CCS Ratings:

DC Plate Voltage	4000 *max*	volts
DC Grid-No.2 (screen-grid) Voltage	500 *max*	volts
DC Grid-No.1 (control-grid) Voltage	-500 *max*	volts
DC Plate Current	350 *max*	ma
Grid-No.2 Input	30 *max*	watts
Grid-No.1 Input	10 *max*	watts
Plate Dissipation	500 *max*	watts

Typical Operation at 110 Mc:

DC Plate Voltage	2500	3000	4000	volts
DC Grid-No.2 Voltage	500	500	500	volts
DC Grid-No.1 Voltage	−150	−150	−150	volts
DC Plate Current	310	310	315	ma
DC Grid-No.2 Current	26	24	22	ma
DC Grid-No.1 Current	15	16	16	ma
Driving Power (Approx.)	5	5	5	watts
Useful Power Output (Approx.)	475	600	835	watts

Key-down conditions per tube without amplitude modulation. Amplitude modulation essentially negative may be used if the positive peak of the audio-frequency envelope does not exceed 115 per cent of the carrier conditions.

OPERATING CONSIDERATIONS

Type 4X500A may be mounted in vertical position only, base up or down. Outline 73, *Outlines* Section.

Adequate forced-air cooling must be provided to limit the temperature of the metal-to-glass seals and the radiator core to 150°C. Forced-air cooling must start before filament voltage is applied, and must be continued until all voltages have been removed from the tube. A minimum air flow of 40 cubic feet per minute is required when the tube is operated at maximum CCS ratings.

BEAM POWER TUBE — 5-125B

See type 4E27A/5-125B.

BEAM POWER TUBE — 5D22

See type 4-250A/5D22.

FULL-WAVE VACUUM RECTIFIER — 5R4-GY

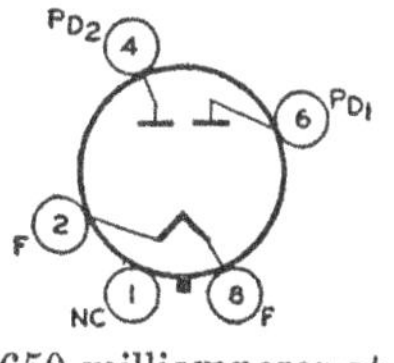

Coated-filament type used in power supply of transmitting and industrial equipment. Rated for a maximum peak inverse plate voltage of 2800 volts and maximum peak plate current of 650 milliamperes at altitudes up to 20,000 feet, it may be used at altitudes up to 40,000 feet with reduced plate voltages. Requires Octal socket and may be mounted in vertical position, base up or down, or in horizontal position with pins 1 and 4 in vertical plane. Outline 28, *Outlines* Section.

Filament Voltage (ac/dc)	5	volts
Filament Current	2	amperes
Tube Voltage Drop (Approx.):		
Measured with applied dc at 250 milliamperes per plate	67	volts

FULL-WAVE RECTIFIER

Maximum Ratings, *Design-Center Values:*	*For Altitudes up to 40000 Feet*		*For Altitudes up to 20000 Feet*	
Peak Inverse Plate Voltage (No load)	2100 *max*	2400 *max*	2800 *max*	volts
Peak Plate Current (Per plate)	650 *max*	650 *max*	650 *max*	ma

DC OUTPUT CURRENT:				
With capacitor input to filter	250 *max*	175 *max*	150 *max*	ma
With choke input to filter	250 *max*	250• *max*	175▲ *max*	ma

Typical Operation with Capacitor-Input Filter:

RMS Plate-to-Plate Supply Voltage:				
Full load	1400	1500	1800	volts
No Load	1500	1700	2000	volts
Filter Input Capacitor	4	4	4	µf
Total Effective Plate-Supply Impedance (Per plate)△	125	500	575	ohms
DC Output Current	250	150	150	ma
DC Output Voltage at Input to Filter (Approx.):				
At Half Load	790	900	1060	volts
At Full Load	700	810	950	volts
Voltage Regulation, Half-Load to Full-Load Current (Approx.)	90	90	110	volts

Typical Operation with Choke-Input Filter:

RMS Plate-to-Plate Supply Voltage:				
Full Load		1500	1900	volts
No Load		1700	2000	volts
Filter Input Choke		5	10	henries
DC Output Current		250	175	ma
DC Output Voltage at Input to Filter (Approx.):				
At Half Load		590	810	volts
At Full Load		550	750	volts
Voltage Regulation, Half-Load to Full-Load Current (Approx.)		40	60	volts

• For choke not less than 5 henries.

▲ For choke not less than 10 henries.

△ Indicated values for conditions shown will limit peak plate current to maximum rated value. When a filter-input capacitor larger than 4 microfarads is used, it may be necessary to use more plate-supply impedance than the value shown to limit the peak plate current to the rated value.

6C24

POWER TRIODE

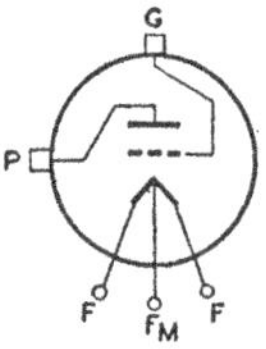

Forced-air-cooled type having integral radiator used as af power amplifier and modulator and as rf power amplifier and oscillator at frequencies up to 160 Mc. Maximum over-all length, 8-23/32 inches; maximum diameter, 1-29/32 inches. Filament volts (ac/dc), 11.0; amperes, 12.1; starting current, 24 *max* amperes. Direct interelectrode capacitances; grid to plate, 4.4 µµf; grid to filament, 4.6 µµf; plate to filament, 3.2 µµf. Maximum CCS ratings as RF POWER AMPLIFIER AND OSCILLATOR: dc plate volts, 3000 *max*; dc grid volts, -500 *max*; dc plate milliamperes, 500 *max;* dc grid milliamperes, 150 *max;* plate input, 1500 *max* watts; plate dissipation 600 *max* watts. The 6C24 is a DISCONTINUED type listed for reference only. As a replacement, the 5786 is a similar type although not directly interchangeable because of either electrical and/or mechanical differences.

6F4

POWER TRIODE

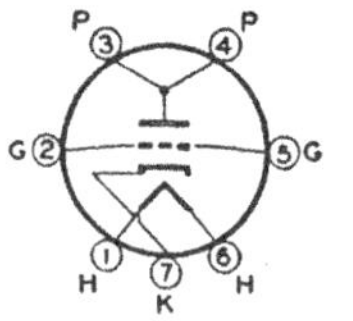

VIEWED FROM SHORT END

Acorn type having heater-cathode used as rf power ampilfier and oscillator at frequencies up to 1200 Mc. Class C Telegraphy maximum plate dissipation (design-center value), 2 watts. Requires Acorn radial 7-contact socket and may be mounted in any position. OUTLINE 1, *Outlines* Section. Plate shows no color when tube is operated at maximum CCS ratings.

HEATER VOLTAGE (AC/DC)	6.3	volts
HEATER CURRENT	0.225	ampere
TRANSCONDUCTANCE*	5800	umhos
AMPLIFICATION FACTOR*	17	
PLATE RESISTANCE (Approx.)*	2900	ohms

DIRECT INTERELECTRODE CAPACITANCES:		
Grid to plate	1.8	μμf
Grid to cathode and heater	1.9	μμf
Plate to cathode and heater	0.6	μμf

* Plate-supply volts, 80; cathode resistor, 150 ohms; plate milliamperes, 13.

RF POWER AMPLIFIER AND OSCILLATOR—Class C Telegraphy and RF POWER AMPLIFIER—Class C FM Telephony

Maximum CCS Ratings, *Design-Center Values:*

DC PLATE VOLTAGE	150 *max*	volts
DC PLATE SUPPLY VOLTAGE	300 *max*	volts
DC GRID VOLTAGE	−50 *max*	volts
DC PLATE CURRENT	20 *max*	ma
DC GRID CURRENT	8 *max*	ma
PLATE DISSIPATION	2 *max*	watts
PEAK HEATER-CATHODE VOLTAGE:		
Heater negative with respect to cathode	80 *max*	volts
Heater positive with respect to cathode	80 *max*	volts

Typical Operation at Moderate Frequencies:♦

DC Plate Voltage	150	volts
DC Grid Voltage■	−15	volts
From a grid resistor of	550	ohms
From a cathode resistor of	2000	ohms
DC Plate Current	20	ma
DC Grid Current (Approx.)	7.5	ma
Driving Power (Approx.)	0.2	watt
Power Output (Approx.)	1.8	watts

Maximum Circuit Values:

Grid-Circuit Resistance	0.5 *max*	megohm

♦ Approximately 45 milliwatts can be obtained when the 6F4 is used at 1200 megacycles per second as an oscillator with 100 volts on plate, maximum rated plate dissipation, and grid resistor of 2000 ohms.

■ Obtained from fixed supply, grid resistor, cathode resistor, or from a combination of grid resistor with either fixed supply or cathode resistor.

POWER TRIODE

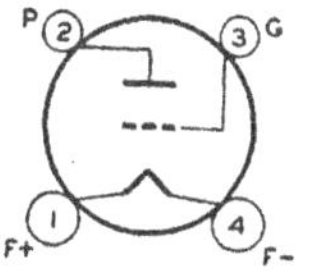

10-Y

Thoriated-tungsten-filament type used as rf power amplifier and oscillator. May be used with full input up to 8 Mc. Requires Small four-contact socket and may be mounted in vertical position only, base down. OUTLINE 29, *Outlines* Section. Filament volts (ac/dc), 7.5; amperes, 1.25. Direct interelectrode capacitances: grid to plate, 7 μμf; grid to filament, 4 μμf; plate to filament, 3 μμf. Maximum CCS ratings as RF POWER AMPLIFIER AND OSCILLATOR, CLASS C TELEGRAPHY: dc plate volts, 450 *max;* dc grid volts, -200 *max;* dc plate milliamperes, 60 *max;* dc grid milliamperes, 15 *max;* plate input, 27 *max* watts; plate dissipation, 15 *max* watts. Characteristics as CLASS A_1 AMPLIFIER: plate volts, 425; grid volts, -35; amplification factor, 8; plate resistance (approx.), 5000 ohms; transconductance, 1600 μmhos. Plate shows no color when tube is operated at maximum CCS ratings. The 10-Y is used principally for renewal purposes.

POWER TRIODE

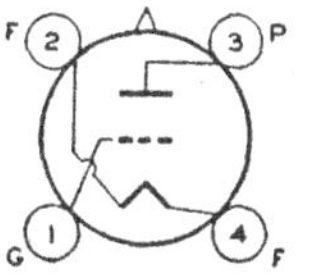

203-A

Thoriated-tungsten-filament type used as af power amplifier and modulator and rf power amplifier and oscillator. May be used with full input up to 15 Mc and with reduced input up to 80 Mc. Requires Jumbo four-contact socket and may be mounted in vertical position only, base down. Maximum over-all length, 7-7/8 inches; maximum diameter, 2-5/16 inches. Filament volts (ac/dc), 10; amperes, 3.25. Direct interelectrode capacitances: grid to plate, 14 μμf; grid to filament, 5.7 μμf; plate to filament, 4.4 μμf. Maximum CCS ratings as RF POWER AMPLIFIER AND OSCILLATOR, CLASS C TELEGRAPHY: dc plate volts, 1250 *max;* dc grid volts, -400 *max;* dc plate milliamperes, 175 *max;* dc grid milliamperes, 60 *max;* plate input, 220 *max* watts; plate dissipation, 100 *max* watts. Plate shows no color when tube is operated at maximum CCS ratings. The 203-A is a DISCONTINUED type listed for reference only. As a replacement, the 8005 is a similar type although not directly interchangeable because of either electrical and/or mechanical differences.

POWER TRIODE

204-A

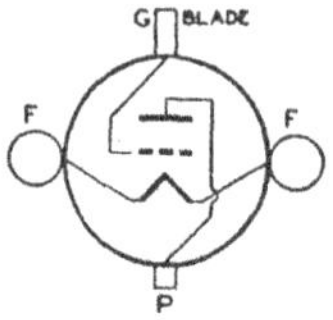

Thoriated-tungsten-filament type used as af power amplifier and modulator and as rf power amplifier and oscillator. May be used with full input up to 3 Mc and with reduced input up to 30 Mc. Requires special end-mounting and may be mounted in vertical position with filament end up, or in horizontal position with plane of plate in vertical plane. Maximum overall length, 14⅝ inches; maximum diameter, 4-1/16 inches. Filament volts (ac/dc), 11; amperes, 3.85. Direct interelectrode capacitances: grid to plate, 15 μμf; grid to filament, 12.5 μμf; plate to filament, 2.3 μμf. Maximum CCS ratings as RF POWER AMPLIFIER AND OSCILLATOR, CLASS C TELEGRAPHY: dc plate volts, 2500 *max;* dc grid volts, -500 *max;* dc plate milliamperes, 275 *max;* dc grid milliamperes, 80 *max;* rf grid amperes, 10 *max;* plate input, 690 *max* watts; plate dissipation, 250 *max* watts. Plate shows a barely perceptible red color when tube is operated at maximum CCS ratings. The 204-A is a DISCONTINUED type listed for reference only.

POWER TRIODE

211

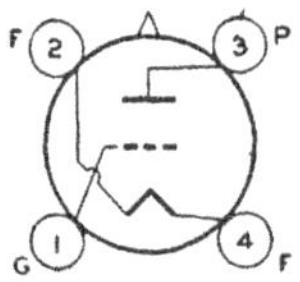

Thoriated-tungsten-filament type used as af power amplifier and modulator and as rf power amplifier and oscillator. May be used with full input up to 15 Mc and with reduced input up to 80 Mc. Requires Jumbo four-contact socket and may be mounted in vertical position, base down, or in horizontal position with pins 1 and 3 in vertical plane. OUTLINE 49, *Outlines* Section. Plate shows a barely perceptible red color when tube is operated at maximum CCS ratings. The 211 is used principally for renewal purposes.

FILAMENT VOLTAGE (AC/DC)	10	volts
FILAMENT CURRENT	3.25	amperes
AMPLIFICATION FACTOR	12	
DIRECT INTERELECTRODE CAPACITANCES:		
Grid to plate	14	μμf
Grid to filament	5.4	μμf
Plate to filament	4.8	μμf

Maximum CCS Ratings:	Class B Modulator	Class C Telegraphy#	
DC PLATE VOLTAGE	1250 *max*	1250 *max*	volts
DC GRID VOLTAGE	—	–400 *max*	volts
DC PLATE CURRENT	175•■ *max*	175 *max*	ma
DC GRID CURRENT	—	50 *max*	ma
PLATE INPUT	220•■ *max*	220 *max*	watts
PLATE DISSIPATION	100■ *max*	100 *max*	watts

Key-down conditions per tube without amplitude modulation. Amplitude modulation essentially negative may be used if the positive peak of the audio-frequency envelope does not exceed 115 per cent of the carrier conditions.

• For maximum-signal conditions.

■ Averaged over any audio-frequency cycle of sine-wave form.

HALF-WAVE VACUUM RECTIFIER

217-C

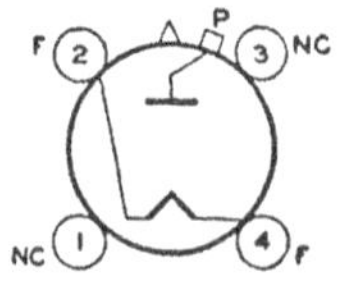

Thoriated-tungsten-filament type used in power supply of transmitting and industrial equipment. Requires Jumbo four-contact socket and may be mounted in vertical position, base down, or in horizontal position with pins 1 and 3 in vertical plane. OUTLINE 51, *Outlines* Section. Filament volts (ac), 10; amperes, 3.25. Maximum ratings: peak inverse plate volts, 7500 *max;* peak plate amperes, 0.6 *max;* average plate amperes, 0.15 *max.* The 217-C is used principally for renewal purposes.

HALF-WAVE MERCURY-VAPOR RECTIFIER

575-A

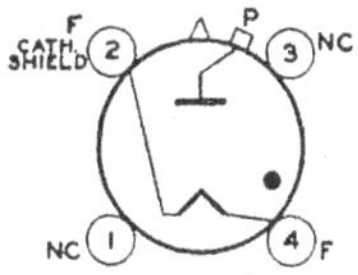

Coated-filament type used in power supply of transmitting and industrial equipment. Maximum peak inverse anode volts, 15000; maximum average anode amperes, 1.5. Requires Jumbo four-contact socket and may be mounted in vertical position only, base down. OUTLINE 60, *Outlines* Section.

FILAMENT VOLTAGE (AC)°	5.0	volts
FILAMENT CURRENT	10.0	amperes
PEAK TUBE VOLTAGE DROP (Approx.)	10	volts

° Filament voltage must be applied at least 30 seconds before application of anode voltage.

HALF-WAVE RECTIFIER—In-Phase Operation

For supply frequency of 60 cps

Maximum Ratings:

PEAK INVERSE ANODE VOLTAGE	10000 *max*	15000 *max*	volts
ANODE CURRENT:			
Peak	7 *max*	6 *max*	amperes
Average∅	1.75 *max*	1.5 *max*	amperes
Fault, for duration of 0.1 second maximum	100 *max*	100 *max*	amperes
CONDENSED-MERCURY-TEMPERATURE RANGE	20 to 60	20 to 50	°C

HALF-WAVE RECTIFIER—Quadrature Operation

For supply frequency of 60 cps

Maximum Ratings:

PEAK INVERSE ANODE VOLTAGE	10000 *max*	15000 *max*	volts
ANODE CURRENT:			
Peak	10 *max*	10 *max*	amperes
Average∅	2.5 *max*	2.5 *max*	amperes
Fault, for duration of 0.1 second maximum	100 *max*	100 *max*	amperes
CONDENSED-MERCURY-TEMPERATURE RANGE	20 to 60	20 to 50	°C

∅ Averaged over any interval of 20 seconds maximum.

Operating Values:

Circuit (For circuit figures, refer to *Rectifier Considerations* Section)	*Fig.*	*Max. Trans. Sec. Volts (RMS)* E	*Approx. DC Output Volts To Filter* Eav	*Max. DC Output Amperes* Iav	*Max. DC Output KW To Filter* Pdc
In-Phase Operation					
		10600•	4800	1.50	7.1
Half-Wave Single-Phase	54	7000▲	3200	1.75	5.5
		5300•	4800	3.00	14.2
Full-Wave Single-Phase	55	3500▲	3200	3.50	11.0
		10600•	9600	3.00	28.4
Series Single-Phase	56	7000▲	6400	3.50	22.0
		6100•	7200	4.50	32.2
Half-Wave Three-Phase	57	4000▲	4800	5.25	25.0
Quadrature Operation					
		6100•	7200	15.0	108
Parallel Three-Phase	58	4000▲	4800	15.0	72
		6100•	14300	7.5	108
Series Three-Phase	59	4000▲	9600	7.5	72
		5300•	6750	9.0* 10.0■	60.8* 67.5■
Half-Wave Four-Phase	60	3500▲	4500	9.0* 10.0■	40.5* 45.0■
		5300•	7200	9.5* 10.0■	68.4* 72.0■
Half-Wave Six-Phase	61	3500▲	4800	9.5* 10.0■	45.6* 48.0■

• For maximum peak inverse anode voltage of 15000 volts and condensed-mercury-temperature range of 20° to 50°C.

▲ For maximum peak inverse anode voltage of 10000 volts and condensed-mercury-temperature range of 20° to 60°C.

* Resistive load. ■ Inductive load.

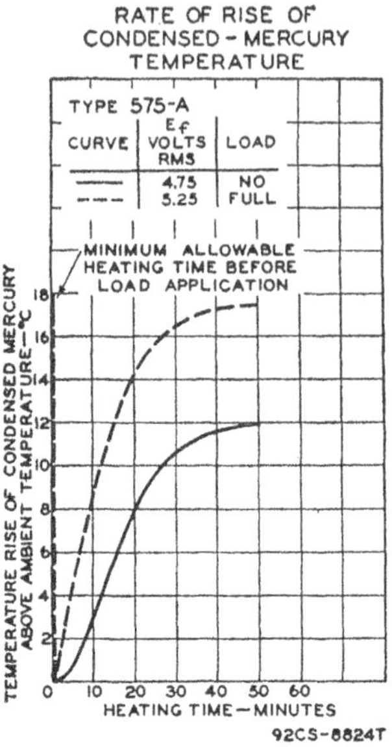

HALF-WAVE MERCURY-VAPOR RECTIFIER

673

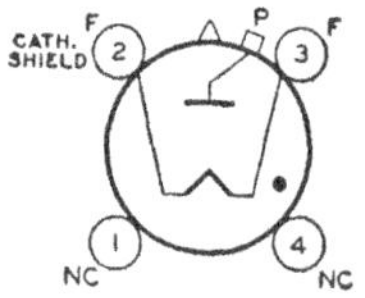

Coated-filament type used in power supply of transmitting and industrial equipment. Maximum peak inverse anode volts, 15000; maximum average anode amperes, 1.5. Requires Super-Jumbo four-contact socket and may be mounted in vertical position only, base down. OUTLINE 62, *Outlines* Section. The 673 is electrically identical with the 575-A.

POWER TRIODE

800

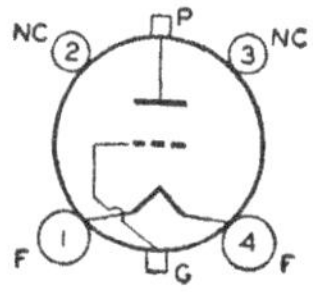

Thoriated-tungsten-filament type used as af power amplifier and modulator and as rf power amplifier and oscillator. May be used with full input up to 60 Mc. Requires Small four-contact socket and may be mounted in vertical position only, base up or down. OUTLINE 38, *Outlines* Section. Filament volts (ac/dc), 7.5; amperes, 3.1. Direct interelectrode capacitances: grid to plate, 2.5 μμf; grid to filament, 2.8 μμf; plate to filament, 2.8 μμf. Maximum CCS ratings as RF POWER AMPLIFIER AND OSCILLATOR: dc plate volts, 1250 *max;* dc grid volts, -400 *max;* dc plate milliamperes, 80 *max;* dc grid milliamperes, 25 *max;* plate input, 100 *max* watts; plate dissipation, 35 *max* watts. Plate shows no color when tube is operated at maximum CCS ratings. The 800 is used principally for renewal purposes.

POWER TRIODE

801-A

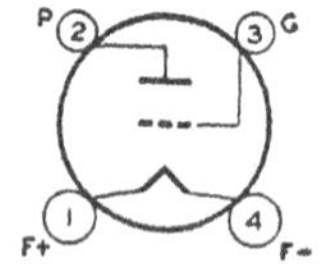

Thoriated-tungsten-filament type used as af power amplifier and modulator and as rf power amplifier and oscillator. May be used with full input up to 60 Mc and with reduced input up to 120 Mc. Requires Small four-contact socket and may be mounted in vertical position with base down, or in horizontal position with pins 1 and 4 in vertical plane. OUTLINE 29, *Outlines* Section. The 801-A is used principally for renewal purposes.

FILAMENT VOLTAGE (AC/DC)	7.5	volts
FILAMENT CURRENT	1.25	amperes
AMPLIFICATION FACTOR	8	
DIRECT INTERELECTRODE CAPACITANCES:		
Grid to plate	6	μμf
Grid to filament	4.5	μμf
Plate to filament	1.5	μμf

Maximum CCS Ratings:	Class B Modulator	Class C Telegraphy#	
DC PLATE VOLTAGE	600 *max*	600 *max*	volts
DC GRID VOLTAGE	–	–200 *max*	volts
DC PLATE CURRENT	70●■ *max*	70 *max*	ma
DC GRID CURRENT	–	15 *max*	ma
PLATE INPUT	42●■ *max*	42 *max*	watts
PLATE DISSIPATION	20■ *max*	20 *max*	watts

Key-down conditions per tube without amplitude modulation. Amplitude modulation essentially negative may be used if the positive peak of the audio-frequency envelope does not exceed 115 per cent of the carrier conditions.
● For maximum-signal conditions.
■ Averaged over any audio-frequency cycle of sine-wave form.

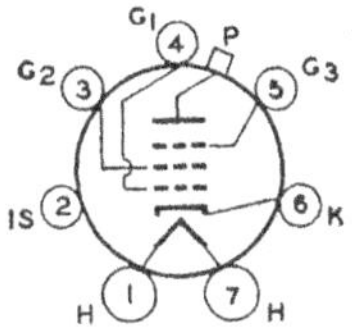

POWER PENTODE

802

Heater-cathode type used as af power amplifier and modulator and as rf power amplifier and oscillator. May be used with full input up to 30 Mc. For operation at 55 Mc, plate voltage and plate input should be reduced to 77 per cent of maximum ratings; at 100 Mc, to 55 per cent. Class C Telegraphy maximum plate dissipation, CCS 10 watts, ICAS 13 watts. Requires Medium seven-contact socket and may be mounted in any position. OUTLINE 31, *Outlines* Section. Plate shows no color when the tube is operated at maximum CCS or ICAS ratings.

HEATER VOLTAGE (AC/DC)	6.3	volts
HEATER CURRENT	0.9	ampere
TRANSCONDUCTANCE (For plate current of 20 milliamperes)	2250	μmhos
DIRECT INTERELECTRODE CAPACITANCES:		
Grid No.1 to plate (With external shielding)	0.15 *max*	μμf
Grid No.1 to cathode, grid No.3, grid No.2, internal shield, and heater	11	μμf
Plate to cathode, grid No.3, grid. No.2, internal shield, and heater	6.8	μμf

AF POWER AMPLIFIER AND MODULATOR—Class A

Maximum Ratings:	*CCS*	*ICAS*	
DC PLATE VOLTAGE	500 *max*	600 *max*	volts
DC GRID-NO.2 (SCREEN-GRID) VOLTAGE	250 *max*	250 *max*	volts
PLATE INPUT	15 *max*	18 *max*	watts
GRID-NO.2 INPUT	3 *max*	3 *max*	watts
PEAK HEATER-CATHODE VOLTAGE:			
Heater negative with respect to cathode	100 *max*	100 *max*	volts
Heater positive with respect to cathode	100 *max*	100 *max*	volts

Typical Operation:■					
DC Plate Voltage	400	500	500	600	volts
DC Grid-No.3 (Suppressor-Grid) Voltage	0*	0*	0*	40	volts
DC Grid-No.2 Voltage	250	175	225	250	volts
DC Grid-No.1 (Control-Grid) Voltage●	–18	–10	–17	–18.5	volts
From cathode resistor of	450	325	530	490	ohms
Peak AF Grid-No.1 Voltage	18	10	17	18.5	volts
DC Plate Current	30	25	25	30	ma
DC Grid-No.2 Current	10	6	7	8	ma
Load Resistance	10000	18000	16000	13200	ohms
Total Harmonic Distortion	8	4	10	9	per cent
Power Output	5.5	4	6.5	7.6	watts

Maximum Circuit Values (CCS or ICAS conditions):

Grid-No.1-Circuit Resistance:		
For fixed-bias operation	0.01 *max*	megohm
For cathode-bias operation	0.5 *max*	megohm

■ Internal shield connected to cathode at socket.
* Connected to cathode at socket.
• Obtained from fixed supply or by cathode resistor of value shown.

RF POWER AMPLIFIER AND OSCILLATOR—Class C Telegraphy#
and
RF POWER AMPLIFIER—Class C FM Telephony

Maximum Ratings:	*CCS*	*ICAS*	
DC PLATE VOLTAGE	500 *max*	600 *max*	volts
DC GRID-NO.3 VOLTAGE	200 *max*	200 *max*	volts
DC GRID-NO.2 VOLTAGE	250 *max*	250 *max*	volts
DC GRID-NO.1 VOLTAGE	–200 *max*	–200 *max*	volts
DC PLATE CURRENT	60 *max*	60 *max*	ma
DC GRID-NO.1 CURRENT	7.5 *max*	7.5 *max*	ma
PLATE INPUT	25 *max*	33 *max*	watts
GRID-NO.3 INPUT	2 *max*	2 *max*	watts
GRID-NO.2 INPUT	6 *max*	6 *max*	watts
PLATE DISSIPATION	10 *max*	13 *max*	watts
PEAK HEATER-CATHODE VOLTAGE:			
Heater negative with respect to cathode	100 *max*	100 *max*	volts
Heater positive with respect to cathode	100 *max*	100 *max*	volts

Typical Operation:■					
DC Plate Voltage	400	500	500	600	volts
DC Grid-No.3 Voltage	0	0	40	40	volts
DC Grid-No.2 Voltage⊕	200	200	250	250	volts
From series resistor of	8000	13600	20800	22060	ohms
DC Grid-No.1 Voltage•	–100	–100	–100	–120	volts
From grid resistor of	14000	17000	50000	50000	ohms
From cathode resistor of	1300	1370	1700	1620	ohms
Peak RF Grid-No.1 Voltage	155	155	155	165	volts
DC Plate Current	45	45	45	55	ma
DC Grid-No.2 Current	25	22	12	16	ma
DC Grid-No.1 Current (Approx.)	7	6	2	2.4	ma
Driving Power (Approx.)	1.1	0.9	0.25	0.3	watt
Power Output (Approx.)	10	14	16	23	watts

Key-down conditions per tube without amplitude modulation. Amplitude modulation essentially negative may be used if the positive peak of the audio-frequency envelope does not exceed 115 per cent of the carrier conditions.
■ Internal shield connected to cathode at socket.
⊕ Obtained preferably from separate source, or from the plate-supply voltage with a voltage divider, or through series resistor. Grid-No.2 voltage must not exceed 500 volts under key-up conditions.
• Obtained from fixed supply, by grid-No.1 resistor, by cathode resistor, or by combination methods.

POWER PENTODE

803

Thoriated-tungsten-filament type used as rf power amplifier and oscillator. May be used with full input up to 20 Mc and with reduced input up to 60 Mc. Requires Giant five-contact socket and may be mounted in vertical position with base up or down, or in horizontal position with pins 2 and 5 in vertical plane. OUTLINE 57, *Outlines* Section. Plate shows a barely perceptible red color when tube is operated at maximum CCS ratings. The 803 is used principally for renewal purposes.

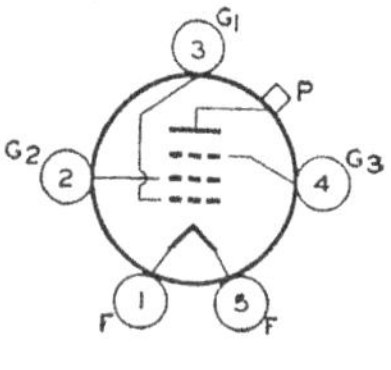

FILAMENT VOLTAGE (AC/DC)	10	volts
FILAMENT CURRENT	5	amperes
TRANSCONDUCTANCE (For plate current of 62.5 milliamperes)	4000	μmhos
DIRECT INTERELECTRODE CAPACITANCES:		
Grid No.1 to plate (With external shielding)	0.15 *max*	μμf
Grid No.1 to filament, grid No.3, and grid No.2	17	μμf
Plate to filament, grid No.3, and grid No.2	29	μμf

RF POWER AMPLIFIER AND OSCILLATOR—Class C Telegraphy#
and
RF POWER AMPLIFIER—Class C FM Telephony

Maximum CCS Ratings:		
DC Plate Voltage	2000 *max*	volts
DC Grid-No.3 (suppressor-grid) Voltage	500 *max*	volts
DC Grid-No.2 (screen-grid) Voltage	600 *max*	volts
DC Grid-No.1 (control-grid) Voltage	−500 *max*	volts
DC Plate Current	175 *max*	ma
DC Grid-No.1 Current	50 *max*	ma
Plate Input	350 *max*	watts
Grid-No.3 Input	10 *max*	watts
Grid-No.2 Input	30 *max*	watts
Plate Dissipation	125 *max*	watts

Key-down conditions per tube without amplitude modulation. Amplitude modulation essentially negative may be used if the positive peak of the audio-frequency envelope does not exceed 115 per cent of the carrier conditions.

POWER PENTODE

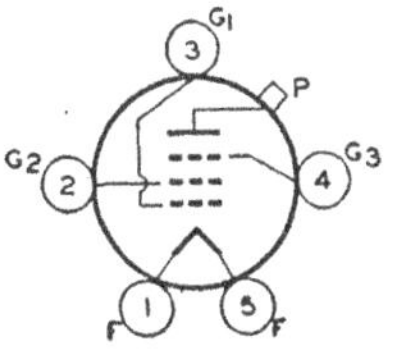

Thoriated-tungsten-filament type used as rf power amplifier and oscillator. May be used with full input up to 15 Mc and with reduced input up to 80 Mc. Requires Small five-contact socket and may be mounted in vertical position with base down, or in horizontal position with pins 2 and 4 in vertical plane. Outline 48, *Outlines* Section. Plate shows no color when tube is operated at maximum CCS or ICAS ratings. The 804 is used principally for renewal purposes.

804

Filament Voltage (ac/dc)	7.5	volts
Filament Current	3.0	amperes
Transconductance (For plate current of 32 milliamperes)	3250	μmhos
Direct Interelectrode Capacitances:		
Grid No.1 to plate (With external shielding)	0.03 *max*	μμf
Grid No.1 to filament, grid No.3, and grid No.2	13	μμf
Plate to filament, grid No.3, and grid No.2	14	μμf

RF POWER AMPLIFIER AND OSCILLATOR—Class C Telegraphy#
and
RF POWER AMPLIFIER—Class C FM Telephony

Maximum Ratings:	*CCS*	*ICAS*	
DC Plate Voltage	1250 *max*	1500 *max*	volts
DC Grid-No.3 (suppressor-grid) Voltage	200 *max*	200 *max*	volts
DC Grid-No.2 (screen-grid) Voltage	300 *max*	300 *max*	volts
DC Grid-No.1 (control-grid) Voltage	−300 *max*	−300 *max*	volts
DC Plate Current	95 *max*	100 *max*	ma
DC Grid-No. 1 Current	15 *max*	15 *max*	ma
Plate Input	120 *max*	150 *max*	watts
Grid-No. 3 Input	5 *max*	5 *max*	watts
Grid-No. 2 Input	15 *max*	15 *max*	watts
Plate Dissipation	40 *max*	50 *max*	watts

Key-down conditions per tube without amplitude modulation. Amplitude modulation essentially negative may be used if the positive peak of the audio-frequency envelope does not exceed 115 per cent of the carrier conditions.

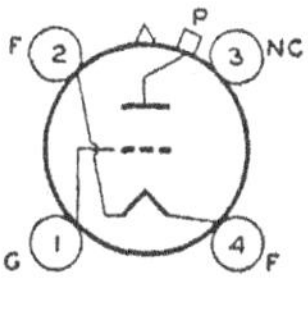

POWER TRIODE

805

Thoriated-tungsten-filament type used as af power amplifier and modulator and as rf power amplifier and oscillator. May be used with full input up to 30 Mc. For operation at 45 Mc, plate voltage and plate input should be reduced to 82 per cent of maximum ratings; at 80 Mc, to 55 per cent. Class C Telegraphy maximum CCS plate dissipation, 125 watts. Requires Jumbo four-contact socket and may be mounted in vertical position with base down, or in horizontal position with pins 1 and 3 in vertical plane. OUTLINE 51, *Outlines* Section. Plate shows no color when tube is operated at maximum CCS ratings.

FILAMENT VOLTAGE (AC/DC)	10	volts
FILAMENT CURRENT	3.25	amperes
DIRECT INTERELECTRODE CAPACITANCES:		
Grid to plate	6.0	μμf
Grid to filament	7.6	μμf
Plate to filament	9.0	μμf

AF POWER AMPLIFIER AND MODULATOR—Class B

Maximum CCS Ratings:

DC PLATE VOLTAGE	1500 *max*	volts
MAXIMUM-SIGNAL DC PLATE CURRENT■	210 *max*	ma
MAXIMUM-SIGNAL PLATE INPUT■	315 *max*	watts
PLATE DISSIPATION■	125 *max*	watts

Typical Operation (*Values are for 2 tubes*):

DC Plate Voltage	1250	1500	volts
DC Grid Voltage	0	–16	volts
Peak AF Grid-to-Grid Voltage	235	280	volts
Zero-Signal DC Plate Current	148	84	ma
Maximum-Signal DC Plate Current	400	400	ma
Effective Load Resistance (Plate to plate)	6700	8200	ohms
Maximum-Signal Driving Power (Approx.)	6	7	watts
Maximum-Signal Power Output (Approx.)	300††	370†	watts

■ Averaged over any audio-frequency cycle of sine-wave form.
†† With 4 per cent harmonic distortion.
† With 3 per cent harmonic distortion.

RF POWER AMPLIFIER AND OSCILLATOR—Class C Telegraphy# and RF POWER AMPLIFIER—Class C FM Telephony

Maximum CCS Ratings:

DC PLATE VOLTAGE	1500 *max*	volts
DC GRID VOLTAGE	–500 *max*	volts
DC PLATE CURRENT	210 *max*	ma
DC GRID CURRENT	70 *max*	ma
PLATE INPUT	315 *max*	watts
PLATE DISSIPATION	125 *max*	watts

Typical Operation:

DC Plate Voltage	1000	1250	1500	volts
DC Grid Voltage	–95	–100	–105	volts
Peak RF Grid Voltage	225	230	235	volts
DC Plate Current	200	200	200	ma
DC Grid Current (Approx.)	40	40	40	ma
Driving Power (Approx.)	8.5	8.5	8.5	watts
Power Output (Approx.)	130	170	215	watts

Key-down conditions per tube without amplitude modulation. Amplitude modulation essentially negative may be used if the positive peak of the audio-frequency envelope does not exceed 115 per cent of the carrier conditions.

POWER TRIODE

806

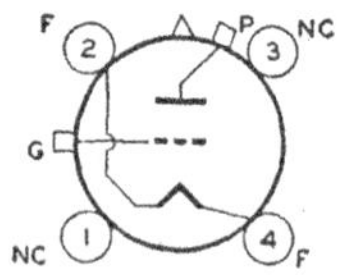

Thoriated-tungsten-filament type used as af power amplifier and modulator and as rf power amplifier and oscillator. May be used with full input up to 30 Mc and with reduced input up to 100 Mc. Requires Jumbo four-contact socket and may be mounted in vertical position only, base down. OUTLINE 59, *Outlines* Section. Filament volts (ac/dc), 5; amperes, 9.5. Direct interelectrode capacitances: grid to plate, 4 μμf; grid to filament, 5.6 μμf; plate to filament, 0.4 μμf. Maximum CCS ratings as AF POWER AMPLIFIER AND MODULATOR: dc plate volts, 3000 *max* (ICAS, 3300 *max*); maximum-signal dc plate milliamperes, 200 *max* (ICAS, 250 *max*); maximum-signal plate input, 500 *max* watts (ICAS, 825 *max* watts); plate dissipation, 150 *max* watts (ICAS, 225 *max* watts). Maximum CCS ratings as RF POWER AMPLIFIER AND OSCILLATOR: dc plate volts, 3000 *max* (ICAS, 3300 *max*); dc grid volts, –1000 *max;* dc plate milliamperes, 200 *max* (ICAS, 305 *max*); dc grid milliamperes, 50 *max;* plate input, 600 *max* watts (ICAS, 1000 *max* watts); plate dissipation, 150 *max* watts (ICAS, 225 *max* watts). Plate shows cherry-red color when tube is operated at maximum CCS ratings, and orange-red color at maximum ICAS ratings. The 806 is used principally for renewal purposes.

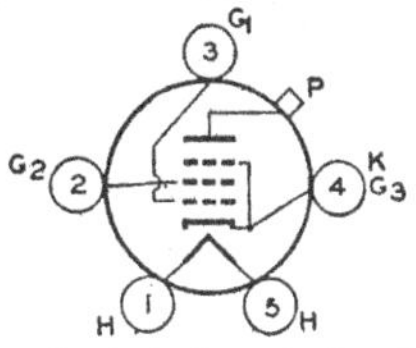

BEAM POWER TUBE

807

Heater-cathode type used as af power amplifier and modulator and as rf power amplifier and oscillator. May be used with full input up to 60 Mc. For operation at 80 Mc, plate voltage and plate input should be reduced to 80 per cent of maximum ratings; at 125 Mc, to 55 per cent. Class C Telegraphy maximum plate dissipation, CCS 25 watts, ICAS 30 watts. Requires Small five-contact socket and may be mounted in any position. OUTLINE 31, *Outlines* Section, except has no bayonet pin. Plate shows no color when tube is operated at maximum CCS or ICAS ratings.

HEATER VOLTAGE (AC/DC)	6.3 ± 0.6	volts
HEATER CURRENT	0.9	ampere
TRANSCONDUCTANCE (Approx.)*	6000	μmhos
MU-FACTOR, Grid No.2 to Grid No.1**	8	
DIRECT INTERELECTRODE CAPACITANCES:		
Grid No.1 to plate (With external shielding)	0.2 *max*	μμf
Grid No.1 to cathode, grid No.3, grid No.2, and heater	12	μμf
Plate to cathode, grid No.3, grid No.2, and heater	7	μμf

* Plate and grid-No.2 volts, 250; grid-No.1 volts, –14.
** Plate and grid-No.2 volts, 250; grid-No.1 volts, –20.

AF POWER AMPLIFIER AND MODULATOR—Class AB2

Maximum Ratings:	CCS	ICAS	
DC PLATE VOLTAGE	600 *max*	750 *max*	volts
DC GRID-NO.2 (SCREEN-GRID) VOLTAGE	300 *max*	300 *max*	volts
MAXIMUM-SIGNAL DC PLATE CURRENT■	120 *max*	120 *max*	ma
MAXIMUM-SIGNAL PLATE INPUT■	60 *max*	90 *max*	watts
MAXIMUM-SIGNAL GRID-NO.2 INPUT■	3.5 *max*	3.5 *max*	watts
PLATE DISSIPATION■	25 *max*	30 *max*	watts
PEAK HEATER-CATHODE VOLTAGE:			
Heater negative with respect to cathode	135 *max*	135 *max*	volts
Heater positive with respect to cathode	135 *max*	135 *max*	volts

Typical Operation (*Values are for 2 tubes*):					
DC Plate Voltage	400	500	600	750	volts
DC Grid-No.2 Voltage‡	300	300	300	300	volts
DC Grid-No.1 (Control-Grid) Voltage	–28	–30	–32	–35	volts
Peak AF Grid-No.1-to-No.1 Voltage	80	86	80	96	volts
Zero-Signal DC Plate Current	72	60	48	30	ma
Maximum-Signal DC Plate Current	240	240	200	240	ma
Zero-Signal DC Grid-No.2 Current	2	0.9	0.7	0.5	ma
Maximum-Signal DC Grid-No.2 Current	20	20	18	20	ma
Effective Load Resistance (Plate to plate)	3700	4600	6900	7300	ohms
Maximum-Signal Driving Power (Approx.)	0.2	0.2	0.1	0.2	watt
Maximum-Signal Power Output (Approx.)▲	55	75	80	120	watts

Maximum Circuit Values (CCS or ICAS conditions):		
Grid-No.1-Circuit Resistance		
For fixed-bias operation	30000 *max*	ohms
For cathode-bias operation	Not recommended	

■ Averaged over any audio-frequency cycle of sine-wave form.
‡ Preferably obtained from a separate source, or from the plate-voltage supply with a voltage divider.
▲ With zero-impedance driver and perfect regulation, plate-circuit distortion does not exceed 2 per cent. In practice, regulation of plate voltage, grid-No.2 voltage, and grid-No.1 voltage should not be greater than 5 per cent, 5 per cent, and 3 per cent, respectively.

PLATE-MODULATED RF POWER AMPLIFIER—Class C Telephony

Carrier conditions per tube for use with a maximum modulation factor of 1.0

Maximum Ratings:	CCS	ICAS	
DC PLATE VOLTAGE	475 *max*	600 *max*	volts
DC GRID-NO.2 VOLTAGE	300 *max*	300 *max*	volts

DC GRID-No.1 VOLTAGE	−200 *max*	−200 *max*	volts
DC PLATE CURRENT	83 *max*	100 *max*	ma
DC GRID-No.1 CURRENT	5 *max*	5 *max*	ma
PLATE INPUT	40 *max*	60 *max*	watts
GRID-No.2 INPUT	2.5 *max*	2.5 *max*	watts
PLATE DISSIPATION	16.5 *max*	25 *max*	watts
PEAK HEATER-CATHODE VOLTAGE:			
Heater negative with respect to cathode	135 *max*	135 *max*	volts
Heater positive with respect to cathode	135 *max*	135 *max*	volts

Typical Operation:

DC Plate Voltage	325	400	475	600	volts
DC Grid-No.2 Voltage♦	250	250	250	300	volts
From series resistor of	12500	25000	28000	37500	ohms
DC Grid-No.1 Voltage♢	−75	−75	−85	−85	volts
From grid-No.1 resistor of	21400	21400	21200	21200	ohms
Peak RF Grid-No.1 Voltage	95	95	108	107	volts
DC Plate Current	80	80	83	100	ma
DC Grid-No.2 Current	6	6	8	8	ma
DC Grid-No.1 Current (Approx.)	3.5	3.5	4	4	ma
Driving Power (Approx.)	0.3	0.3	0.4	0.4	watt
Power Output (Approx.)	17	22	28	44	watts

Maximum Circuit Values (CCS or ICAS conditions):

Grid-No.1-Circuit Resistance	30000 *max*	ohms

♦ Obtained preferably from separate source modulated along with the plate supply, or from the modulated plate supply through series resistor of value shown.

♢ Obtained from grid-No.1 resistor of value shown or from a combination of grid-No.1 resistor with either fixed supply or cathode resistor.

RF POWER AMPLIFIER AND OSCILLATOR—Class C Telegraphy#
and
RF POWER AMPLIFIER—Class C FM Telephony

Maximum Ratings:

	CCS	*ICAS*	
DC PLATE VOLTAGE	600 *max*	750 *max*	volts
DC GRID-No.2 VOLTAGE	300 *max*	300 *max*	volts
DC GRID-No.1 VOLTAGE	−200 *max*	−200 *max*	volts
DC PLATE CURRENT	100 *max*	100 *max*	ma
DC GRID-No.1 CURRENT	5 *max*	5 *max*	ma
PLATE INPUT	60 *max*	75 *max*	watts
GRID-No.2 INPUT	3.5 *max*	3.5 *max*	watts
PLATE DISSIPATION	25 *max*	30 *max*	watts
PEAK HEATER-CATHODE VOLTAGE:			
Heater negative with respect to cathode	135 *max*	135 *max*	volts
Heater positive with respect to cathode	135 *max*	135 *max*	volts

AVERAGE PLATE CHARACTERISTICS

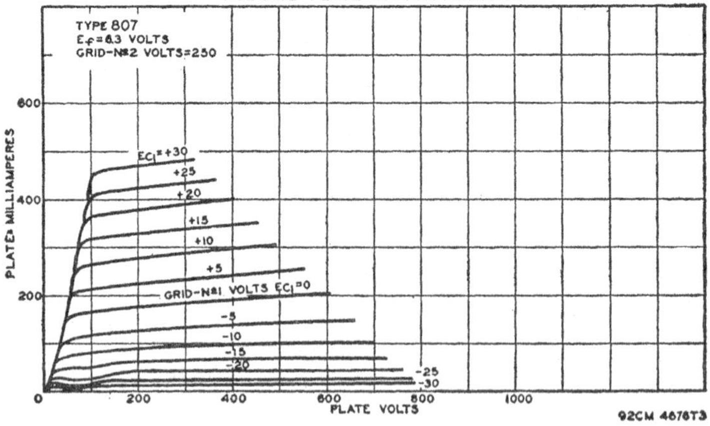

Typical Operation:

DC Plate Voltage	400	500	600	750	volts
DC Grid-No.2 Voltage⊕	250	250	250	250	volts
From series resistor of	19000	31000	44000	62000	ohms
DC Grid-No.1 Voltage •	-45	-45	-45	-45	volts
From grid-No.1 resistor of	11200	11200	11200	11200	ohms
From cathode resistor of	400	400	400	400	ohms
Peak RF Grid-No.1 Voltage	65	65	65	65	volts
DC Plate Current	100	100	100	100	ma
DC Grid-No.2 Current	8	8	8	8	ma
DC Grid-No.1 Current (Approx.)	4	4	4	4	ma
Driving Power (Approx.)	0.3	0.3	0.3	0.3	watt
Power Output (Approx.)	25	32	40	54	watts

Maximum Circuit Values (CCS or ICAS conditions):

Grid-No.1-Circuit Resistance	30000 *max*	ohms

Key-down conditions per tube without amplitude modulation. Amplitude modulation essentially negative may be used if the positive peak of the audio-frequency envelope does not exceed 115 per cent of the carrier conditions.

⊕ Obtained preferably from a separate source, from plate-voltage supply with a voltage divider, or through series resistor of value shown. Grid-No.2 voltage must not exceed 400 volts under key-up conditions.

• Obtained from fixed supply, by grid-No.1 resistor, by cathode resistor, or by combination methods.

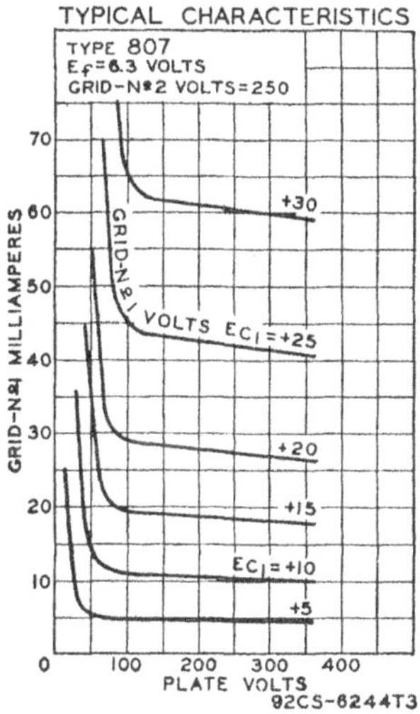

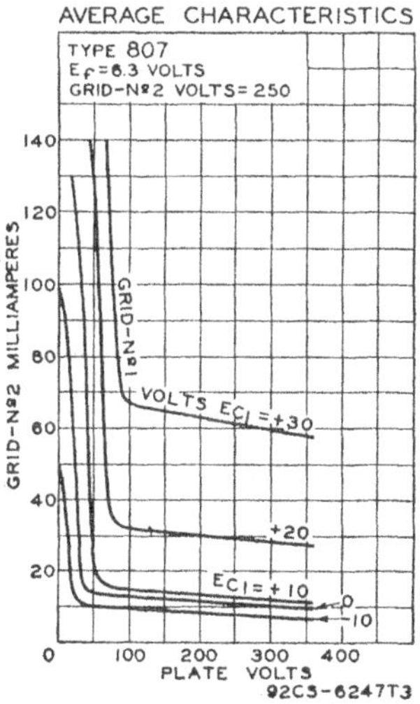

POWER TRIODE

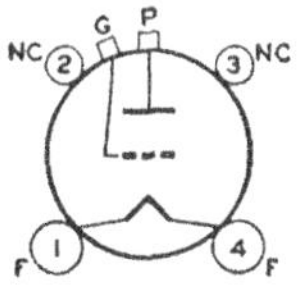

Thoriated-tungsten-filament type used as rf power amplifier and oscillator. May be used with full input up to 30 Mc and with reduced input up to 130 Mc. Class C Telegraphy maximum plate dissipation, CCS 50 watts, ICAS 75 watts.

808

Filament Voltage (ac/dc)	.5	volts
Filament Current	4.0	amperes
Amplification Factor	47	
Direct Interelectrode Capacitances:		
Grid to plate	2.8	μμf
Grid to filament	5.3	μμf
Plate to filament	0.25	μμf

RF POWER AMPLIFIER AND OSCILLATOR—Class C Telegraphy#
and
RF POWER AMPLIFIER—Class C FM Telephony

Maximum Ratings:	CCS		ICAS	
DC PLATE VOLTAGE	1500 *max*		2000 *max*	volts
DC GRID VOLTAGE	-400 *max*		-400 *max*	volts
DC PLATE CURRENT	150 *max*		150 *max*	ma
DC GRID CURRENT	35 *max*		40 *max*	ma
PLATE INPUT	200 *max*		300 *max*	watts
PLATE DISSIPATION	50 *max*		75 *max*	watts
Typical Operation:				
DC Plate Voltage	1250	1500	2000	volts
DC Grid Voltage♠	-150	-150	-150	volts
From grid resistor of	4300	4300	4200	ohms
From cathode resistor of	880	940	800	ohms
Peak RF Grid Voltage	290	300	280	volts
DC Plate Current	135	125	150	ma
DC Grid Current (Approx.)	35	35	36	ma
Driving Power (Approx.)	9	9.5	9	watts
Power Output (Approx.)	125	140	225	watts

Key-down conditions per tube without amplitude modulation. Amplitude modulation essentially negative may be used if the positive peak of the audio-frequency envelope does not exceed 115 per cent of the carrier conditions.

♠ Obtained from fixed supply, by grid resistor, by cathode resistor, or by combination methods.

OPERATING CONSIDERATIONS

Type 808 requires Small four-contact socket and may be mounted in vertical position only, base down. OUTLINE 32, *Outlines* Section.

For operation at 60 Mc, plate voltage and plate input should be reduced to 75 per cent of maximum ratings; at 130 Mc, to 50 per cent. Plate shows cherry-red color when tube is operated at maximum CCS ratings, and orange-red color at maximum ICAS ratings.

When the 808 is used in the final amplifier or a preceding stage of a transmitter designed for break-in operation and oscillator keying, a small amount of fixed bias must be used to maintain plate current at a safe value. With a plate voltage of 2000 volts, a fixed bias of at least −30 volts should be used.

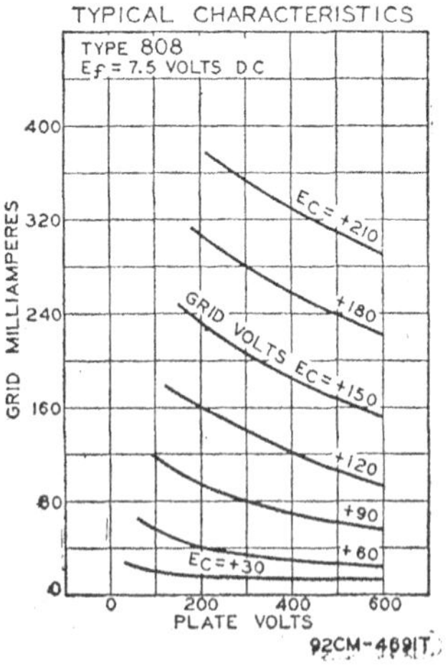

AVERAGE PLATE CHARACTERISTICS

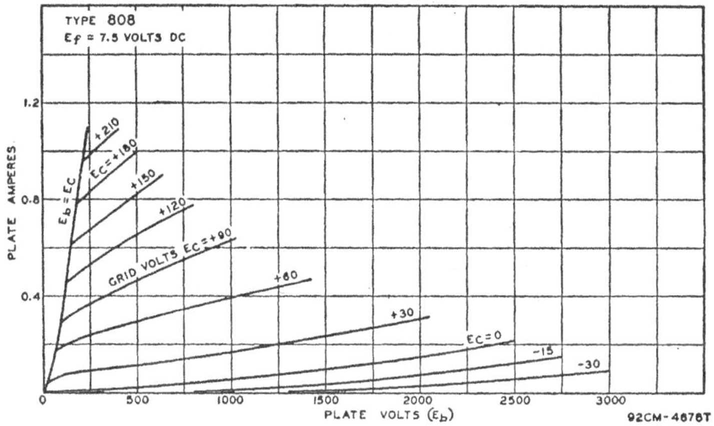

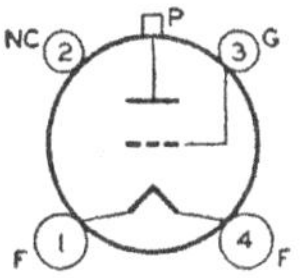

POWER TRIODE

809

Thoriated-tungsten-filament type used as rf power amplifier and oscillator. May be used with full input up to 60 Mc and with reduced input up to 120 Mc. Class C Telegraphy maximum plate dissipation, CCS 25 watts, ICAS 30 watts.

Filament Voltage (ac/dc)	6.3	volts
Filament Current	2.5	amperes
Amplification Factor	50	
Direct Interelectrode Capacitances:		
Grid to plate	6.7	μμf
Grid to filament	5.7	μμf
Plate to filament	0.9	μμf

RF POWER AMPLIFIER AND OSCILLATOR—Class C Telegraphy#
and
RF POWER AMPLIFIER—Class C FM Telephony

Maximum Ratings:	*CCS*	*ICAS*	
DC Plate Voltage	750 *max*	1000 *max*	volts
DC Grid Voltage	-200 *max*	-200 *max*	volts
DC Plate Current	100 *max*	100 *max*	ma
DC Grid Current	35 *max*	35 *max*	ma
Plate Input	75 *max*	100 *max*	watts
Plate Dissipation	25 *max*	30 *max*	watts

Typical Operation:				
DC Plate Voltage	500	750	1000	volts
DC Grid Voltage ♦	-50	-60	-75	volts
From grid resistor of	2500	3000	3000	ohms
From cathode resistor of	420	500	600	ohms
Peak RF Grid Voltage	135	140	160	volts
DC Plate Current	100	100	100	ma
DC Grid Current (Approx.)	20	20	25	ma
Driving Power (Approx.)	2.5	2.5	3.8	watts
Power Output (Approx.)	35	55	75	watts

Key-down conditions per tube without amplitude modulation. Amplitude modulation essentially negative may be used if the positive peak of the audio-frequency envelope does not exceed 115 per cent of the carrier conditions.

♦ Obtained from fixed supply, by grid resistor, by cathode resistor, or by combination methods.

OPERATING CONSIDERATIONS

Type 809 requires Small four-contact socket and may be mounted in vertical position with base down, or in horizontal position with pins 1 and 4 in vertical plane. OUTLINE 40, *Outlines* Section.

For operation at 70 Mc, plate voltage and plate input should be reduced to 88 per cent of maximum ratings; at 120 Mc, to 50 per cent. Plate shows no color when tube is operated at maximum CCS ratings, and shows a barely perceptible red color at maximum ICAS ratings.

When the 809 is used in the final amplifier or a preceding stage of a transmitter designed for break-in operation and oscillator keying, a small amount of fixed bias must be used to maintain the plate current at a safe value. With a plate voltage of 1000 volts, a fixed bias of at least −10 volts should be used.

810

POWER TRIODE

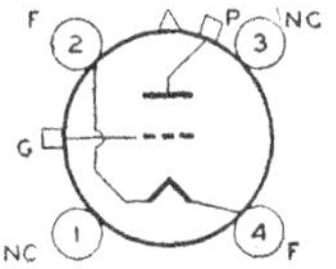

Thoriated-tungsten-filament type used as af power amplifier and modulator and as rf power amplifier and oscillator. May be used with full input up to 30 Mc and with reduced input up to 100 Mc. Class C Telegraphy maximum plate dissipation, CCS 125 watts, ICAS 175 watts.

FILAMENT VOLTAGE (AC/DC)	10	volts
FILAMENT CURRENT	4.5	amperes
AMPLIFICATION FACTOR	36	
DIRECT INTERELECTRODE CAPACITANCES:		
Grid to plate	4.8	μμf
Grid to filament	8.7	μμf
Plate to filament	12	μμf

AF POWER AMPLIFIER AND MODULATOR—Class B

Maximum Ratings:	*CCS*	*ICAS*	
DC PLATE VOLTAGE	2500 *max*	2750 *max*	volts
MAXIMUM-SIGNAL DC PLATE CURRENT■	250 *max*	250 *max*	ma
MAXIMUM-SIGNAL PLATE INPUT■	425 *max*	510 *max*	watts
PLATE DISSIPATION■	125 *max*	175 *max*	watts

Typical Operation (*Values are for 2 tubes*):			
DC Plate Voltage	2000	2250	volts
DC Grid Voltage†	−50	−60	volts
Peak AF Grid-to-Grid Voltage	345	380	volts
Zero-Signal DC Plate Current	60	70	ma
Maximum-Signal DC Plate Current	420	450	ma
Effective Load Resistance (Plate to plate)	11000	11600	ohms
Maximum-Signal Driving Power (Approx.)	10	13	watts
Maximum-Signal Power Output (Approx.)	590	725	watts

■ Averaged over any audio-frequency cycle of sine-wave form.
† For ac filament supply.

PLATE-MODULATED RF POWER AMPLIFIER—Class C Telephony

Carrier conditions per tube for use with a maximum modulation factor of 1.0

Maximum Ratings:	*CCS*	*ICAS*	
DC PLATE VOLTAGE	1600 *max*	2000 *max*	volts
DC GRID VOLTAGE	−500 *max*	−500 *max*	volts
DC PLATE CURRENT	210 *max*	250 *max*	ma
DC GRID CURRENT	70 *max*	75 *max*	ma
PLATE INPUT	335 *max*	500 *max*	watts
PLATE DISSIPATION	85 *max*	125 *max*	watts

Typical Operation:				
DC Plate Voltage	1250	1600	2000	volts
DC Grid Voltage ◊	−200	−200	−350	volts
From grid resistor of	4000	4000	5000	ohms
Peak RF Grid Voltage	370	370	550	volts
DC Plate Current	210	210	250	ma

DC Grid Current (Approx.)	50	50	70	ma
Driving Power (Approx.)	17	17	35	watts
Power Output (Approx.)	180	250	380	watts

♁ Obtained from grid resistor of value shown or from a combination of grid resistor with either fixed supply or cathode resistor.

RF POWER AMPLIFIER AND OSCILLATOR—Class C Telegraphy#
and
RF POWER AMPLIFIER—Class C FM Telephony

Maximum Ratings:	*CCS*	*ICAS*	
DC PLATE VOLTAGE	2000 *max*	2500 *max*	volts
DC GRID VOLTAGE	–500 *max*	–500 *max*	volts
DC PLATE CURRENT	250 *max*	300 *max*	ma
DC GRID CURRENT	70 *max*	75 *max*	ma
PLATE INPUT	500 *max*	750 *max*	watts
PLATE DISSIPATION	125 *max*	175 *max*	watts

Typical Operation:				
DC Plate Voltage	1500	2000	2500	volts
DC Grid Voltage♣	–120	–160	–180	volts
From grid resistor of	3000	4000	3000	ohms
From cathode resistor of	415	550	500	ohms
Peak RF Grid Voltage	280	330	350	volts
DC Plate Current	250	250	300	ma
DC Grid Current (Approx.)	40	40	60	ma
Driving Power (Approx.)	10	12	19	watts
Power Output (Approx.)	275	375	575	watts

Key-down conditions per tube without amplitude modulation. Amplitude modulation essentially negative may be used if the positive peak of the audio-frequency envelope does not exceed 115 per cent of the carrier conditions.

♣ Obtained from fixed supply, by grid resistor, by cathode resistor, or by combination methods.

OPERATING CONSIDERATIONS

Type 810 requires Jumbo four-contact socket and may be mounted in vertical position with base down, or in horizontal position with pins 1 and 2 in vertical plane. OUTLINE 53, *Outlines* Section.

For operation at 60 Mc, plate voltage and plate input should be reduced to 70 per cent of maximum ratings; at 100 Mc, to 50 per cent. Plate shows a barely perceptible red color when tube is operated at maximum CCS ratings, and shows a cherry-red color at maximum ICAS ratings.

When the 810 is used in the final amplifier or a preceding stage of a transmitter designed for break-in operation and oscillator keying, a small amount of fixed bias must be used to maintain the plate current at a safe value. With a plate voltage of 2500 volts, a fixed bias of at least –40 volts should be used.

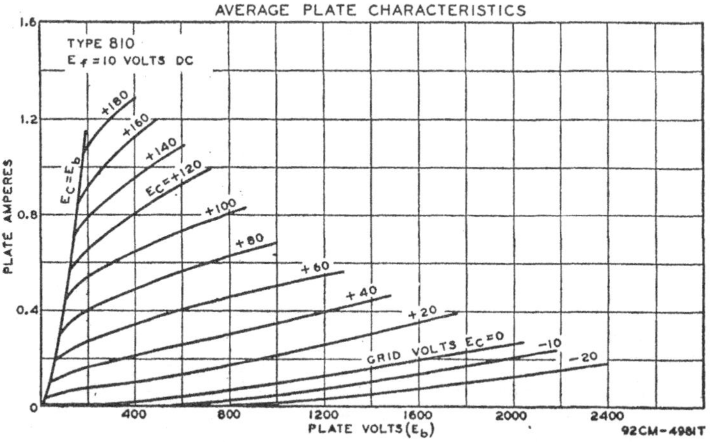

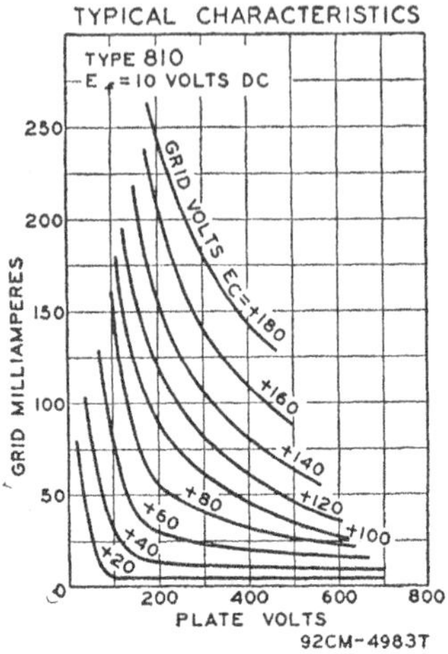

POWER TRIODE

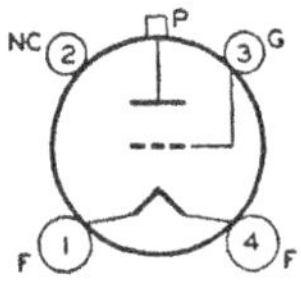

811-A

Thoriated-tungsten-filament type used as af power amplifier and modulator and as rf power amplifier and oscillator. May be used with full input up to 30 Mc. For operation at 60 Mc, plate voltage and plate input should be reduced to 89 per cent of maximum ratings; at 80 Mc, to 70 per cent; at 100 Mc, to 55 per cent. Class C Telegraphy maximum plate dissipation, CCS 45 watts, ICAS 65 watts. Requires Small four-contact socket and may be mounted in vertical position with base down, or in horizontal position with pins 1 and 4 in vertical plane. OUTLINE 39, *Outlines* Section. Plate shows no color when tube is operated at maximum CCS ratings, and shows a barely perceptible red color at maximum ICAS ratings.

FILAMENT VOLTAGE (AC/DC)	6.3	volts
FILAMENT CURRENT	4	amperes
AMPLIFICATION FACTOR*	160	
DIRECT INTERELECTRODE CAPACITANCES:		
Grid to plate	5.6	μμf
Grid to filament	5.9	μμf
Plate to filament	0.7	μμf

* Grid volts,−1; plate milliamperes, 20.

AF POWER AMPLIFIER AND MODULATOR—Class B

Maximum Ratings:	*CCS*	*ICAS*	
DC PLATE VOLTAGE	1250 *max*	1500 *max*	volts
MAXIMUM-SIGNAL DC PLATE CURRENT■	175 *max*	175 *max*	ma
MAXIMUM-SIGNAL PLATE INPUT■	165 *max*	235 *max*	watts
PLATE DISSIPATION■	45 *max*	65 *max*	watts

Typical Operation (*Values are for 2 tubes*):						
DC Plate Voltage	750	1250	1000	1250	1500	volts
DC Grid Voltage†	0	0	0	0	−4.5	volts
Peak AF Grid-to-Grid Voltage	197	145	185	175	170	volts
Zero-Signal DC Plate Current	32	50	44	54	32	ma
Maximum-Signal DC Plate Current	350	260	350	350	318	ma
Effective Load Resistance (Plate to plate)	5100	12400	7400	9200	12400	ohms
Maximum-Signal Driving Power (Approx.)	9.7	3.8	7.5	6	4.4	watts
Maximum-Signal Power Output (Approx.)	178	235	248	310	340	watts

■ Averaged over any audio-frequency cycle of sine-wave form.
† For ac filament supply.

PLATE-MODULATED RF POWER AMPLIFIER—Class C Telephony

Carrier conditions per tube for use with a maximum modulation factor of 1.0

Maximum Ratings:	CCS	ICAS	
DC PLATE VOLTAGE	1000 *max*	1250 *max*	volts
DC GRID VOLTAGE	−200 *max*	−200 *max*	volts
DC PLATE CURRENT	125 *max*	150 *max*	ma
DC GRID CURRENT	50 *max*	50 *max*	ma
PLATE INPUT	115 *max*	175 *max*	watts
PLATE DISSIPATION	30 *max*	45 *max*	watts

Typical Operation:			
DC Plate Voltage	1000	1250	volts
DC Grid Voltage♁	−55	−120	volts
From grid resistor of	1200	2700	ohms
Peak RF Grid Voltage	150	250	volts
DC Plate Current	115	140	ma
DC Grid Current (Approx.)	45	45	ma
Driving Power (Approx.)	6.1	10	watts
Power Output (Approx.)	88	135	watts

♁Obtained from grid resistor of value shown or from a combination of grid resistor with either fixed supply or cathode resistor.

RF POWER AMPLIFIER AND OSCILLATOR—Class C Telegraphy#
and
RF POWER AMPLIFIER—Class C FM Telephony

Maximum Ratings:	CCS	ICAS	
DC PLATE VOLTAGE	1250 *max*	1500 *max*	volts
DC GRID VOLTAGE	−200 *max*	−200 *max*	volts
DC PLATE CURRENT	175 *max*	175 *max*	ma
DC GRID CURRENT	50 *max*	50 *max*	ma
PLATE INPUT	175 *max*	260 *max*	watts
PLATE DISSIPATION	45 *max*	65 *max*	watts

Typical Operation:			
DC Plate Voltage	1250	1500	volts
DC Grid Voltage♣	−50	−70	volts
From grid resistor of	1100	1750	ohms
From cathode resistor of	270	330	ohms
Peak RF Grid Voltage	140	175	volts
DC Plate Current	140	173	ma
DC Grid Current (Approx.)	45	40	ma
Driving Power (Approx.)	5.7	7.1	watts
Power Output (Approx.)	135	200	watts

Key-down conditions per tube without amplitude modulation. Amplitude modulation essentially negative may be used if the positive peak of the audio-frequency envelope does not exceed 115 per cent of the carrier conditions.

♣Obtained from fixed supply, by grid resistor, by cathode resistor, or by combination methods.

SELF-RECTIFYING AMPLIFIER®—Class C

Maximum CCS Ratings:		
RMS PLATE VOLTAGE	1750 *max*	volts
DC GRID VOLTAGE	−125 *max*	volts
DC PLATE CURRENT	65 *max*	ma
DC GRID CURRENT	25 *max*	ma
PLATE INPUT	125 *max*	watts
PLATE DISSIPATION	45 *max*	watts

Typical Push-Pull Operation at 27 Mc (*Values are for 2 tubes*):		
RMS Plate Voltage	1750	volts
DC Grid Voltage♁	−70	volts
From grid resistor of	1500	ohms
DC Plate Current	130	ma
DC Grid Current (Approx.)	46	ma
Driving Power (Approx.)●	12	watts
Power Output (Approx.)	175	watts
Useful Power Output (Approx.)–75-per-cent circuit efficiency	130	watts

♁Obtained from grid resistor of value shown or from a combination of grid resistor with either fixed supply or cathode resistor.

●From a self-rectifying driver.

AMPLIFIER[⊕] —Class C

With separate, rectified, unfiltered, single-phase, full-wave plate supply

Maximum CCS Ratings:

DC Plate Voltage	1125 *max*	volts
DC Grid Voltage	–125 *max*	volts
DC Plate Current	160 *max*	ma
DC Grid Current	45 *max*	ma
Plate Input▲	175 *max*	watts
Plate Dissipation	45 *max*	watts

Typical Operation:

DC Plate Voltage	1125	volts
DC Grid Voltage ♁	–85	volts
From grid resistor of	1400	ohms
DC Plate Current	125	ma
DC Grid Current (Approx.)	25	ma
Driving Power (Approx.)▲▲	3	watts
Power Output (Approx.)	135	watts

▲ Power input is 1.23 times the product of dc plate voltage and dc plate current.

♁ Obtained from grid resistor of value shown or from a combination of grid resistor with either fixed supply or cathode resistor.

▲▲From a driver having a rectified, unfiltered, single-phase, full-wave plate supply.

⊕The 811-A is not recommended for oscillator service in applications involving wide variations of load. For such applications, the 812-A having a lower amplification factor is preferred for its ability to oscillate over a wide variation of load.

AVERAGE CHARACTERISTICS

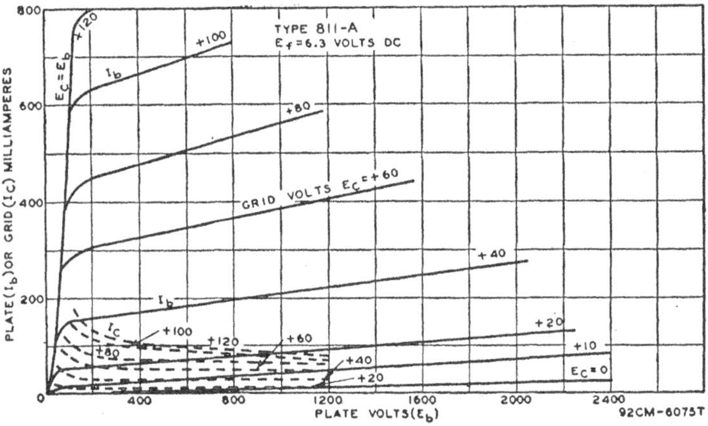

POWER TRIODE

812-A

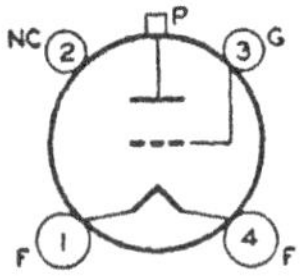

Thoriated-tungsten-filament type used as af power amplifier and modulator and as rf power amplifier and oscillator. May be used with full input up to 30 Mc and with reduced input up to 100 Mc. Class C Telegraphy maximum plate dissipation, CCS 45 watts, ICAS 65 watts.

Filament Voltage (ac/dc)	6.3	volts
Filament Current	4	amperes
Amplification Factor*	29	
Direct Interelectrode Capacitances:		
Grid to plate	5.5	μμf
Grid to filament	5.4	μμf
Plate to filament	0.77	μμf

*Grid volts, –30; plate milliamperes, 30.

AF POWER AMPLIFIER AND MODULATOR—Class B

Maximum Ratings:	CCS	ICAS	
DC PLATE VOLTAGE	1250 *max*	1500 *max*	volts
MAXIMUM-SIGNAL DC PLATE CURRENT■	175 *max*	175 *max*	ma
MAXIMUM-SIGNAL PLATE INPUT■	165 *max*	235 *max*	watts
PLATE DISSIPATION■	45 *max*	65 *max*	watts

Typical Operation (*Values are for 2 tubes*):			
DC Plate Voltage	1250	1500	volts
DC Grid Voltage†	-40	-48	volts
Peak AF Grid-to-Grid Voltage	225	270	volts
Zero-Signal DC Plate Current	22	28	ma
Maximum-Signal DC Plate Current	260	310	ma
Effective Load Resistance (Plate to plate)	12200	13200	ohms
Maximum-Signal Driving Power (Approx.)	3.5	5	watts
Maximum-Signal Power Output (Approx.)	235	340	watts

■ Averaged over any audio-frequency cycle of sine-wave form.
† For ac filament supply.

PLATE-MODULATED RF POWER AMPLIFIER—Class C Telephony

Carrier conditions per tube for use with a maximum modulation factor of 1.0

Maximum Ratings:	CCS	ICAS	
DC PLATE VOLTAGE	1000 *max*	1250 *max*	volts
DC GRID VOLTAGE	-200 *max*	-200 *max*	volts
DC PLATE CURRENT	125 *max*	150 *max*	ma
DC GRID CURRENT	35 *max*	35 *max*	ma
PLATE INPUT	115 *max*	175 *max*	watts
PLATE DISSIPATION	30 *max*	45 *max*	watts

Typical Operation:			
DC Plate Voltage	1000	1250	volts
DC Grid Voltage♁	-110	-115	volts
From grid resistor of	3400	3300	ohms
Peak RF Grid Voltage	220	240	volts
DC Plate Current	115	140	ma
DC Grid Current (Approx.)	33	35	ma
Driving Power (Approx.)	6.6	7.6	watts
Power Output (Approx.)	85	130	watts

♁ Obtained from grid resistor of value shown or from a combination of grid resistor with either fixed supply or cathode resistor.

RF POWER AMPLIFIER AND OSCILLATOR—Class C Telegraphy# and RF POWER AMPLIFIER—Class C FM Telephony

Maximum Ratings:	CCS	ICAS	
DC PLATE VOLTAGE	1250 *max*	1500 *max*	volts
DC GRID VOLTAGE	-200 *max*	-200 *max*	volts
DC PLATE CURRENT	175 *max*	175 *max*	ma
DC GRID CURRENT	35 *max*	35 *max*	ma
PLATE INPUT	175 *max*	260 *max*	watts
PLATE DISSIPATION	45 *max*	65 *max*	watts

Typical Operation:			
DC Plate Voltage	1250	1500	volts
DC Grid Voltage♦	-90	-120	volts
From grid resistor of	3000	4000	ohms
From cathode resistor of	530	590	ohms
Peak RF Grid Voltage	200	240	volts
DC Plate Current	140	173	ma
DC Grid Current (Approx.)	30	30	ma
Driving Power (Approx.)	5.4	6.5	watts
Power Output (Approx.)	130	190	watts

Key-down conditions per tube without amplitude modulation. Amplitude modulation essentially negative may be used if the positive peak of the audio-frequency envelope does not exceed 115 per cent of the carrier conditions.
♦ Obtained from fixed supply, by grid resistor, by cathode resistor, or by combination methods.

SELF-RECTIFYING OSCILLATOR OR AMPLIFIER—Class C

Maximum CCS Ratings:

RMS PLATE VOLTAGE	1750 *max*	volts
DC GRID VOLTAGE	−125 *max*	volts
DC PLATE CURRENT	75 *max*	ma
DC GRID CURRENT	20 *max*	ma
PLATE INPUT	145 *max*	watts
PLATE DISSIPATION	45 *max*	watts

Typical Push-Pull Operation at 27 Mc (*Values are for 2 tubes*):

RMS Plate Voltage	1740	volts
DC Grid Voltage ♁	−100	volts
From grid resistor of	3500	ohms
DC Plate Current	150	ma
DC Grid Current (At full load)	29	ma
Driving Power (Approx.)●	12	watts
Power Output (Approx.)	200	watts
Useful Power Output (Approx.)–75-per-cent circuit efficiency	150	watts

♁ Obtained from grid resistor of value shown or from a combination of grid resistor with either fixed supply or cathode resistor.

● From a self-rectified driver.

AMPLIFIER OR OSCILLATOR—Class C

With separate, rectified, unfiltered, single-phase, full-wave plate supply

Maximum CCS Ratings:

DC PLATE VOLTAGE	1125 *max*	volts
DC GRID VOLTAGE	−125 *max*	volts
DC PLATE CURRENT	160 *max*	ma
DC GRID CURRENT	32 *max*	ma
PLATE INPUT▲	175 *max*	watts
PLATE DISSIPATION	45 *max*	watts

Typical Operation:

DC Plate Voltage	1125	volts
DC Grid Voltage ♁	−65	volts
From grid resistor of	2200	ohms
DC Plate Current	125	ma
DC Grid Current (Approx.)	30	ma
Driving Power (Approx.)▲▲	5	watts
Power Output (Approx.)	135	watts

▲ Power input is 1.23 times the product of dc plate voltage and dc plate current.

♁ Obtained from grid resistor of value shown or from a combination of grid resistor with either fixed supply or cathode resistor.

▲▲ From a driver having a rectified, unfiltered, single-phase, full-wave plate supply.

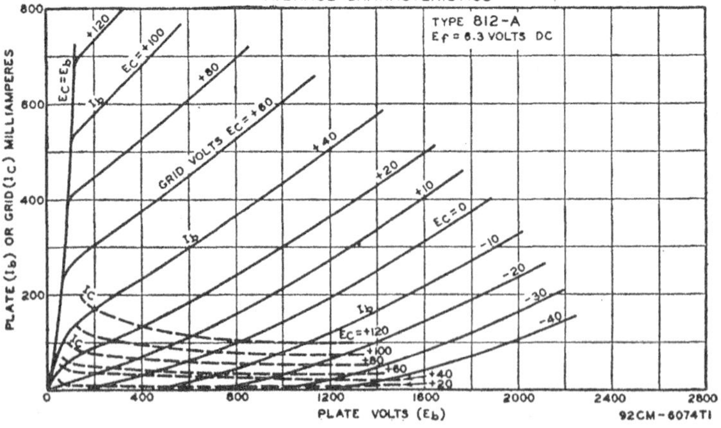

OPERATING CONSIDERATIONS

Type 812-A requires Small four-contact socket and may be mounted in vertical position with base down, or in horizontal position with pins 1 and 4 in vertical plane. OUTLINE 39, *Outlines* Section.

For operation at 60 Mc, plate voltage and plate input should be reduced to 89 per cent of maximum ratings; at 80 Mc, to 70 per cent; at 100 Mc, to 55 per cent. Plate shows no color when tube is operated at maximum CCS ratings, and shows a barely perceptible red color at maximum ICAS ratings.

When the 812-A is used in the final amplifier or a preceding stage of a transmitter designed for break-in operation and oscillator keying, a small amount of fixed bias must be used to maintain the plate current at a safe value. With a plate voltage of 1500 volts, a fixed bias of at least −45 volts should be used.

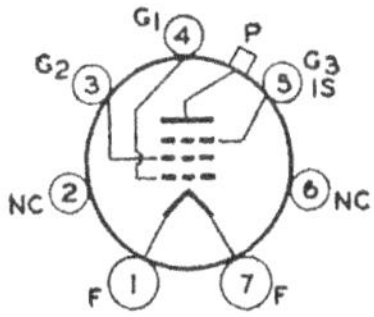

BEAM POWER TUBE

813

Thoriated-tungsten-filament type used as af power amplifier and modulator and as rf power amplifier and oscillator. May be used with full input up to 30 Mc. For operation at 45 Mc, plate voltage and plate input should be reduced to 87 per cent of maximum ratings; at 60 Mc, to 75 per cent; at 120 Mc, to 50 per cent. Class C Telegraphy maximum plate dissipation, CCS 100 watts, ICAS 125 watts. Requires Giant seven-contact socket and may be mounted in vertical position with base up or down, or in horizontal position with pins 2 and 6 in vertical plane. OUTLINE 47, *Outlines* Section. Plate shows no color when tube is operated at maximum CCS or ICAS ratings.

FILAMENT VOLTAGE (AC/DC)	10	volts
FILAMENT CURRENT	5	amperes
TRANSCONDUCTANCE*	3750	μmhos
MU-FACTOR, Grid No.2 to Grid No.1*	8.5	
DIRECT INTERELECTRODE CAPACITANCES:		
Grid No.1 to plate	0.25 *max*	μμf
Grid No.1 to filament, grid No.3, internal shield, grid No.2, and base shell	16.3	μμf
Plate to filament, grid No.3, internal shield, grid No.2, and base shell	14	μμf

* Plate volts, 2000; grid-No.2 volts, 400; plate milliamperes, 50.

AF POWER AMPLIFIER AND MODULATOR—Class AB1

Maximum Ratings:

	CCS	*ICAS*	
DC PLATE VOLTAGE	2250 *max*	2500 *max*	volts
DC GRID-No.2 (SCREEN-GRID) VOLTAGE	1100 *max*	1100 *max*	volts
MAXIMUM-SIGNAL DC PLATE CURRENT■	180 *max*	225 *max*	ma
MAXIMUM-SIGNAL PLATE INPUT■	360 *max*	450 *max*	watts
MAXIMUM-SIGNAL DC GRID-No.2 INPUT■	22 *max*	22 *max*	watts
PLATE DISSIPATION■	100 *max*	125 *max*	watts

Typical Operation (*Values are for 2 tubes*):

DC Plate Voltage	1500	2000	2250	2500	volts
DC Grid-No.3 (Suppressor-Grid) Voltage▲	0	0	0	0	volts
DC Grid-No.2 Voltage‡	750	750	750	750	volts
DC Grid-No.1 (Control-Grid) Voltage†	−85	−90	−95	−95	volts
Peak AF Grid-No.1-to-Grid-No.1 Voltage	160	160	170	180	volts
Zero-Signal DC Plate Current	50	50	50	50	ma
Maximum-Signal DC Plate Current	305	265	255	290	ma
Zero-Signal DC Grid-No.2 Current	2	2	2	2	ma
Maximum-Signal DC Grid-No.2 Current	45	43	53	54	ma
Effective Load Resistance (Plate to plate)	9300	16000	20000	19000	ohms
Maximum-Signal Driving Power (Approx.)	0	0	0	0	watts
Maximum-Signal Power Output (Approx.)	260	335	380	490	watts

Maximum Circuit Values (CCS or ICAS conditions):

Grid-No.1-Circuit Resistance	30000 *max*	ohms

■ Averaged over any audio-frequency cycle of sine-wave form.
▲ Grid No.3 should be connected to the mid-tap on the filament-transformer secondary winding or to the negative end of a filament operated on dc.
‡ Preferably obtained from a separate source or from the plate-voltage supply with a voltage divider.
† For ac filament supply.

PLATE-MODULATED PUSH-PULL RF POWER AMPLIFIER—Class C Telephony

Carrier conditions per tube for use with a maximum modulation factor of 1.0

Maximum Ratings:

	CCS	*ICAS*	
DC PLATE VOLTAGE	1600 *max*	2000 *max*	volts
DC GRID-NO.2 VOLTAGE	400 *max*	400 *max*	volts
DC GRID-NO.1 VOLTAGE	-300 *max*	-300 *max*	volts
DC PLATE CURRENT	150 *max*	200 *max*	ma
DC GRID-NO.1 CURRENT	25 *max*	30 *max*	ma
PLATE INPUT	240 *max*	400 *max*	watts
GRID-NO.2 INPUT	15 *max*	20 *max*	watts
PLATE DISSIPATION	67 *max*	100 *max*	watts

Typical Operation:

DC Plate Voltage	1250	1600	2000	volts
DC Grid-No.3 Voltage▲	0	0	0	volts
DC Grid-No.2 Voltage♣	300	300	350	volts
From series resistor of	27000	43000	41000	ohms
DC Grid-No.1 Voltage♁	–160	–160	–175	volts
From grid-No.1 resistor of	12500	13500	11000	ohms
Peak RF Grid-No.1 Voltage	250	250	300	volts
DC Plate Current	150	150	200	ma
DC Grid-No.2 Current	35	30	40	ma
DC Grid-No.1 Current (Approx.)	13	12	16	ma
Driving Power (Approx.)	2.9	2.7	4.3	watts
Power Output (Approx.)	140	180	300	watts

Maximum Circuit Values (CCS or ICAS conditions):

Grid-No.1-Circuit Resistance	30000 *max*	ohms

▲ Grid No.3 should be connected to the mid-tap on the filament-transformer secondary winding or to the negative end of a filament operated on dc.
♣ Obtained preferably from separate source modulated along with the plate supply, or from the modulated plate supply through series resistor of value shown for each operating condition.
♁ Obtained from a grid-No.1 resistor of value shown or from a combination of grid-No.1 resistor with either fixed supply or cathode resistor.

RF POWER AMPLIFIER AND OSCILLATOR—Class C Telegraphy#
and
RF POWER AMPLIFIER—Class C FM Telephony

Maximum Ratings:

	CCS	*ICAS*	
DC PLATE VOLTAGE	2000 *max*	2250 *max*	volts
DC GRID-NO.2 VOLTAGE	400 *max*	400 *max*	volts
DC GRID-NO.1 VOLTAGE	-300 *max*	-300 *max*	volts
DC PLATE CURRENT	180 *max*	225 *max*	ma
DC GRID-NO.1 CURRENT	25 *max*	30 *max*	ma
PLATE INPUT	360 *max*	500 *max*	watts
GRID-NO.2 INPUT	22 *max*	22 *max*	watts
PLATE DISSIPATION	100 *max*	125 *max*	watts

Typical Operation:

DC Plate Voltage	1250	1500	2000	2250	volts
DC Grid-No.3 Voltage▲	0	0	0	0	volts
DC Grid-No.2 Voltage♣	300	300	400	400	volts
From series resistor of	27000	40000	36000	46000	ohms
DC Grid-No.1 Voltage†®	–75	–90	–120	–155	volts
From grid-No.1 resistor of	6000	7500	12000	10000	ohms
From cathode resistor of	330	400	520	565	ohms
Peak RF Grid-No.1 Voltage	160	175	205	275	volts
DC Plate Current	180	180	180	220	ma
DC Grid-No.2 Current	35	30	45	40	ma
DC Grid-No.1 Current (Approx.)	12	12	10	15	ma
Driving Power (Approx.)	1.7	1.9	1.9	4.0	watts
Power Output (Approx.)	170	210	275	375	watts

Maximum Circuit Values:

Grid-No.1-Circuit Resistance	30000 *max*	ohms

‡ Key-down conditions per tube without amplitude modulation. Amplitude modulation essentially negative may be used if the positive peak of the audio-frequency envelope does not exceed 115 per cent of the carrier conditions.

▲ Grid No.3 should be connected to the mid-tap on the filament-transformer secondary winding or to the negative end of a filament operated on dc.

♦ Obtained from separate source, from plate-voltage supply with a voltage divider, or through series resistor of value shown for each operating condition. Grid-No. 2 voltage must not exceed 800 volts under key-up conditions.

† For ac filament supply.

° Obtained from a grid-No.1 resistor, from cathode resistor, or from a combination of grid-No.1 resistor with either fixed supply or cathode resistor. If preceding stage is keyed, bias must be obtained partially from a fixed supply to limit the plate current and plate dissipation to a safe value.

SELF-RECTIFYING OSCILLATOR OR AMPLIFIER—Class C

Maximum CCS Ratings:

RMS PLATE VOLTAGE	2800 *max*	volts
RMS GRID-NO.2 VOLTAGE	550 *max*	volts
DC GRID-NO.1 VOLTAGE	–100 *max*	volts
DC PLATE CURRENT	95 *max*	ma
DC GRID-NO.1 CURRENT	10 *max*	ma
PLATE INPUT•	295 *max*	watts
GRID-NO.2 INPUT•	22 *max*	watts
PLATE DISSIPATION	100 *max*	watts

Typical Operation:

RMS Plate Voltage	2800	volts
DC Grid-No.3 Voltage▲	0	volts
RMS Grid-No.2 Voltage□	530	volts
DC Grid-No.1 Voltage♁	–37	volts
From grid-No.1 resistor of	37000	ohms
DC Plate Current	95	ma
DC Grid-No.2 Current	12	ma
DC Grid-No.1 Current (Approx.)	1	ma
Driving Power (Approx.)♦	1	watt
Useful Power Output (Approx.)–75-per-cent circuit efficiency	170	watts

• Power input is 1.11 times the product of the rms voltage and the dc current.

▲ Grid No.3 should be connected to the mid-tap on the filament-transformer secondary winding or to the negative end of filament operated on dc.

□ Obtained from a separate ac supply in phase with the plate supply or from a low-voltage tap on the plate transformer. Use of a grid-No.2 series voltage-dropping resistor is not recommended.

♁ Obtained from a grid-No.1 resistor of value shown or from a combination of grid-No.1 resistor and cathode resistor. Fixed-bias operation is not recommended. The bias resistors should not be bypassed for the plate and grid-No.2 voltage supply frequency.

♦ From a self-rectified driver.

AMPLIFIER OR OSCILLATOR—Class C

With separate, rectified, unfiltered, single-phase, full-wave plate and grid-No.2 supply

Maximum CCS Ratings:

DC PLATE VOLTAGE	1800 *max*	volts
DC GRID-NO.2 VOLTAGE	360 *max*	volts
DC GRID-NO.1 VOLTAGE	–200 *max*	volts
DC PLATE CURRENT	190 *max*	ma
DC GRID-NO.1 CURRENT	22 *max*	ma
PLATE INPUT•	360 *max*	watts
GRID-NO.2 INPUT•	22 *max*	watts
PLATE DISSIPATION	100 *max*	watts

Typical Operation:

DC Plate Voltage	1800	volts
DC Grid-No.3 Voltage▲	0	volts
DC Grid-No.2 Voltage△	250	volts
DC Grid-No.1 Voltage♁	–120	volts
From grid-No.1 resistor of	10000	ohms
DC Plate Current	160	ma
DC Grid-No.2 Current	37	ma
DC Grid-No.1 Current (Approx.)	12	ma

Driving Power (Approx.)▲▲	2	watts
Useful Power Output (Approx.)–75-per-cent circuit efficiency	210	watts

● Power input is 1.23 times the product of dc plate voltage and dc plate current.
▲ Grid No.3 should be connected to the mid-tap on the filament-transformer secondary winding or the negative end of a filament operated on dc.
△ Obtained from a separate, rectified, unfiltered, single-phase, full-wave supply in phase with the plate supply, or from the rectified, unfiltered, single-phase, full-wave supply by means of taps on the plate transformer.
♁ Obtained from grid-No.1 resistor of value shown or from a combination of grid-No.1 resistor and cathode resistor. Fixed-bias operation is not recommended. The bias resistors should not be bypassed for the plate and grid-No.2 voltage supply frequency.
▲▲ From a driver having a rectified, unfiltered, single-phase, full-wave plate supply.

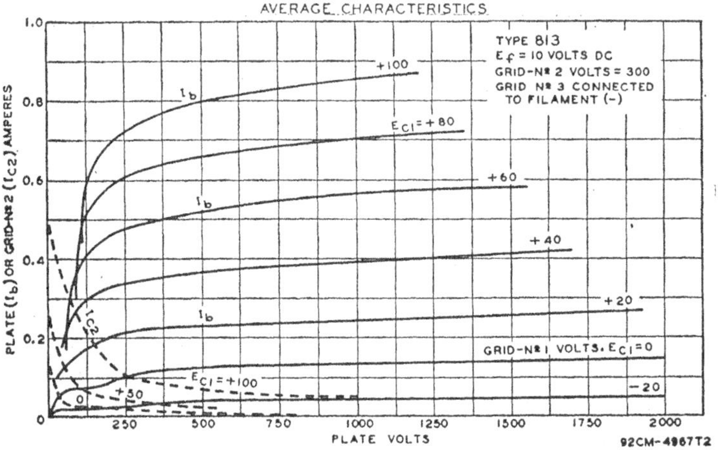

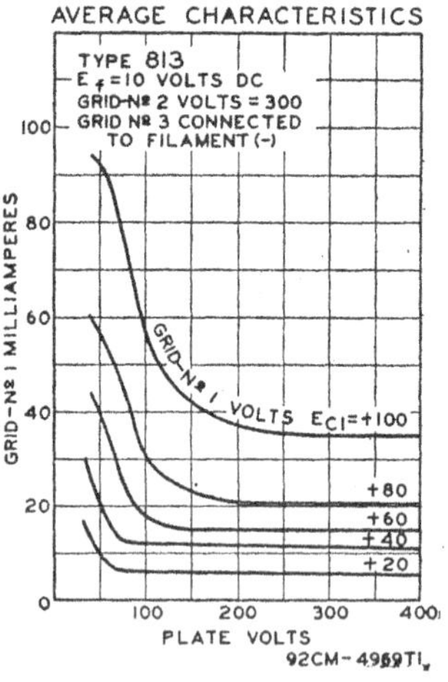

BEAM POWER TUBE

814

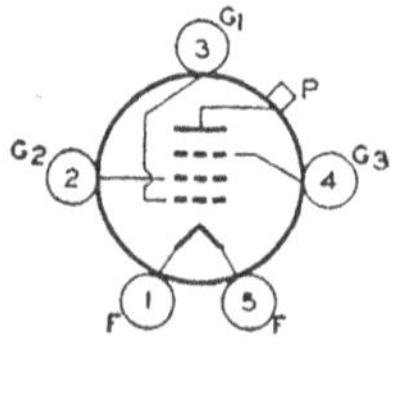

Thoriated-tungsten-filament type used as rf power amplifier and oscillator. May be used with full input up to 30 Mc. For operation at 50 Mc, plate voltage and plate input should be reduced to 80 per cent; at 75 Mc, to 64 per cent. Class C Telegraphy maximum plate dissipation, CCS 50 watts, ICAS 65 watts. Requires Small five-contact socket and may be mounted in vertical position with base down, or in horizontal position with pins 2 and 4 in vertical plane. OUTLINE 48, *Outlines* Section. Plate shows no color when tube is operated at maximum CCS ratings, and shows a barely perceptible red color at maximum ICAS ratings.

FILAMENT VOLTAGE (AC/DC)	10	volts
FILAMENT CURRENT	3.25	amperes
TRANSCONDUCTANCE (For plate current of 39 milliamperes)	3300	μmhos
DIRECT INTERELECTRODE CAPACITANCES:		
Grid No.1 to plate	0.15 *max*	μμf
Grid No.1 to filament, grid No.3, and grid No.2	13.5	μμf
Plate to filament, grid No.3, and grid No.2	13.5	μμf

RF POWER AMPLIFIER AND OSCILLATOR—Class C Telegraphy#
and
RF POWER AMPLIFIER—Class C FM Telephony

Maximum Ratings:	*CCS*	*ICAS*	
DC PLATE VOLTAGE	1250 *max*	1500 *max*	volts
DC GRID-NO.2 (SCREEN-GRID) VOLTAGE	400 *max*	400 *max*	volts
DC GRID-NO.1 (CONTROL-GRID) VOLTAGE	-300 *max*	-300 *max*	volts
DC PLATE CURRENT	150 *max*	150 *max*	ma
DC GRID-NO.1 CURRENT	15 *max*	15 *max*	ma
PLATE INPUT	180 *max*	225 *max*	watts
GRID-NO.2 INPUT	10 *max*	10 *max*	watts
PLATE DISSIPATION	50 *max*	65 *max*	watts

Typical Operation:				
DC Plate Voltage	1000	1250	1500	volts
DC Grid-No.3 (Suppressor-Grid) Voltage^△	0	0	0	volts
DC Grid-No.2 Voltage^⊕	300	300	300	volts
From series resistor of	40000	42000	50000	ohms
DC Grid-No.1 Voltage †♁	-70	-80	-90	volts
From grid-No.1 resistor of	7000	8000	9000	ohms
From cathode resistor of	395	455	490	ohms
Peak RF Grid-No.1 Voltage	150	165	170	volts
DC Plate Current	150	144	150	ma
DC Grid-No.2 Current	17.5	22.5	24	ma
DC Grid-No.1 Current (Approx.)	10	10	10	ma
Driving Power (Approx.)	1.35	1.5	1.5	watts
Power Output (Approx.)	100	130	160	watts

Key-down conditions per tube without amplitude modulation. Amplitude modulation essentially negative may be used if the positive peak of the audio-frequency envelope does not exceed 115 per cent of the carrier conditions.

△ Grid-No.3 should be connected to the mid-tap on the filament-transformer secondary winding or to the negative end of filament operated on dc.

⊕ Obtained from separate source, from plate-voltage supply with a voltage divider, or through series resistor of value shown. If preceding stage is keyed, partial fixed bias is required.

† For ac filament supply.

♁ Obtained preferably from grid-No.1 resistor or from a combination of grid-No.1 resistor with either fixed supply or cathode resistor.

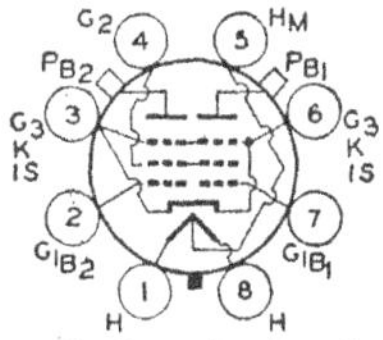

TWIN BEAM POWER TUBE

815

Heater-cathode type used as af power amplifier and modulator and as rf power amplifier and oscillator. May be used with full input up to 125 Mc. For operation at 175 Mc, plate voltage and plate input should be reduced to 80 per cent of maximum ratings; at 200 Mc, to 70 per cent. Class C Telegraphy maximum plate dissipation (per tube), CCS 20 watts, ICAS 25 watts. Requires Octal socket and may be mounted in any position. OUTLINE 25, *Outlines* Section. Plates show no color when tube is operated at maximum CCS or ICAS ratings.

HEATER ARRANGEMENT	*Series*	*Parallel*	
HEATER VOLTAGE (AC/DC)	12.6 ± 10%	6.3 ± 10%	volts
HEATER CURRENT	0.8	1.6	amperes
TRANSCONDUCTANCE (Each unit, for plate current of 25 milliamperes.)	4000		μmhos
MU-FACTOR, Grid No.2 to Grid No.1., (Each unit)	6.5		
DIRECT INTERELECTRODE CAPACITANCES (Each unit):			
Grid No.1 to plate	0.22 *max*		μμf
Grid No.1 to cathode, grid No.3, internal shield, grid No.2, and heater mid-tap	14.0		μμf
Plate to cathode, grid No.3, internal shield, grid No.2, and heater mid-tap	8.5		μμf

PUSH-PULL AF POWER AMPLIFIER AND MODULATOR—Class AB2

Values are on a per-tube basis

Maximum Ratings:	CCS	ICAS	
DC PLATE VOLTAGE	400 *max*	500 *max*	volts
DC GRID-NO.2 (SCREEN-GRID) VOLTAGE	225 *max*	225 *max*	volts
MAXIMUM-SIGNAL DC PLATE CURRENT■	150 *max*	150 *max*	ma
MAXIMUM-SIGNAL PLATE INPUT■	60 *max*	75 *max*	watts
MAXIMUM-SIGNAL GRID-NO.2 INPUT■	4.5 *max*	4.5 *max*	watts
PLATE DISSIPATION■	20 *max*	25 *max*	watts
PEAK HEATER-CATHODE VOLTAGE:			
Heater negative with respect to cathode	100 *max*	100 *max*	volts
Heater positive with respect to cathode	100 *max*	100 *max*	volts

Typical Operation:			
DC Plate Voltage	400	500	volts
DC Grid-No.2 Voltage♦♠	125	125	volts
DC Grid-No.1 (Control-Grid) Voltage	–15	–15	volts
Peak AF Grid-No.1-to-Grid-No.1 Voltage	60	60	volts
Zero-Signal DC Plate Current	20	22	ma
Maximum-Signal DC Plate Current	150	150	ma
Maximum-Signal DC Grid-No.2 Current	32	32	ma
Effective Load Resistance (Plate to plate)	6200	8000	ohms
Maximum-Signal Driving Power (Approx.)	0.36	0.36	watt
Maximum-Signal Power Output (Approx.)	42	54	watts

■ Averaged over any audio-frequency cycle of sine-wave form.

♦ In applications requiring the use of grid-No.2 voltages above 135 volts, provision should be made for the adjustment of grid-No.1 bias for each unit separately. The necessity for this adjustment at the lower grid-No.2 voltages depends on the distortion requirements and on whether the plate-dissipation rating is exceeded at zero-signal plate current.

♠ Obtained preferably from a separate source, or from the plate-voltage supply with a voltage divider.

PLATE-MODULATED PUSH-PULL RF POWER AMPLIFIER—Class C Telephony

Carrier conditions per tube for use with a maximum modulation factor of 1.0

Maximum Ratings:	CCS	ICAS	
DC PLATE VOLTAGE	325 *max*	400 *max*	volts
DC GRID-NO.2 VOLTAGE	225 *max*	225 *max*	volts
DC GRID-NO.1 VOLTAGE	–175 *max*	–175 *max*	volts
DC PLATE CURRENT	125 *max*	150 *max*	ma
DC GRID-NO.1 CURRENT	7 *max*	7 *max*	ma
PLATE INPUT	40 *max*	60 *max*	watts
GRID-NO.2 INPUT	4 *max*	4 *max*	watts
PLATE DISSIPATION	13.5 *max*	20 *max*	watts
PEAK HEATER-CATHODE VOLTAGE:			
Heater negative with respect to cathode	100 *max*	100 *max*	volts
Heater positive with respect to cathode	100 *max*	100 *max*	volts

Typical Operation:			
DC Plate Voltage	325	400	volts
DC Grid-No.2 Voltage♠♦	165	175	volts
From series resistor of	10000	15000	ohms
DC Grid-No.1 Voltage ◊	–45	–45	volts
From grid-No.1 resistor of	11250	15000	ohms
Peak RF Grid-No.1-to-Grid-No.1 Voltage	112	116	volts
DC Plate Current	123	150	ma
DC Grid-No.2 Current	16	15	ma
DC Grid-No.1 Current (Approx.)	4	3	ma
Driving Power (Approx.)	0.2	0.16	watt
Power Output (Approx.)	30	45	watts

Maximum Circuit Values:

Grid-No.1-Circuit Resistance	15000 *max*	ohms

♣ Obtained preferably from separate source modulated along with the plate supply, or from the modulated plate supply through series resistor of value shown.
♦ In applications requiring the use of grid-No.2 voltage above 135 volts, provision should be made for the adjustment of grid-No.1 bias for each unit separately. The necessity for this adjustment at lower grid-No.2 voltages depends on the distortion requirements and on whether the plate-dissipation rating is exceeded at zero-signal plate current.
⁶ Obtained from grid-No.1 resistor or from a combination of grid-No.1 resistor with either fixed supply or cathode resistor.

PUSH-PULL RF POWER AMPLIFIER AND OSCILLATOR—Class C Telegraphy# and PUSH-PULL RF POWER AMPLIFIER—Class C FM Telephony

Maximum Ratings:	*CCS*	*ICAS*	
DC Plate Voltage	400 *max*	500 max	volts
DC Grid-No.2 Voltage	225 *max*	225 *max*	volts
DC Grid-No.1 Voltage	–175 *max*	–175 *max*	volts
DC Plate Current	150 *max*	150 *max*	ma
DC Grid-No.1 Current	7 *max*	7 *max*	ma
Plate Input	60 *max*	75 *max*	watts
Grid-No.2 Input	4.5 *max*	4.5 *max*	watts
Plate Dissipation	20 *max*	25 *max*	watts
Peak Heater-Cathode Voltage:			
Heater negative with respect to cathode	100 *max*	100 *max*	volts
Heater positive with respect to cathode	100 *max*	100 *max*	volts

Typical Operation:			
DC Plate Voltage	400	500	volts
DC Grid-No.2 Voltage ♣♦	145	200	volts
From a series resistor of	15000	17500	ohms
DC Grid-No.1 Voltage ⁶	–45	–45	volts
From grid-No.1 resistor of	10000	13000	ohms
From cathode resistor of	260	265	ohms
Peak RF Grid-No.1-to-Grid-No.1 Voltage	116	112	volts
DC Plate Current	150	150	ma
DC Grid-No.2 Current	17	17	ma
DC Grid-No.1 Current (Approx.)	4.5	3.5	ma
Driving Power (Approx.)	0.23	0.18	watt
Power Output (Approx.)	44	56	watts

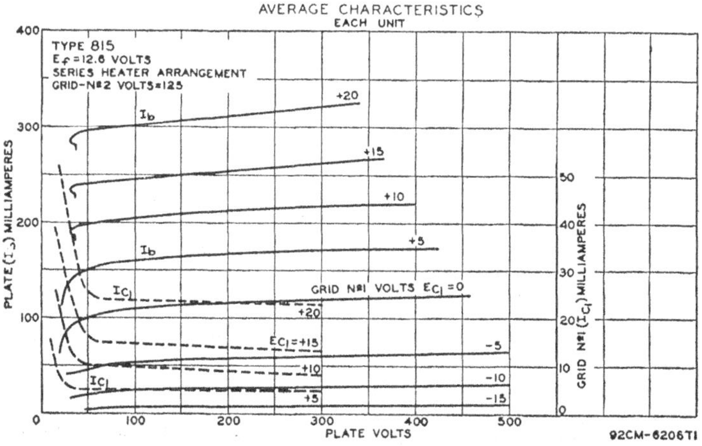

Maximum Circuit Values:

Grid-No.1-Circuit Resistance	15000 *max*	ohms

ƒ Key-down conditions per tube without amplitude modulation. Amplitude modulation essentially negative may be used if the positive peak of the audio-frequency envelope does not exceed 115 per cent of the carrier conditions.

♦ Obtained from separate source, from plate-voltage supply with a voltage divider, or through series resistor of value shown. Grid-No.2 voltage must not exceed 600 volts under key-up conditions.

♦ In applications requiring the use of grid-No.2 voltages above 135 volts, provision should be made for adjustment of grid-No.1 bias for each unit separately. The necessity for this adjustment at lower grid-No.2 voltages depends on the distortion requirements and on whether the plate-dissipation rating is exceeded at zero-signal plate current.

△ Obtained from fixed supply, by grid-No.1 resistor, by cathode resistor, or by combination methods.

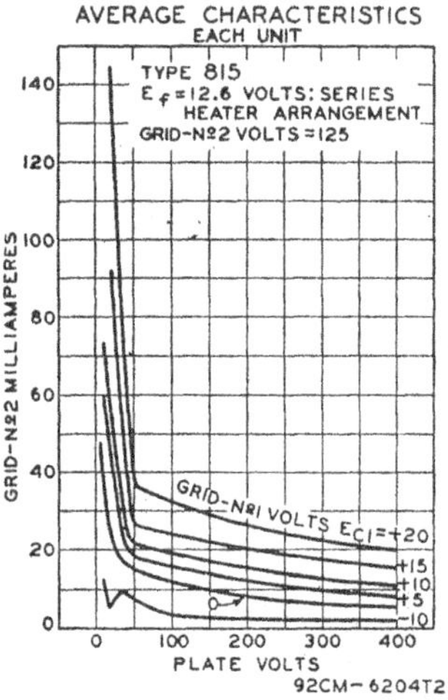

816

HALF-WAVE MERCURY-VAPOR RECTIFIER

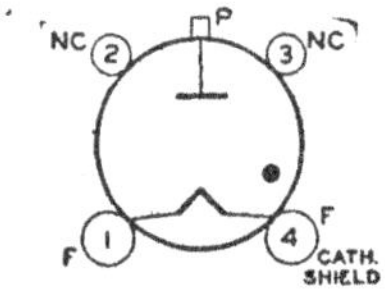

Coated-filament type used in power supply of transmitting and industrial equipment. Maximum peak inverse anode volts, 7500; maximum average anode amperes, 125. Requires Small four-contact socket and may be mounted in vertical position only, base down. OUTLINE 26, *Outlines* Section.

FILAMENT VOLTAGE (AC)°	2.5 ± 10%	volts
FILAMENT CURRENT	2.0	amperes
TUBE VOLTAGE DROP (Approx.)	15	volts

° Filament voltage must be applied at least 10 seconds before the application of anode voltage.

HALF-WAVE RECTIFIER

Maximum Ratings *(For power-supply frequency of 60 cps)*:

PEAK INVERSE ANODE VOLTAGE	7500 *max*	volts
ANODE CURRENT:		
Peak	500 *max*	ma
Average ♦	125 *max*	ma
Fault, for duration of 0.1 second maximum	5 *max*	amperes
CONDENSED-MERCURY-TEMPERATURE RANGE	20 to 60	°C

♦ Averaged over any interval of 30 seconds maximum.

Operating Values:

Circuit (For circuit figures, refer to *Rectifier Considerations* Section)	*Fig.*	*Max. Trans. Sec. Volts (RMS) E*	*Approx. DC Output Volts To Filter Eav*	*Max. DC Output Amperes Iav*	*Max. DC Output KW To Filter Pdc*
		In-Phase Operation			
Half-Wave Single-Phase...	54	5300	2400	0.125	0.3
Full-Wave Single-Phase...	55	2600	2400	0.250	0.6
Series Single-Phase.......	56	5300	4800	0.250	1.2
Half-Wave Three-Phase...	57	3000	3600	0.750	2.7
		Quadrature Operation			
Parallel Three-Phase......	58	3000	3600	1.5	5.4
Series Three-Phase........	59	3000	7200	0.75	5.4
Half-Wave Four-Phase....	60	2600	3500	0.45* 0.5□	1.55* 1.75□
Half-Wave Six-Phase.....	61	2600	3600	0.47* 0.5□	1.70* 1.80□

* Resistive load. □ Inductive load.

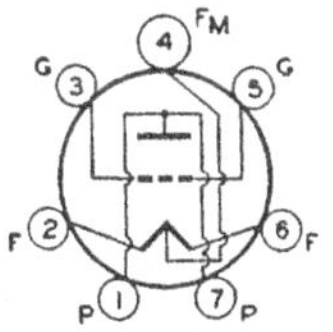

POWER TRIODE

826

Thoriated-tungsten-filament type used as rf power amplifier and oscillator. May be used with full input up to 250 Mc and with reduced input up to 300 Mc. Class C Telegraphy maximum plate dissipation, with natural cooling, CCS 45 watts, ICAS 55 watts; with forced-air cooling, CCS 60 watts, ICAS 75 watts.

Filament Voltage (ac/dc)	7.5	volts
Filament Current	4	amperes
Amplification Factor	31	
Direct Interelectrode Capacitances:		
Grid to plate	3	μμf
Grid to filament mid-tap	3	μμf
Plate to filament mid-tap	1.1	μμf

PLATE-MODULATED RF POWER AMPLIFIER—Class C Telephony

Carrier conditions per tube for use with a maximum modulation factor of 1.0

	Natural Cooling		Forced-Air Cooling		
Maximum Ratings:	*CCS*	*ICAS*	*CCS*	*ICAS*	
DC Plate Voltage	800 *max*	1000 *max*	800 *max*	1000 *max*	volts
DC Grid Voltage	−600 *max*	−600 *max*	−600 *max*	−600 *max*	volts
DC Plate Current	95 *max*	125 *max*	95 *max*	125 *max*	ma
DC Grid Current	40 *max*	40 *max*	40 *max*	40 *max*	ma
Plate Input	60 *max*	95 *max*	75 *max*	125 *max*	watts
Plate Dissipation	30 *max*	45 *max*	40 *max*	60 *max*	watts
Typical Operation:					
DC Plate Voltage	—	1000	800	1000	volts
DC Grid Voltage† ♂	—	−160	−100	−100	volts
From grid resistor of	—	4000	2800	2800	ohms
Peak RF Grid Voltage	—	320	198	210•	volts
DC Plate Current	—	95	94	125	ma
DC Grid Current (Approx.)	—	40	35	35	ma
Driving Power (Approx.)	—	11.5	6.3	6.6	watts
Power Output (Approx.)	—	70	53	90	watts

† For ac filament supply.

♂ Obtained by grid resistor of value shown. Fixed-bias operation is not recommended for linear modulation.

• To obtain linear modulation to 100 per cent, the driver stage should be modulated approximately 10 per cent.

RF POWER AMPLIFIER AND OSCILLATOR—Class C Telegraphy#

and

RF POWER AMPLIFIER—Class C FM Telephony

	Natural Cooling		Forced-Air Cooling		
Maximum Ratings:	*CCS*	*ICAS*	*CCS*	*ICAS*	
DC Plate Voltage	1000 *max*	1000 *max*	1000 *max*	1250 *max*	volts
DC Grid Voltage	−600 *max*	−600 *max*	−600 *max*	−600 *max*	volts

DC PLATE CURRENT	125 *max*	140 *max*	125 *max*	140 *max*	ma
DC GRID CURRENT	40 *max*	40 *max*	40 *max*	40 *max*	ma
PLATE INPUT	95 *max*	130 *max*	125 *max*	175 *max*	watts
PLATE DISSIPATION	45 *max*	55 *max*	60 *max*	75 *max*	watts
Typical Operation:					
DC Plate Voltage	—	1000	1000	1250	volts
DC Grid Voltage†△	—	-70	-70	-125	volts
From grid resistor of	—	2000	2000	3600	ohms
From cathode resistor of	—	425	440	780	ohms
Peak RF Grid Voltage	—	183	183	245	volts
DC Plate Current	—	130	125	125	ma
DC Grid Current (Approx.)	—	35	35	35	ma
Driving Power (Approx.)	—	5.8	5.8	7.7	watts
Power Output (Approx.)	—	90	86	120	watts

♯ Key-down conditions per tube without amplitude modulation. Amplitude modulation essentially negative may be used if the positive peak of the audio-frequency envelope does not exceed 115 per cent of the carrier conditions.

† For ac filament supply.

△ Obtained from fixed supply, by grid resistor, by cathode resistor, or by combination methods.

OPERATING CONSIDERATIONS

Type 826 requires Septar seven-contact socket and may be mounted in vertical position only, base up or down. OUTLINE 16, *Outlines* Section.

For operation at 300 Mc, plate voltage and plate input should be reduced to 80 per cent of maximum ratings. Plate shows an orange-red color when tube is operated at maximum CCS ratings, and shows a bright orange-red color at maximum ICAS ratings.

When the 826 is used in the final amplifier or a preceding stage of a transmitter designed for break-in operation and oscillator keying, a small amount of fixed bias must be used to maintain the plate current at a safe value. With plate voltage of 1250 volts, a fixed bias of at least -22.5 volts should be used.

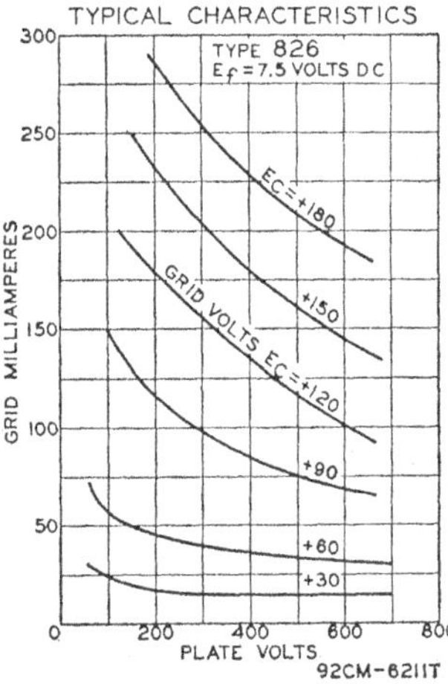

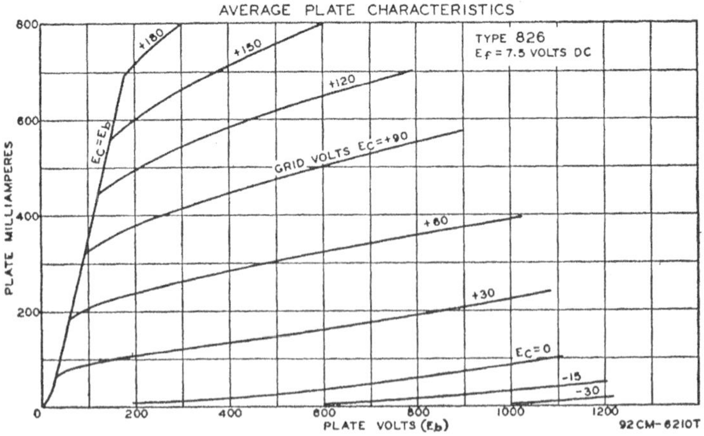

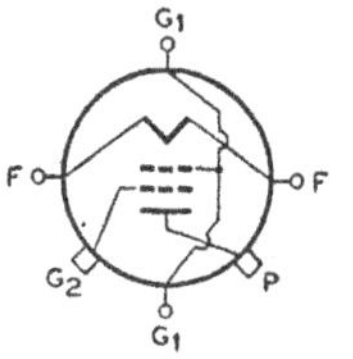

BEAM POWER TUBE

827-R

Forced-air-cooled type having thoriated-tungsten filament and integral radiator used as rf power amplifier and oscillator at frequencies up to 110 Mc. Class C Telegraphy maximum CCS plate dissipation, 800 watts.

Filament Voltage (ac/dc)	7.5	volts
Filament Current	25	amperes
Filament Starting Current	50 *max*	amperes
Mu-Factor, Grid No.2 to Grid No.1*	16	
Direct Interelectrode Capacitances (With external shielding):		
Grid No.1 to plate	0.22 *max*	μμf
Grid No.1 to filament and grid No.2	18.5	μμf
Plate to filament and grid No.2	11.0	μμf

* Plate volts, 2000; grid-No.2 volts, 1100; plate milliamperes, 350.

RF POWER AMPLIFIER AND OSCILLATOR—Class C Telegraphy#
and
RF POWER AMPLIFIER—Class C FM Telephony

Maximum CCS Ratings:

DC Plate Voltage	3500 *max*	volts
DC Grid-No.2 (screen-grid) Voltage	1000 *max*	volts
DC Grid-No.1 (control-grid) Voltage	–500 *max*	volts
DC Plate Current	500 *max*	ma
DC Grid-No.1 Current	150 *max*	ma
Plate Input	1500 *max*	watts
Grid-No.2 Input	150 *max*	watts
Plate Dissipation	800 *max*	watts
Ambient Temperature	45 *max*	°C

Key-down conditions per tube without amplitude modulation. Amplitude modulation essentially negative may be used if the positive peak of the audio-frequency envelope does not exceed 115 per cent of the carrier conditions.

OPERATING CONSIDERATIONS

Type 827-R requires special mounting and may be mounted in vertical position only with grid-No.1 and filament terminals up. Outline 76, *Outlines* Section.

At maximum CCS ratings, 100 cubic feet of forced air per minute from plate to seal end are required. Also, flow of 10 cubic feet per minute from 1-inch diameter nozzle should be directed into header. Air flow must start before any voltages are applied to the 827-R. Maximum temperatures: incoming air, 45°C; radiator, 150°C; glass, 150°C; filament seals, 175°C.

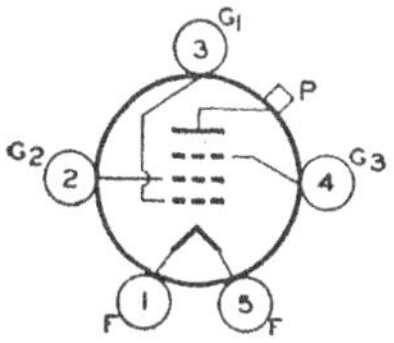

BEAM POWER TUBE

828

Thoriated-tungsten-filament type used as af power amplifier and modulator and as rf power amplifier and oscillator. May be used with full input up to 30 Mc. For operation at 50 Mc, plate voltage and plate input should be reduced to 80 per cent of maximum ratings; at 75 Mc, to 65 per cent. Class C Telegraphy maximum plate dissipation, CCS 70 watts, ICAS 80 watts. Requires Small five-contact socket and may be mounted in vertical position with base down, or in horizontal position with pins 2 and 4 in vertical plane. Outline 48, *Outlines* Section. Plate shows no color when tube is operated at maximum CCS ratings, and shows a barely perceptible red color at maximum ICAS ratings.

Filament Voltage (ac/dc)	10	volts
Filament Current	3.25	amperes
Transconductance (For plate current of 43 milliamperes)	2700	μmhos

Direct Interelectrode Capacitances:		
Grid No.1 to plate	0.07 *max*	μμf
Grid No.1 to filament, grid No.3, and grid No.2	12	μμf
Plate to filament, grid No.3, and grid No.2	14	μμf

AF POWER AMPLIFIER AND MODULATOR—Class AB1

Maximum Ratings:	*CCS*	*ICAS*	
DC Plate Voltage	1750 *max*	2000 *max*	volts
DC Grid-No.3 (suppressor-grid) Voltage	100 *max*	100 *max*	volts
DC Grid-No.2 (screen-grid) Voltage	750 *max*	750 *max*	volts
Maximum-Signal DC Plate Current■	150 *max*	150 *max*	ma
Maximum-Signal Plate Input■	225 *max*	270 *max*	watts
Maximum-Signal DC Grid-No.2 Input■	16 *max*	23 *max*	watts
Plate Dissipation■	70 *max*	80 *max*	watts

Typical Operation *(Values are for 2 tubes):*			
DC Plate Voltage	1700	2000	volts
DC Grid-No.3 Voltage	60	60	volts
DC Grid-No.2 Voltage●	750	750	volts
DC Grid-No.1 (Control-Grid) Voltage†	-120	-120	volts
Peak AF Grid-No.1-to-Grid-No.1 Voltage	240	240	volts
Zero-Signal DC Plate Current	50	50	ma
Maximum-Signal DC Plate Current	248	270	ma
DC Grid-No.3 Current	9	9	ma
Zero-Signal DC Grid-No.2 Current	4	2	ma
Maximum-Signal DC Grid-No.2 Current	43	60	ma
Effective Load Resistance (Plate to plate)	16200	18500	ohms
Maximum-Signal Driving Power (Approx.)	0	0	watts
Maximum-Signal Power Output (Approx.)	300▲	385	watts

Maximum Circuit Values (CCS or ICAS conditions):		
Grid-No.1-Circuit Resistance:		
For fixed-bias operation	0.22 *max*	megohm
For cathode-bias operation	Not recommended	

■ Averaged over any audio-frequency cycle of sine-wave form.
● Zero-signal grid-No.2 voltage must not exceed 775 volts.
† For ac filament supply.
▲ Distortion only one per cent with 20 db of feedback to grid of driver.

PLATE-MODULATED RF POWER AMPLIFIER—Class C Telephony

Carrier conditions per tube for use with a maximum modulation factor of 1.0

Maximum Ratings:	*CCS*	*ICAS*	
DC Plate Voltage	1000 *max*	1250 *max*	volts
DC Grid-No.3 Voltage	100 *max*	100 *max*	volts
DC Grid-No.2 Voltage	400 *max*	400 *max*	volts
DC Grid-No.1 Voltage	-300 *max*	-300 *max*	volts
DC Plate Current	135 *max*	160 *max*	ma
DC Grid-No.1 Current	15 *max*	15 *max*	ma
Plate Input	135 *max*	200 *max*	watts
Grid-No.3 Input	5 *max*	5 *max*	watts
Grid-No.2 Input	11 *max*	11 *max*	watts
Plate Dissipation	47 *max*	70 *max*	watts

Typical Operation:			
DC Plate Voltage	1000	1250	volts
DC Grid-No.3 Voltage	75	75	volts
DC Grid-No.2 Voltage♦	400	400	volts
From series resistor of	26000	30000	ohms
DC Grid-No.1 Voltage†♢	-140	-140	volts
From grid-No.1 resistor of	14000	11700	ohms
Peak RF Grid-No.1 Voltage	230	250	volts
DC Plate Current	135	160	ma
DC Grid-No.3 Current	13	15	ma
DC Grid-No.2 Current	23	28	ma
DC Grid-No.1 Current (Approx.)	10	12	ma
Driving Power (Approx.)	2.1	2.7	watts
Power Output (Approx.)	100	150	watts

♦ Obtained preferably from separate source modulated along with the plate supply, or from the modulated plate supply through series resistor of value shown.
† For ac filament supply.
♢ Obtained from grid-No.1 resistor of value shown or from a combination of grid-No.1 resistor with either fixed supply or cathode resistor.

RF POWER AMPLIFIER AND OSCILLATOR—Class C Telegraphy#
and
RF POWER AMPLIFIER—Class C FM Telephony

Maximum Ratings:	*CCS*	*ICAS*	
DC Plate Voltage	1250 *max*	1500 *max*	volts
DC Grid-No.3 Voltage	100 *max*	100 *max*	volts
DC Grid-No.2 Voltage	400 *max*	400 *max*	volts
DC Grid-No.1 Voltage	-300 *max*	-300 *max*	volts
DC Plate Current	160 *max*	180 *max*	ma
DC Grid-No.1 Current	15 *max*	15 *max*	ma
Plate Input	200 *max*	270 *max*	watts
Grid-No.3 Input	5 *max*	5 *max*	watts
Grid-No.2 Input	16 *max*	16 *max*	watts
Plate Dissipation	70 *max*	80 *max*	watts

Typical Operation:			
DC Plate Voltage	1250	1500	volts
DC Grid-No.3 Voltage	75	75	volts
DC Grid-No.2 Voltage⊕	400	400	volts
From series resistor of	24300	39300	ohms
DC Grid-No.1 Voltage†□	-95	-100	volts
From grid-No.1 resistor of	7900	8300	ohms
From cathode resistor of	415	430	ohms
Peak RF Grid-No.1 Voltage	195	205	volts
DC Plate Current	160	180	ma
DC Grid-No.3 Current	22	14	ma
Grid-No.2 Current	35	28	ma
Grid-No.1 Current (Approx.)	12	12	ma
Driving Power (Approx.)	2.1	2.2	watts
Power Output (Approx.)	150	200	watts

Key-down conditions per tube without amplitude modulation. Amplitude modulation essentially negative may be used if the positive peak of the audio-frequency envelope does not exceed 115 per cent of the carrier conditions.

⊕ Obtained from separate source, from plate-voltage supply with a voltage divider, or through series resistor of value shown. Grid-No.2 voltage must not exceed 800 volts under key-up conditions.

† For ac filament supply.

□ Obtained from fixed supply, by grid-No.1 resistor, by cathode resistor, or by combination methods. If preceding stage is keyed, partial fixed bias is required.

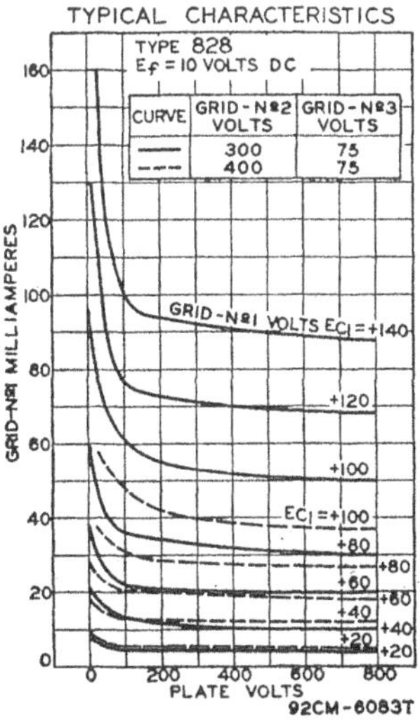

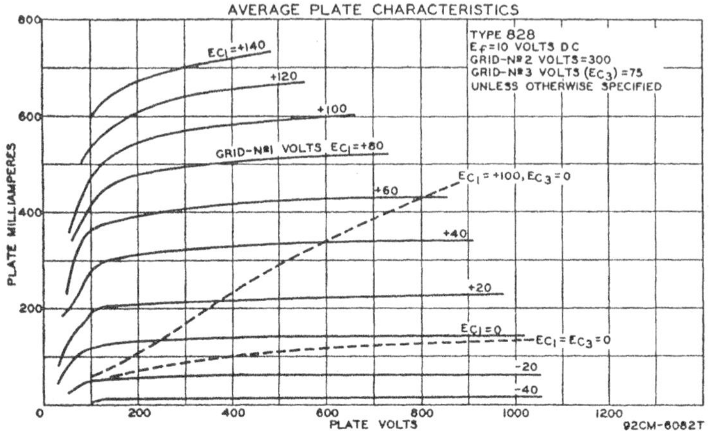

TWIN BEAM POWER TUBE

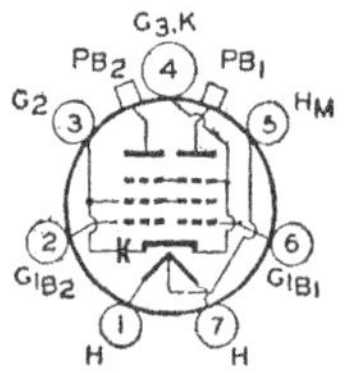

829-B

Heater-cathode type having mid-tapped heater used as af power amplifier and modulator and as rf power amplifier and oscillator. May be used with full input up to 200 Mc. For operation at 250 Mc, plate voltage and plate input should be reduced to 89 per cent of maximum ratings. Class C Telegraphy maximum plate dissipation (per tube) with natural cooling, CCS 30 watts, ICAS 40 watts; with forced-air cooling, CCS 40 watts, ICAS 45 watts. Requires Septar seven-contact socket and may be mounted in vertical position with base up or down, or in horizontal position with pins 2 and 6 in vertical plane. OUTLINE 22, *Outlines* Section. Plates show no color when tube is operated at maximum CCS or ICAS ratings.

HEATER ARRANGEMENT	*Series*	*Parallel*	
HEATER VOLTAGE (AC/DC)	12.6 ± 10%	6.3 ± 10%	volts
HEATER CURRENT	1.125	2.25	amperes
TRANSCONDUCTANCE (Each unit)*	8500		μmhos
MU-FACTOR, Grid No.2 to Grid No.1 (Each unit)**	9		
DIRECT INTERELECTRODE CAPACITANCES (Each unit):■			
Grid No.1 to plate	0.12 *max*		μμf
Grid No.1 to cathode, grid No.3, grid No.2, and heater mid-tap	14.5		μμf
Plate to cathode, grid No.3, grid No.2, and heater mid-tap	7		μμf

* Plate volts, 250; grid-No.2 volts, 175; plate milliamperes, 60.
** Plate and grid-No.2 volts, 225; plate milliamperes, 60.
■ With external shield up to flange seal.

PUSH-PULL AF POWER AMPLIFIER AND MODULATOR—Class AB1

Values are on a per-tube basis

Maximum CCS Ratings:	Natural Cooling	
DC PLATE VOLTAGE	750 *max*	volts
DC GRID-NO.2 (SCREEN-GRID) VOLTAGE	225 *max*	volts
MAXIMUM-SIGNAL DC PLATE CURRENT■	250 *max*	ma
MAXIMUM-SIGNAL PLATE INPUT■	100 *max*	watts
MAXIMUM-SIGNAL GRID-NO.2 INPUT■	7 *max*	watts
PLATE DISSIPATION■	30 *max*	watts
PEAK HEATER-CATHODE VOLTAGE:		
Heater negative with respect to cathode	100 *max*	volts
Heater positive with respect to cathode	100 *max*	volts
BULB TEMPERATURE	235 *max*	°C

Typical Operation:

DC Plate Voltage	600	volts
DC Grid-No.2 Voltage♦	200	volts
DC Grid-No.1 (Control-Grid) Voltage	-18	volts
Peak AF Grid-No.1-to-Grid-No.1 Voltage	36	volts
Zero-Signal DC Plate Current	40	ma
Maximum-Signal DC Plate Current	110	ma
Zero-Signal DC Grid-No.2 Current	4	ma
Maximum-Signal DC Grid-No.2 Current	26	ma
Effective Load Resistance (Plate to plate)	13750	ohms
Maximum-Signal Driving Power	0	watts
Maximum-Signal Power Output	44	watts

Maximum Circuit Values:

Grid-No.1-Circuit Resistance:		
For fixed-bias operation	0.1 *max*	megohm
For cathode-bias operation	Not recommended	

■ Averaged over any audio-frequency cycle of sine-wave form.

♦ Obtained preferably from a separate source, or from the plate-voltage supply with a voltage divider.

PLATE-MODULATED PUSH-PULL RF POWER AMPLIFIER—Class C Telephony

Carrier conditions per tube for use with a maximum modulation factor of 1.0

Maximum Ratings:	Natural Cooling CCS	Natural Cooling ICAS	Forced-Air Cooling CCS	Forced-Air Cooling ICAS	
DC PLATE VOLTAGE	600 *max*	600 *max*	600 *max*	600 *max*	volts
DC GRID-NO.2 VOLTAGE	225 *max*	225 *max*	225 *max*	250 *max*	volts
DC GRID-NO.1 VOLTAGE	-175 *max*	-175 *max*	-175 *max*	-175 *max*	volts
DC PLATE CURRENT	212 *max*	212 *max*	212 *max*	240 *max*	ma
DC GRID-NO.1 CURRENT	15 *max*	15 *max*	15 *max*	20 *max*	ma
PLATE INPUT	67.5 *max*	90 *max*	90 *max*	120 *max*	watts
GRID-NO.2 INPUT	7 *max*	7 *max*	7 *max*	8 *max*▲	watts
PLATE DISSIPATION	21 *max*	28 *max*	28 *max*	40 *max*	watts
PEAK HEATER-CATHODE VOLTAGE:					
Heater negative with respect to cathode	100 *max*	100 *max*	100 *max*	100 *max*	volts
Heater positive with respect to cathode	100 *max*	100 *max*	100 *max*	100 *max*	volts
BULB TEMPERATURE	235 *max*	235 *max*	235 *max*	235 *max*	°C

Typical Operation:

DC Plate Voltage	600	425	600	425	600	600	volts
DC Grid-No.2 Voltage♦	190	200	200	200	200	200	volts
From series resistor of	32000	11000	25000	11000	25000	20000	ohms
DC Grid-No.1 Voltage♢	-60	-60	-60	-60	-60	-70	volts
From grid-No.1 resistor of	15000	4300	8600	4300	8600	5400	ohms

AVERAGE PLATE CHARACTERISTICS
EACH UNIT

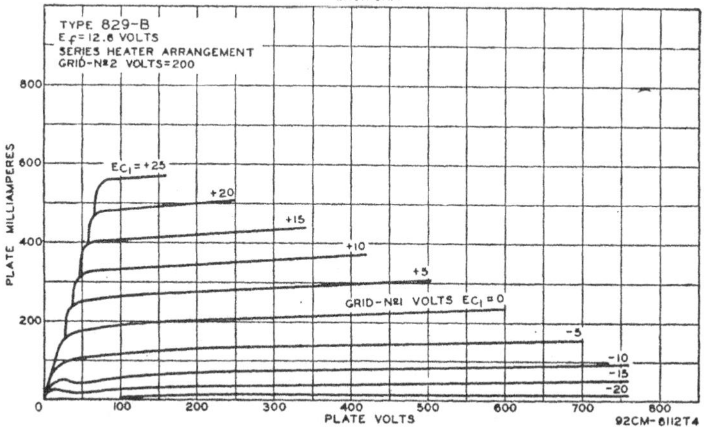

Peak RF Grid-No.1-to-Grid-No.1 Voltage	138	160	144	160	144	180	volts
DC Plate Current	112	212	150	212	150	200	ma
DC Grid-No.2 Current	13	21	16	21	16	20	ma
DC Grid-No.1 Current (Approx.)	4	14	7	14	7	13	ma
Driving Power (Approx.)	0.3	1	0.5	1	0.5	1.1	watts
Power Output (Approx.)	50	63	70	63	70	90	watts

Maximum Circuit Values (CCS or ICAS conditions):

Grid-No.1-Circuit Resistance .... 15000 *max* ohms

▲ In ICAS applications, at frequencies less than 20 Mc, where the duty factor does not exceed 0.2, maximum "on" period does not exceed 30 seconds, and average modulation factor does not exceed 0.25, maximum grid-No.2 input of 12 watts is permitted.

♦ Obtained preferably from separate source modulated along with the plate supply, or from the modulated plate supply through series resistor of value shown.

♁ Obtained from grid-No.1 resistor of value shown or from a combination of grid-No.1 resistor with either fixed supply or cathode resistor.

PUSH-PULL RF POWER AMPLIFIER AND OSCILLATOR—Class C Telegraphy#
and
PUSH-PULL RF POWER AMPLIFIER—Class C FM Telephony

Values are on a per-tube basis

Maximum Ratings:	Natural Cooling		Forced-Air Cooling		
	CCS	*ICAS*	*CCS*	*ICAS*	
DC Plate Voltage	750 *max*	750 *max*	750 *max*	750 *max*	volts
DC Grid-No.2 Voltage	225 *max*	225 *max*	225 *max*	250 *max*	volts
DC Grid-No.1 Voltage	-175 *max*	-175 *max*	-175 *max*	-175 *max*	volts
DC Plate Current	240 *max*	240 *max*	240 *max*	240 *max*	ma
DC Grid-No.1 Current	15 *max*	15 *max*	15 *max*	20 *max*	ma
Plate Input	90 *max*	120 *max*	120 *max*	150 *max*	watts
Grid-No.2 Input	7 *max*	7 *max*	7 *max*	8 *max*	watts
Plate Dissipation	30 *max*	40 *max*	40 *max*	45 *max*	watts
Peak Heater-Cathode Voltage:					
Heater negative with respect to cathode	100 *max*	100 *max*	100 *max*	100 *max*	volts
Heater positive with respect to cathode	100 *max*	100 *max*	100 *max*	100 *max*	volts
Bulb Temperature	265 *max*	265 *max*	235 *max*	235 *max*	°C

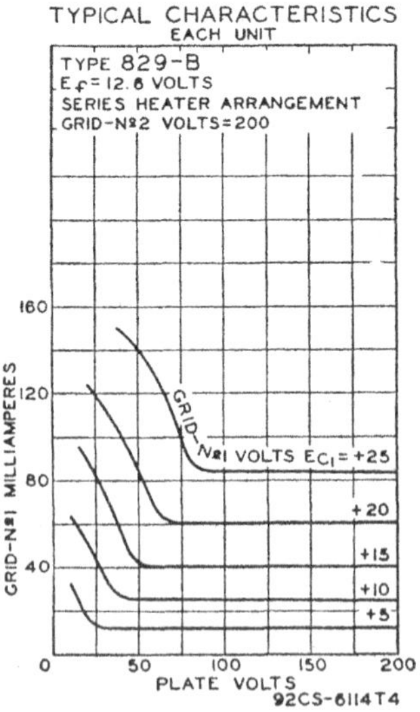

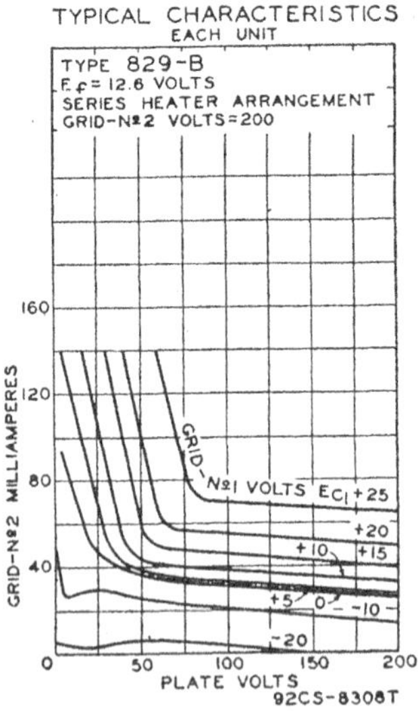

Typical Operation:

DC Plate Voltage	750	500	750	500	750	750	volts
DC Grid-No.2 Voltage®	190	200	200	200	200	200	volts
From series resistor of	40000	13000	32000	13000	32000	27500	ohms
DC Grid-No.1 Voltage△	-50	-45	-50	-45	-50	-50	volts
From grid-No.1 resistor of	12500	3000	7200	3000	7200	4200	ohms
From cathode resistor of	360	170	270	170	270	200	ohms
Peak RF Grid-No.1-to-Grid-No.1 Voltage	116	128	124	128	124	134	volts
DC Plate Current	120	230	160	230	160	200	ma
DC Grid-No.2 Current	14	23	17	23	17	20	ma
DC Grid-No.1 Current (Approx.)	4	15	7	15	7	12	ma
Driving Power (Approx.)	0.3	0.9	0.4	0.9	0.4	0.8	watt
Power Output (Approx.)	70	83	90	83	90	115	watts

Maximum Circuit Values (CCS or ICAS conditions):

Grid-No.1-Circuit Resistance . . . . . . 15000 *max* ohms

♯ Key-down conditions per tube without amplitide modulation. Amplitude modulation essentially negative may be used if the positive peak of the audio-frequency envelope does not exceed 115 per cent of the carrier conditions.

® Obtained preferably from separate source, from plate-voltage supply with a voltage divider, or through series resistor of value shown. The grid-No.2 voltage must not exceed 600 volts under key-up conditions.

△ Obtained from fixed supply, by grid-No.1 resistor, by cathode resistor, or by combination methods.

POWER TRIODE

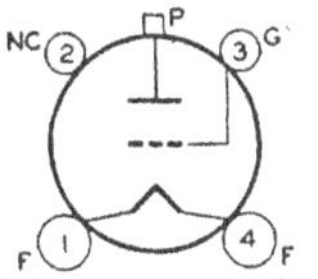

Thoriated-tungsten-filament type used as af power amplifier and modulator and as rf power amplifier and oscillator. May be used with full input up to 15 Mc and with reduced input up to 60 Mc. Requires Small four-contact socket and may be mounted in vertical position with base down, or in horizontal position with pins 1 and 4 in vertical plane. OUTLINE 43, *Outlines* Section. Plate shows no color when tube is operated at maximum CCS ratings. The 830-B is used principally for renewal purposes.

830-B

FILAMENT VOLTAGE (AC/DC)	10	volts
FILAMENT CURRENT	2	amperes
AMPLIFICATION FACTOR	25	
DIRECT INTERELECTRODE CAPACITANCES:		
Grid to plate	11	μμf
Grid to filament	5	μμf
Plate to filament	1.8	μμf

Maximum CCS Ratings:	Class B Modulator	Class C Telegraphy♯	
DC PLATE VOLTAGE	1000 *max*	1000 *max*	volts
DC GRID VOLTAGE	—	-300 *max*	volts
DC PLATE CURRENT	150•■ *max*	150 *max*	ma
DC GRID CURRENT	—	30 *max*	ma
PLATE INPUT	150•■ *max*	150 *max*	watts
PLATE DISSIPATION	60 *max*	60 *max*	watts

♯ Key-down conditions per tube without amplitude modulation. Amplitude modulation essentially negative may be used if the positive peak of the audio-frequency envelope does not exceed 115 per cent of the carrier conditions.

• For maximum-signal conditions.

■ Averaged over any audio-frequency cycle of sine-wave form.

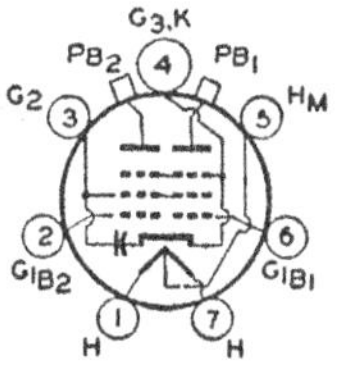

TWIN BEAM POWER TUBE

832-A

Heater-cathode type having mid-tapped heater used as rf power amplifier and oscillator. May be used with full input up to 200 Mc. For operation at 250 Mc, plate voltage and plate input should be reduced to 89 per cent of maximum ratings. Class C Telegraphy maximum plate dissipation (per tube), CCS 15 watts, ICAS 20 watts. Requires

Septar seven-contact socket and may be mounted in any position. OUTLINE 12, *Outlines* Section. Plates show no color when tube is operated at maximum CCS or ICAS ratings.

HEATER ARRANGEMENT	*Series*	*Parallel*	
HEATER VOLTAGE (AC/DC)	12.6 ± 10%	6.3 ± 10%	volts
HEATER CURRENT	0.8	1.6	amperes
TRANSCONDUCTANCE (Each unit)*	3500		μmhos
MU-FACTOR, Grid No.2 to Grid No.1 (Each unit)**	6.5		
DIRECT INTERELECTRODE CAPACITANCES (Each unit):□			
Grid No.1 to plate	0.07 *max*		μμf
Grid No.1 to cathode, grid No.3, grid No.2, and heater mid-tap	8.0		μμf
Plate to cathode, grid No.3, grid No.2, and heater mid-tap	3.8		μμf
Grid No. 2 to cathode (including internal Grid-No. 2 by-pass capacitor)	65		μμf

* Plate volts, 250; grid-No.2 volts, 135; plate milliamperes, 30.
** Plate and grid-No.2 volts, 250; plate milliamperes, 30.
□ With external shield in plane of seal flange.

PLATE-MODULATED PUSH-PULL RF POWER AMPLIFIER—Class C Telephony

Carrier conditions per tube for use with a maximum modulation factor of 1.0

Maximum Ratings:	*CCS*	*ICAS*	
DC PLATE VOLTAGE	600 *max*	600 *max*	volts
DC GRID-NO.2 (SCREEN-GRID) VOLTAGE	250 *max*	250 *max*	volts
DC GRID-NO.1 (CONTROL-GRID) VOLTAGE	-175 *max*	-175 *max*	volts
DC PLATE CURRENT	75 *max*	95 *max*	ma
DC GRID-NO.1 CURRENT	6 *max*	6 *max*	ma
PLATE INPUT	22 *max*	36 *max*	watts
GRID-NO.2 INPUT	3.4 *max*	5 *max*	watts
PLATE DISSIPATION	10 *max*	15 *max*	watts
PEAK HEATER-CATHODE VOLTAGE:			
Heater negative with respect to cathode	100 *max*	100 *max*	volts
Heater positive with respect to cathode	100 *max*	100 *max*	volts

Typical Operation:				
DC Plate Voltage	425	600	600	volts
DC Grid-No.2 Voltage♦	200	200	200	volts
From series resistor of	14000	25000	20000	ohms
DC Grid-No.1 Voltage♢	-60	-65	-70	volts
From grid-No.1 resistor of	25000	25000	23000	ohms
Peak RF Grid-No.1-to-Grid-No.1 Voltage	140	150	160	volts
DC Plate Current	52	36	60	ma
DC Grid-No.2 Current	16	16	20	ma
DC Grid-No.1 Current (Approx.)	2.4	2.6	3	ma
Driving Power (Approx.)	0.15	0.18	0.21	watt
Power Output (Approx.)	16	17	26	watts

Maximum Circuit Values (CCS or ICAS conditions):		
Grid-No.1-Circuit Resistance	25000 *max*	ohms

♦ Obtained preferably from separate source modulated along with the plate supply or from the modulated plate supply through series resistor of value shown.
♢ Obtained from grid-No.1 resistor of value shown or from a combination of grid-No.1 resistor with either fixed supply or cathode resistor.

PUSH-PULL RF POWER AMPLIFIER AND OSCILLATOR—Class C Telegraphy#
and
PUSH-PULL RF POWER AMPLIFIER—Class C FM Telephony

Maximum Ratings:	*CCS*	*ICAS*	
DC PLATE VOLTAGE	750 *max*	750 *max*	volts
DC GRID-NO.2 VOLTAGE	250 *max*	250 *max*	volts
DC GRID-NO.1 VOLTAGE	-175 *max*	-175 *max*	volts
DC PLATE CURRENT	90 *max*	115 *max*	ma
DC GRID-NO.1 CURRENT	6 *max*	6 *max*	ma
PLATE INPUT	36 *max*	50 *max*	watts
GRID-NO.2 INPUT	5 *max*	5 *max*	watts
PLATE DISSIPATION	15 *max*	20 *max*	watts
PEAK HEATER-CATHODE VOLTAGE:			
Heater negative with respect to cathode	100 *max*	100 *max*	volts
Heater positive with respect to cathode	100 *max*	100 *max*	volts

Typical Operation:

DC Plate Voltage	500	750	750	volts
DC Grid-No.2 Voltage⊕	200	200	200	volts
From series resistor of	21000	37000	25000	ohms
DC Grid-No.1 Voltage△	-65	-65	-50	volts
From grid-No.1 resistor of	25000	23000	12500	ohms
From cathode resistor of	730	1000	550	ohms
Peak RF Grid-No.1-to-Grid-No.1 Voltage	150	150	130	volts
DC Plate Current	72	48	65	ma
DC Grid-No.2 Current	14	15	22	ma
DC Grid-No.1 Current (Approx.)	2.6	2.8	4.0	ma
Driving Power (Approx.)	0.18	0.19	0.24	watt
Power Output (Approx.)	26	26	35	watts

Maximum Circuit Values (CCS or ICAS conditions):

Grid-No.1-Circuit Resistance . . . 25000 *max* ohms

ƒ Key-down conditions per tube without amplitude modulation. Amplitude modulation essentially negative may be used if the positive peak of the audio-frequency envelope does not exceed 115 per cent of the carrier conditions.

⊕ Obtained from separate source, from plate-voltage supply with a voltage divider, or from series resistor of value shown. The grid-No.2 voltage must not exceed 600 volts under key-up conditions.

△ Obtained from fixed supply, by grid-No. 1 resistor, by cathode resistor, or by combination methods.

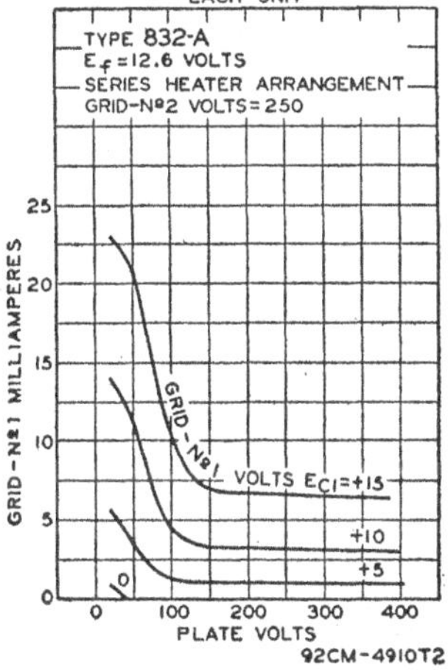

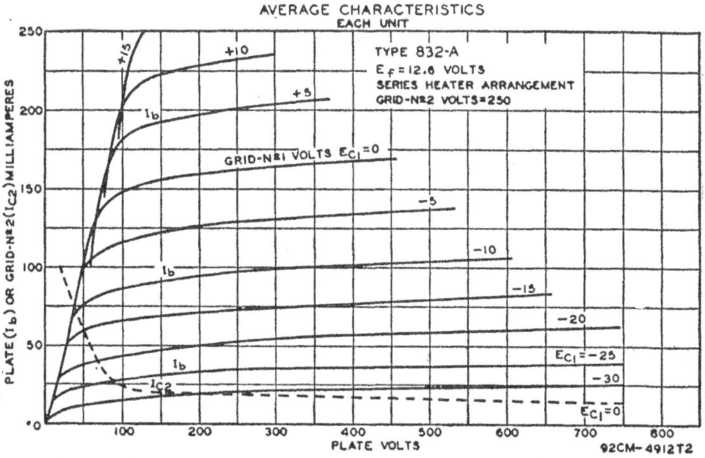

833-A

POWER TRIODE

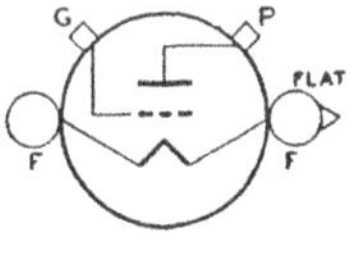

Thoriated-tungsten-filament type used as af power amplifier and modulator and as rf power amplifier and oscillator. May be used with full input up to 30 Mc with natural cooling (20 Mc with forced-air cooling), and with reduced input up to 75 Mc. Class C Telegraphy maximum plate dissipation with natural cooling, CCS 300 watts, ICAS 350 watts; with forced-air cooling, CCS 400 watts, ICAS 450 watts.

Filament Voltage (ac/dc)	10	volts
Filament Current	10	amperes
Amplification Factor*	35	
Direct Interelectrode Capacitances:		
Grid to plate	6.3	μμf
Grid to filament	12.3	μμf
Plate to filament	8.5	μμf

* Grid volts, −10; plate milliamperes, 200.

AF POWER AMPLIFIER AND MODULATOR—Class B

	Natural Cooling		Forced-Air Cooling		
Maximum Ratings:	*CCS*	*ICAS*	*CCS*	*ICAS*	
DC Plate Voltage	3000 *max*	3300 *max*	4000 *max*	4000 *max*	volts
Maximum-Signal DC Plate Current■	500 *max*	500 *max*	500 *max*	500 *max*	ma
Maximum-Signal Plate Input■	1125 *max*	1300 *max*	1600 *max*	1800 *max*	watts
Plate Dissipation■	300 *max*	350 *max*	400 *max*	450 *max*	watts
Typical Operation (*Values are for two tubes*):					
DC Plate Voltage	3000	3300	4000	4000	volts
DC Grid Voltage†	−70	−80	−100	−100	volts
Peak AF Grid-to-Grid Voltage	400	440	480	510	volts
Zero-Signal DC Plate Current	100	100	100	100	ma
Maximum-Signal DC Plate Current	750	780	800	900	ma
Effective Load Resistance (Plate to plate)	9500	10500	12000	11000	ohms
Maximum-Signal Driving Power (Approx.)	20	30	29	38	watts
Maximum-Signal Power Output (Approx.)	1650	1900	2400	2700	watts

■ Averaged over any audio-frequency cycle of sine-wave form.
† For ac filament supply.

AVERAGE PLATE CHARACTERISTICS

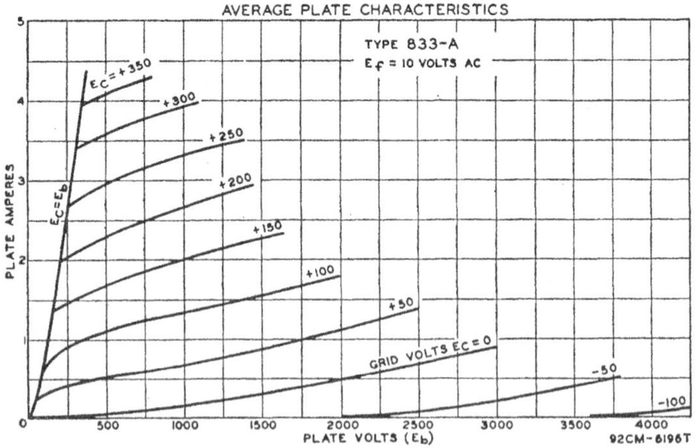

PLATE-MODULATED RF POWER AMPLIFIER—Class C Telephony

Carrier conditions per tube for use with a maximum modulation factor of 1.0

	Natural Cooling		Forced-Air Cooling		
Maximum Ratings:	*CCS*	*ICAS*	*CCS*	*ICAS*	
DC Plate Voltage	2500 *max*	3000 *max*	3000 *max*	4000 *max*	volts
DC Grid Voltage	-500 *max*	-500 *max*	-500 *max*	-500 *max*	volts
DC Plate Current	400 *max*	400 *max*	450 *max*	450 *max*	ma
DC Grid Current	100 *max*	100 *max*	100 *max*	100 *max*	ma
Plate Input	835 *max*	1000 *max*	1250 *max*	1800 *max*	watts
Plate Dissipation	200 *max*	250 *max*	270 *max*	350 *max*	watts
Typical Operation:					
DC Plate Voltage	2500	3000	3000	4000	volts
DC Grid Voltage δ	-300	-240	-300	-325	volts
From grid resistor of	4000	3400	3600	3600	ohms
Peak RF Grid Voltage	460	410	490	520	volts
DC Plate Current	335	335	415	450	ma
DC Grid Current (Approx.)	75	70	85	90	ma
Driving Power (Approx.)	30	26	37	42	watts
Power Output (Approx.)	635	800	1000	1500	watts

δ Obtained from grid resistor of value shown or from a combination of grid resistor with either fixed supply or cathode resistor.

RF POWER AMPLIFIER AND OSCILLATOR—Class C Telegraphy#
and
RF POWER AMPLIFIER—Class C FM Telephony

	Natural Cooling		Forced-Air Cooling		
Maximum Ratings:	*CCS*	*ICAS*	*CCS*	*ICAS*	
DC Plate Voltage	3000 *max*	3300 *max*	4000 *max*	4000 *max*	volts
DC Grid Voltage	-500 *max*	-500 *max*	-500 *max*	-500 *max*	volts
DC Plate Current	500 *max*	500 *max*	500 *max*	500 *max*	ma
DC Grid Current	100 *max*	100 *max*	100 *max*	100 *max*	ma
Plate Input	1250 *max*	1500 *max*	1800 *max*	2000 *max*	watts
Plate Dissipation	300 *max*	350 *max*	400 *max*	450 *max*	watts

Typical Operation:						
DC Plate Voltage	2250	3000	3000	4000	4000	volts
DC Grid Voltage△	-125	-200	-155	-200	-225	volts
From grid resistor of	1500	3600	2150	2650	2400	ohms
From cathode resistor of	235	425	270	380	380	ohms
Peak RF Grid Voltage	300	360	350	375	415	volts
DC Plate Current	445	415	500	450	500	ma
DC Grid Current (Approx.)	85	55	70	75	95	ma
Driving Power (Approx.)	23	20	25	26	35	watts
Power Output (Approx.)	780	1000	1150	1440	1600	watts

Key-down conditions per tube without amplitude modulation. Amplitude modulation essentially negative may be used if the positive peak of the audio-frequency envelope does not exceed 115 per cent of the carrier conditions.

△ Obtained from fixed supply, by grid resistor, by cathode resistor, or by combination methods.

OPERATING CONSIDERATIONS

Type 833-A requires special mounting and may be mounted in vertical position with filament end up or down, or in horizontal position with all terminals in same vertical plane. Outline 56, *Outlines* Section.

For operation with natural cooling at 50 Mc, plate voltage and plate input should be reduced to 90 per cent of maximum ratings; at 75 Mc, to 72 per cent. For operation with forced-air cooling at 50 Mc, plate voltage and plate input should be reduced to 83 per cent of maximum ratings; at 75 Mc, to 65 per cent.

With forced-air cooling, an air flow of 40 cubic feet per minute from a 2-inch-diameter nozzle directed vertically on the bulb between grid and plate seals is required to limit the temperature between these seals to 145°C.

When the 833-A is used in the final amplifier or a preceding stage of a transmitter designed for break-in operation and oscillator keying, a small amount of fixed bias must be used to maintain the plate current at a safe value. With a plate voltage of 4000 volts, a fixed bias of at least -90 volts should be used.

Plate shows an orange-red color when tube is operated at maximum CCS or ICAS ratings.

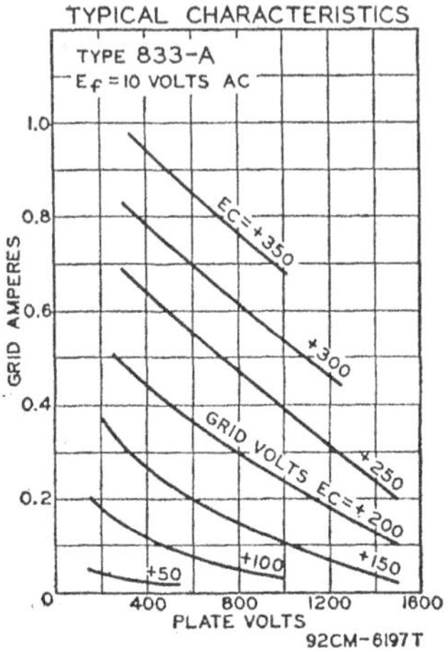

POWER TRIODE

834

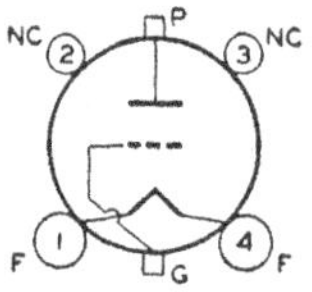

Thoriated-tungsten-filament type used as rf power amplifier and oscillator. May be used with full input up to 100 Mc. For operation at 170 Mc, plate voltage and plate input should be reduced to 80 per cent of maximum ratings; at 350 Mc, to 53 per cent. Class C Telegraphy maximum CCS plate dissipation, 50 watts. Requires Small four-contact socket and may be mounted in vertical position only, base up or down. OUTLINE 44, *Outlines* Section. Plate shows an orange-red color when tube is operated at maximum CCS ratings.

FILAMENT VOLTAGE (AC/DC)	7.5	volts
FILAMENT CURRENT	3.1	amperes
AMPLIFICATION FACTOR	10.5	
DIRECT INTERELECTRODE CAPACITANCES:		
Grid to plate	2.4	μμf
Grid to filament	2.2	μμf
Plate to filament	0.6	μμf

RF POWER AMPLIFIER AND OSCILLATOR—Class C Telegraphy#
and
RF POWER AMPLIFIER—Class C FM Telephony

Maximum CCS Ratings:

DC PLATE VOLTAGE	1250 *max*	volts
DC GRID VOLTAGE	-400 *max*	volts
DC PLATE CURRENT	100 *max*	ma
DC GRID CURRENT	20 *max*	ma
PLATE INPUT	125 *max*	watts
PLATE DISSIPATION	50 *max*	watts

Key-down conditions per tube without amplitude modulation. Amplitude modulation essentially negative may be used if the positive peak of the audio-frequency envelope does not exceed 115 per cent of the carrier conditions.

POWER TRIODE

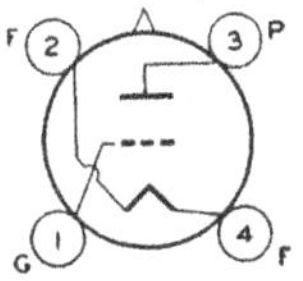

835

Thoriated-tungsten-filament type used as af power amplifier and modulator and as rf power amplifier and oscillator. May be used with full input up to 20 Mc and with reduced input up to 100 Mc. Requires Jumbo four-contact socket and may be mounted in vertical position with base down, or in horizontal position with pins 1 and 3 in vertical plane. OUTLINE 49, *Outlines* Section. Filament volts (ac/dc), 10; amperes, 3.25. Direct interelectrode capacitances: grid to plate, 9.25 μμf; grid to filament, 6 μμf; plate to filament, 5 μμf. Plate shows a barely perceptible red color when tube is operated at maximum CCS ratings. Except for interelectrode capacitances, the 835 is identical with type 211. The 835 is a DISCONTINUED type listed for reference only.

HALF-WAVE VACUUM RECTIFIER

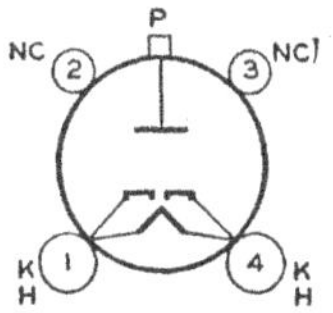

836

Heater-cathode type having two cathodes used in power supply of transmitting and industrial equipment. Maximum peak inverse plate volts, 5000; maximum average plate amperes, 0.25. Requires Small four-contact socket and may be mounted in any position. OUTLINE 40, *Outlines* Section. The 836 has two separate cathodes, each of which is connected to its respective heater. Plate-circuit return should be made to the mid-tap of the heater transformer.

HEATER VOLTAGE (AC)°	2.5	volts
HEATER CURRENT	5.0	amperes

HALF-WAVE RECTIFIER

Maximum Ratings:

PEAK INVERSE PLATE VOLTAGE	5000 *max*	volts
PLATE CURRENT:		
Peak	1 *max*	ampere
Average	0.25 *max*	ampere
Fault, for duration of 0.1 second maximum	5 *max*	amperes

° Heater voltage should be applied approximately 40 seconds before the application of plate voltage.

BEAM POWER TUBE

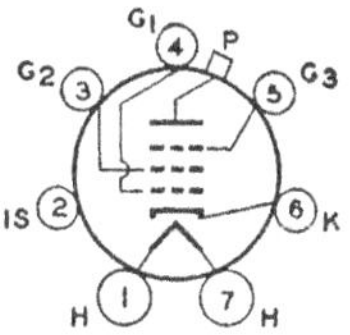

837

Heater-cathode type used as rf power amplifier and oscillator. May be used with full input up to 20 Mc. For operation at 40 Mc, plate voltage and plate input should be reduced to 76 per cent of maximum ratings; at 60 Mc, to 62 per cent. Class C Telegraphy maximum CCS plate dissipation, 12 watts. Requires Medium seven-contact socket and may be mounted in any position. OUTLINE 31, *Outlines* Section, except has no bayonet pin. Plate shows no color when tube is operated at maximum CCS ratings.

HEATER VOLTAGE (AC/DC)	12.6 ± 10%	volts
HEATER CURRENT	0.7	ampere
TRANSCONDUCTANCE (For plate current of 24 milliamperes)	3400	μmhos
DIRECT INTERELECTRODE CAPACITANCES:		
Grid-No.1 to plate (With external shielding)	0.20 *max*	μμf
Grid No.1 to cathode, grid No.3, grid No.2, internal shield, and heater	16	μμf
Plate to cathode, grid No.3, grid No.2, internal shield, and heater	10	μμf

RF POWER AMPLIFIER AND OSCILLATOR—Class C Telegraphy#
and
RF POWER AMPLIFIER—Class C FM Telephony

Maximum CCS Ratings:

DC PLATE VOLTAGE	500 *max*	volts

DC Grid-No.3 (suppressor-grid) Voltage	200 *max*	volts
DC Grid-No.2 (screen-grid) Voltage	200 *max*	volts
DC Grid-No.1 (control-grid) Voltage	–200 *max*	volts
DC Plate Current	80 *max*	ma
DC Grid-No.1 Current	8 *max*	ma
Plate Input	32 *max*	watts
Grid-No.3 Input	5 *max*	watts
Grid-No.2 Input	8 *max*	watts
Plate Dissipation	12 *max*	watts
Peak Heater-Cathode Voltage:		
Heater negative with respect to cathode	100 *max*	volts
Heater positive with respect to cathode	100 *max*	volts

Key-down conditions per tube without amplitude modulation. Amplitude modulation essentially negative may be used provided the positive peak of the audio-frequency envelope does not exceed 115 per cent of the carrier conditions.

POWER TRIODE

838

Thoriated-tungsten-filament type used as af power amplifier and modulator and as rf power amplifier and oscillator. May be used with full input up to 30 Mc and with reduced input up to 120 Mc. Requires Jumbo four-contact socket and may be mounted in vertical position with base down, or in horizontal position with pins 1 and 3 in vertical plane. Outline 49, *Outlines* Section. Plate shows no color when tube is operated at maximum CCS ratings. The 838 is used principally for renewal purposes.

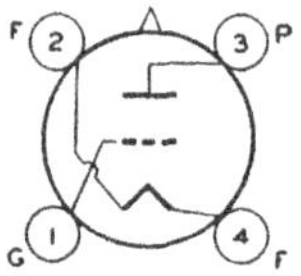

Filament Voltage (ac/dc)	10	volts
Filament Current	3.25	amperes
Direct Interelectrode Capacitances:		
Grid to plate	7.8	μμf
Grid to filament	6.0	μμf
Plate to filament	4.0	μμf

Maximum CCS Ratings:	Class B Modulator	Class C Telegraphy#	
DC Plate Voltage	1250 *max*	1250 *max*	volts
DC Grid Voltage	—	–400 *max*	volts
DC Plate Current	175•■ *max*	175 *max*	ma
DC Grid Current	—	70 *max*	ma
Plate Input	220•■ *max*	220 *max*	watts
Plate Dissipation	100• *max*	100 *max*	watts

Key-down conditions per tube without amplitude modulation. Amplitude modulation essentially negative may be used if the positive peak of the audio-frequency envelope does not exceed 115 per cent of the carrier conditions.
• For maximum-signal conditions.
■ Averaged over any audio-frequency cycle of sine-wave form.

POWER TRIODE

841

Thoriated-tungsten-filament type used as af power amplifier and modulator and as rf power amplifier and oscillator. May be used with full input up to 6 Mc and with reduced input up to 30 Mc. Requires Small four-contact socket and may be mounted in vertical position with base down, or in horizontal position with pins 1 and 4 in vertical plane. Outline 29, *Outlines* Section. Plate shows no color when tube is operated at maximum CCS ratings. The 841 is used principally for renewal purposes.

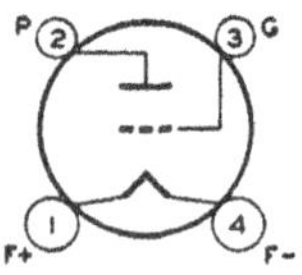

Filament Voltage (ac/dc)	7.5	volts
Filament Current	1.25	amperes
Direct Interelectrode Capacitances:		
Grid to plate	7.5	μμf
Grid to filament	4.0	μμf
Plate to filament	2.6	μμf

Maximum CCS Ratings:	Class B Modulator	Class C Telegraphy#	
DC Plate Voltage	425 *max*	450 *max*	volts
DC Grid Voltage	—	-200• *max*	volts
DC Plate Current	60•■ *max*	60 *max*	ma
DC Grid Current	—	20 *max*	ma
Plate Input	25•■ *max*	27 *max*	watts
Plate Dissipation	15■ *max*	15 *max*	watts

Key-down conditions per tube without amplitude modulation. Amplitude modulation essentially negative may be used if the positive peak of the audio-frequency envelope does not exceed 115 per cent of the carrier conditions.

• For maximum-signal conditions.

■ Averaged over any audio-frequency cycle of sine-wave form.

POWER TRIODE

842

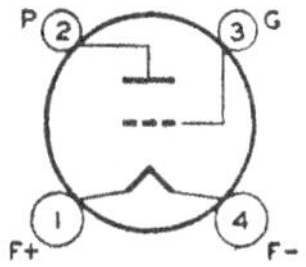

Thoriated-tungsten-filament type used as af power amplifier and modulator. Requires Small four-contact socket and may be mounted in vertical position with base down, or in horizontal position with pins 1 and 4 in vertical plane. OUTLINE 29, *Outlines* Section. Filament volts (ac/dc), 7.5; amperes, 1.25. Direct interelectrode capacitances: grid to plate, 6.4 μμf; grid to filament, 3.2 μμf; plate to filament, 2.6 μμf. Maximum CCS ratings as CLASS A AF POWER AMPLIFIER AND MODULATOR: dc plate volts, 425 *max;* plate dissipation, 12 *max* watts. Plate shows no color when tube is operated at maximum CCS ratings. The 842 is used principally for renewal purposes.

POWER TRIODE

843

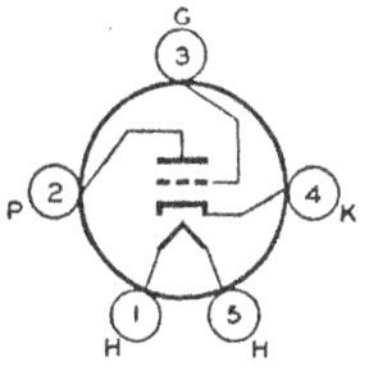

Heater-cathode type used as rf power amplifier and oscillator. May be used with full input up to 6 Mc and with reduced input up to 30 Mc. Requires Small five-contact socket and may be mounted in any position. OUTLINE 29, *Outlines* Section. Heater volts (ac/dc), 2.5; amperes, 2.5. Direct interelectrode capacitances: grid to plate, 3.9 μμf; grid to cathode and heater, 4 μμf; plate to cathode and heater, 2.5 μμf. Maximum CCS ratings as RF POWER AMPLIFIER AND OSCILLATOR: dc plate volts, 450 *max;* dc grid volts, -200 *max;* dc plate milliamperes, 40 *max;* dc grid milliamperes, 7.5 *max;* plate input, 18 *max* watts; plate dissipation, 15 *max* watts, peak heater-cathode volts, ± 45 *max*. Plate shows no color when tube is operated at maximum CCS ratings. The 843 is a DISCONTINUED type listed for reference only.

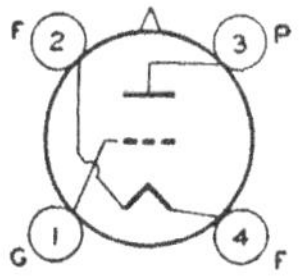

POWER TRIODE

845

Thoriated-tungsten-filament type used as af power amplifier and modulator. Class AB_1 maximum CCS plate dissipation, 100 watts. Requires Jumbo four-contact socket and may be mounted in vertical position with base down, or in horizontal position with pins 1 and 3 in vertical plane. OUTLINE 49, *Outlines* Section. Plate shows no color when tube is operated at maximum CCS ratings.

Filament Voltage (ac/dc)	10	volts
Filament Current	3.25	amperes
Amplification Factor	5.3	
Direct Interelectrode Capacitances:		
Grid to plate	12.1	μμf
Grid to filament	5.0	μμf
Plate to filament	5.0	μμf

AF POWER AMPLIFIER AND MODULATOR—Class AB_1

Maximum CCS Ratings:		
DC Plate Voltage	1250 *max*	volts

DC Grid Voltage	-400 *max*	volts
DC Plate Current	120 *max*	ma
Plate Input	150 *max*	watts
Plate Dissipation	100 *max*	watts

POWER TRIODE

849

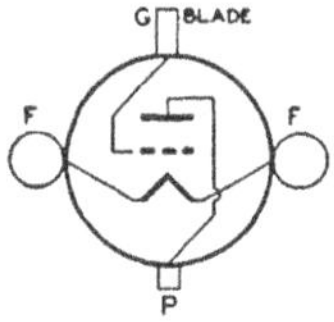

Thoriated-tungsten-filament type used as af power amplifier and modulator and as rf power amplifier and oscillator. May be used with full input up to 3 Mc and with reduced input up to 30 Mc. Tube may be mounted in vertical position with filament end up, or in horizontal position with plate in vertical plane. Maximum over-all length, 14⅜ inches; maximum diameter, 4$\frac{1}{16}$ inches. Filament volts (ac/dc), 11; amperes, 5. Direct interelectrode capacitances: grid to plate, 34 μμf; grid to filament, 17 μμf; plate to filament, 3 μμf. Maximum CCS ratings as RF POWER AMPLIFIER AND OSCILLATOR: dc plate volts, 2500 *max;* dc grid volts, -500 *max;* dc plate amperes, 0.35 *max;* dc grid amperes, 0.125 *max;* plate input, 875 *max* watts; plate dissipation, 400 *max* watts. Plate shows cherry-red color when tube is operated at maximum CCS ratings. The 849 is a DISCONTINUED type listed for reference only.

POWER TETRODE

850

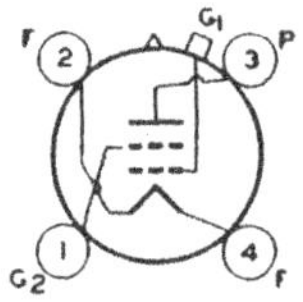

Thoriated-tungsten-filament type used as rf power amplifier and oscillator at frequencies up to 15 Mc. Requires Jumbo four-contact socket and may be mounted in vertical position with base up or down, or in horizontal position with pins 1 and 3 in vertical plane. Outline 51, *Outlines* Section. Filament volts (ac/dc), 10; amperes, 3.25. Direct interelectrode capacitances: grid No.1 to plate (with external shielding), 0.25 *max* μμf; grid No.1 to filament and grid No.2, 17 μμf; plate to filament and grid No.2, 25 μμf. Maximum CCS ratings as RF POWER AMPLIFIER AND OSCILLATOR: dc plate volts, 1250 *max;* dc grid-No.2 volts, 400 *max;* dc grid-No. 1 volts, -400 *max;* dc plate milliamperes, 175 *max;* dc grid-No. 1 milliamperes, 40 *max;* plate input, 220 *max* watts; grid-No.2 input, 10 *max* watts; plate dissipation, 100 *max* watts. Plate shows a barely perceptible red color when tube is operated at maximum CCS ratings. The 850 is a DISCONTINUED type listed for reference only.

POWER TRIODE

851

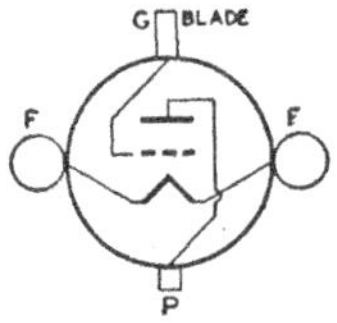

Thoriated-tungsten-filament type used as af power amplifier and modulator and as rf power amplifier and oscillator. May be used with full input up to 3 Mc. For operation at 7 Mc, plate voltage and plate input should be reduced to 75 per cent of maximum ratings; at 15 Mc, to 50 per cent. Tube may be mounted in vertical position with filament end up, or in horizontal position with plate in vertical plane. Outline 64, *Outlines* Section. The 851 is used principally for renewal purposes.

Filament Voltage (ac/dc)	11.0	volts
Filament Current	15.5	amperes
Amplification Factor	20.5	
Direct Interelectrode Capacitances:		
Grid to plate	47	μμf
Grid to filament	25.5	μμf
Plate to filament	4.5	μμf

Maximum CCS Ratings:	Class B Modulator	Class C Telegraphy#	
DC Plate Voltage	3000 *max*	2500 *max*	volts
DC Grid Voltage	—	-500 *max*	volts
DC Plate Current	1•■ *max*	1 *max*	ampere
DC Grid Current	—	0.2 *max*	ampere
Plate Input	2250•■ *max*	2500 *max*	watts
Plate Dissipation	750■ *max*	750 *max*	watts

Key-down conditions per tube without amplitude modulation. Amplitude modulation essentially neg-

ative may be used if the positive peak of the audio-frequency envelope does not exceed 115 per cent of the carrier conditions.
• For maximum-signal conditions.
▪ Averaged over any audio-frequency cycle of sine-wave form.

POWER TETRODE

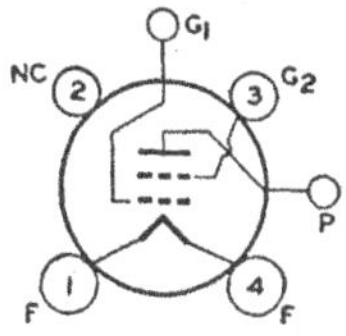

Thoriated-tungsten-filament type used as rf power amplifier and oscillator. May be used with full input up to 30 Mc. For operation at 60 Mc, plate voltage and plate input should be reduced to 75 per cent of maximum ratings; at 120 Mc, to 50 per cent. Requires Small four-contact socket and may be mounted in vertical position only, base down. OUTLINE 55, *Outlines* Section. Plate shows no color when tube is operated at maximum CCS ratings. The 860 is used principally for renewal purposes.

860

FILAMENT VOLTAGE (AC/DC)	10	volts
FILAMENT CURRENT	3.25	amperes
TRANSCONDUCTANCE (For plate current of 50 milliamperes)	1100	μmhos
AMPLIFICATION FACTOR	200	
DIRECT INTERELECTRODE CAPACITANCES:		
Grid No.1 to plate (With external shielding)	0.08 *max*	μμf
Grid No.1 to filament and grid No.2	7.75	μμf
Plate to filament and grid No.2	7.5	μμf

RF POWER AMPLIFIER AND OSCILLATOR—Class C Telegraphy#
and
RF POWER AMPLIFIER—Class C FM Telephony

Maximum CCS Ratings:

DC PLATE VOLTAGE	3000 *max*	volts
DC GRID-NO.2 (SCREEN-GRID) VOLTAGE	500 *max*	volts
DC GRID-NO.1 (CONTROL-GRID) VOLTAGE	-800 *max*	volts
DC PLATE CURRENT	150 *max*	ma
DC GRID-NO.1 CURRENT	40 *max*	ma
PLATE INPUT	300 *max*	watts
GRID-NO.2 INPUT	10 *max*	watts
PLATE DISSIPATION	100 *max*	watts

Key-down conditions per tube without amplitude modulation. Amplitude modulation essentially negative may be used if the positive peak of the audio-frequency envelope does not exceed 115 per cent of the carrier conditions.

POWER TETRODE

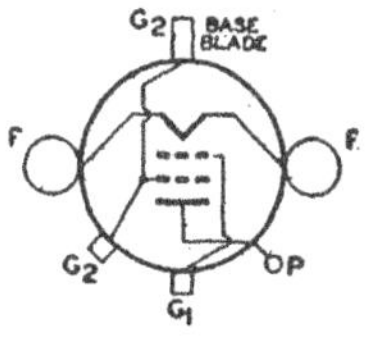

Thoriated-tungsten-filament type used as rf power amplifier and oscillator. May be used with full input up to 20 Mc. For operation at 30 Mc, plate voltage and plate input should be reduced to 82 per cent of maximum ratings; at 68 Mc, to 53 per cent. Tube may be mounted in vertical position only, filament end up. OUTLINE 63, *Outlines* section. Plate shows an orange-red color when tube is operated at maximum CCS ratings. The 861 is used principally for renewal purposes.

861

FILAMENT VOLTAGE (AC/DC)	11	volts
FILAMENT CURRENT	10	amperes
TRANSCONDUCTANCE (For plate current of 130 milliamperes)	2400	μmhos
AMPLIFICATION FACTOR	300	
DIRECT INTERELECTRODE CAPACITANCES:		
Grid No.1 to plate (With external shielding)	0.10 *max*	μμf
Grid No.1 to filament and grid No.2	14	μμf
Plate to filament and grid No.2	11	μμf

RF POWER AMPLIFIER AND OSCILLATOR—Class C Telegraphy#
and
RF POWER AMPLIFIER—Class C FM Telephony

Maximum CCS Ratings:

DC PLATE VOLTAGE	3500 *max*	volts

DC Grid-No.2 (screen-grid) Voltage●	750 *max*	volts
DC Grid-No.1 (control-grid) Voltage	-1000 *max*	volts
DC Plate Current	350 *max*	ma
DC Grid-No.1 Current	75 *max*	ma
Plate Input	1200 *max*	watts
Grid-No.2 Input	35 *max*	watts
Plate Dissipation	400 *max*	watts

Key-down conditions per tube without amplitude modulation. Amplitude modulation essentially negative may be used if the positive peak of the audio-frequency envelope does not exceed 115 per cent of the carrier conditions.

● Grid-No.2 voltage must not exceed 2000 volts under key-up conditions.

POWER TETRODE

865

Thoriated-tungsten-filament type used as rf power amplifier and oscillator. May be used with full input up to 15 Mc. For operation at 30 Mc, plate voltage and plate input should be reduced to 78 per cent of maximum ratings; at 60 Mc, to 55 per cent. Requires Small four-contact socket and may be mounted in vertical position only, base up or down. Outline 31, *Outlines* Section. Plate shows no color when tube is operated at maximum CCS ratings. The 865 is used principally for renewal purposes.

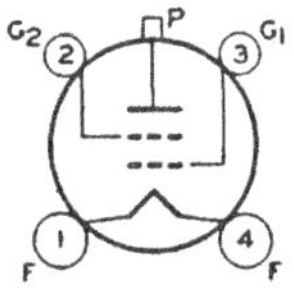

Filament Voltage (ac/dc)	7.5	volts
Filament Current	2.0	amperes
Transconductance (For plate current of 18 milliamperes)	750	μmhos
Amplification Factor (Approx.)	150	
Direct Interelectrode Capacitances:		
Grid No.1 to plate (With external shielding)	0.10 *max*	μμf
Grid No.1 to filament and grid No.2	8.5	μμf
Plate to filament and grid No.2	8.0	μμf

RF POWER AMPLIFIER AND OSCILLATOR—Class C Telegraphy#
and
RF POWER AMPLIFIER—Class C FM Telephony

Maximum CCS Ratings:

DC Plate Voltage	750 *max*	volts
DC Grid-No.2 (screen-grid) Voltage	175 *max*	volts
DC Grid-No.1 (control-grid) Voltage	-200 *max*	volts
DC Plate Current	60 *max*	ma
DC Grid-No.1 Current	15 *max*	ma
Plate Input	45 *max*	watts
Grid-No.2 Input	3 *max*	watts
Plate Dissipation	15 *max*	watts

Key-down conditions per tube without amplitude modulation. Amplitude modulation essentially negative may be used if the positive peak of the audio-frequency envelope does not exceed 115 per cent of the carrier conditions.

HALF-WAVE MERCURY-VAPOR RECTIFIER

866-A

Coated-filament type used in power supply of transmitting and industrial equipment. Maximum peak inverse anode volts, 10,000; maximum average anode amperes, 0.25. Requires Small four-contact socket and may be mounted in vertical position only, base down. Outline 41, *Outlines* Section.

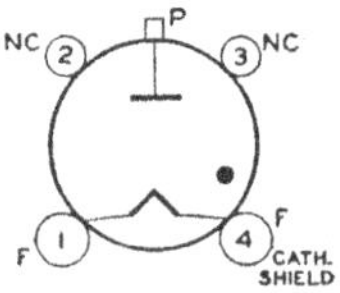

Filament Voltage (ac)°	2.5	volts
Filament Current	5.0	amperes
Peak Tube Voltage Drop (Approx.)	15	volts

° Filament voltage must be applied at least 15 seconds before the application of anode voltage.

HALF-WAVE RECTIFIER

Maximum Ratings: *(For power-supply frequency of 60 cps):*

Peak Inverse Anode Voltage	2500 *max*	5000 *max*	10000 *max*	volts
Anode Current:				
Peak	2 *max*	1 *max*	1 *max*	amperes
Average*	0.5 *max*	0.25 *max*	0.25 *max*	ampere
Fault, for duration of 0.1 second maximum	20 *max*	20 *max*	20 *max*	amperes
Condensed-Mercury-Temperature Range●	20 to 80	20 to 70	20 to 60	°C

* Averaged over any interval of 30 seconds maximum.
● Operation at 40° ± 5°C is recommended.

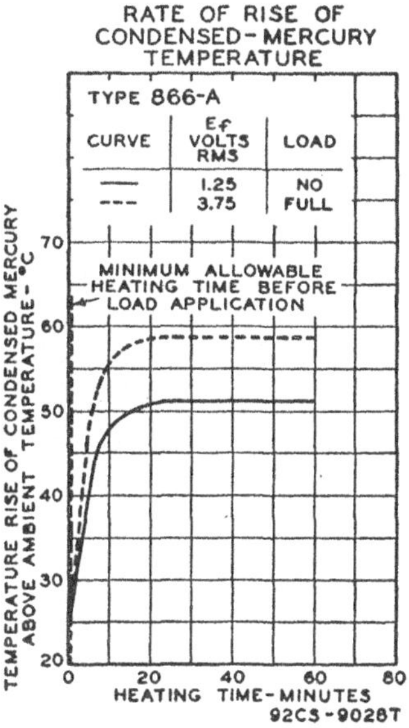

Operating Values:

Circuit (For circuit figures, refer to *Rectifier Considerations* Section)	*Fig.*	*Max. Trans. Sec. Volts (RMS) E*	*Approx. DC Output Volts To Filter Eav*	*Max. DC Output Amperes Iav*	*Max. DC Output KW To Filter Pdc*
		In-Phase Operation			
Half-Wave Single-Phase	54	7000●	3200	0.25	0.8
		3500▲	1600	0.25	0.4
		1700□	800	0.50	0.4
Full-Wave Single-Phase	55	3500●	3200	0.5	1.6
		1700▲	1600	0.5	0.8
		800□	800	1.0	0.8
Series Single-Phase	56	7000●	6400	0.5	3.2
		3500▲	3200	0.5	1.6
		1700□	1600	1.0	1.6
Half-Wave Three-Phase	57	4000●	4800	0.75	3.6
		2000▲	2400	0.75	1.8
		1000□	1200	1.5	1.8
		Quadrature Operation			
Parallel Three-Phase	58	4000●	4800	1.5	7.2
		2000▲	2400	1.5	3.6
		1000□	1200	3.0	3.6

Circuit (For circuit figures, refer to *Rectifier Considerations* Section)	Fig.	*Max. Trans. Sec. Volts (RMS)* E	*Approx. DC Output Volts To Filter* Eav	*Max. DC Output Amperes* Iav	*Max. DC Output KW To Filter* Pdc
		4000•	9600	0.75	7.2
Series Three-Phase........	59	2000▲	4800	0.75	3.6
		1000□	2400	1.5	3.6
		3500•	4500	0.91* 1.0■	4.05* 4.5■
Half-Wave Four-Phase....	60	1700▲	2300	0.91* 1.0■	2.07* 2.3■
		800□	1100	1.82* 2.0■	1.98* 2.2■
		3500•	4800	0.95* 1.0■	4.60* 4.8■
Half-Wave Six-Phase.....	61	1700▲	2400	0.95* 1.0■	2.30* 2.4■
		800□	1200	1.90* 2.0■	2.28* 2.4■

• For maximum peak inverse anode voltage of 10000 volts and maximum average anode current of 0.25 ampere.
▲ For maximum peak inverse anode voltage of 5000 volts and maximum average anode current of 0.25 ampere.
□ For maximum peak inverse anode voltage of 2500 volts and maximum average anode current of 0.5 ampere.
* Resistive load. ■ Inductive load.

872-A

HALF-WAVE MERCURY-VAPOR RECTIFIER

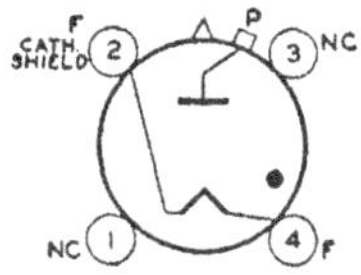

Coated-filament type used in power supply of transmitting and industrial equipment. Maximum peak inverse anode volts, 10,000; maximum average anode amperes, 1.25. Requires Jumbo four-contact socket and may be mounted in vertical position only, base down. OUTLINE 52, *Outlines* Section.

FILAMENT VOLTAGE (AC)°	5.0	volts
FILAMENT CURRENT	7.5	amperes
PEAK TUBE VOLTAGE DROP (Approx.)	10	volts

° Filament voltage must be applied at least 30 seconds before the application of anode voltage.

HALF-WAVE RECTIFIER

Maximum Ratings *(For power-supply frequency of 60 cps)*:

PEAK INVERSE ANODE VOLTAGE	5000 *max*	10000 *max*	volts
ANODE CURRENT:			
Peak	5 *max*	5 *max*	amperes
Average ♁	1.25 *max*	1.25 *max*	amperes
Fault, for duration of 0.2 second maximum	50 *max*	50 *max*	amperes
CONDENSED-MERCURY-TEMPERATURE RANGE•	20 to 70	20 to 60	°C

♁ Averaged over any interval of 15 seconds maximum.
• Operation at 40° ± 5°C is recommended.

Operating Values:

Circuit (For circuit figures, refer to *Rectifier Considerations* Section)	Fig.	*Max. Trans. Sec. Volts (RMS)* E	*Approx. DC Output Volts To Filter* Eav	*Max. DC Output Amperes* Iav	*Max. DC Output KW To Filter* Pdc
		In-Phase Operation			
		7000•	3200	1.25	4.0
Half-Wave Single-Phase...	54	3500▲	1600	1.25	2.0
		3500•	3200	2.5	8.0
Full-Wave Single-Phase...	55	1700▲	1600	2.5	4.0
		7000•	6400	2.5	16.0
Series Single-Phase.......	56	3500▲	3200	2.5	8.0
		4000•	4800	3.75	18.0
Half-Wave Three-Phase...	57	2000▲	2400	3.75	9.0

Circuit (For circuit figures, refer to *Rectifier Considerations* Section)	Fig.	Max. Trans. Sec. Volts (RMS) E	Approx. DC Output Volts To Filter Eav	Max. DC Output Amperes Iav	Max. DC Output KW To Filter Pdc
Quadrature Operation					
Parallel Three-Phase......	58	4000•	4800	7.5	36.0
		2000▲	2400	7.5	18.0
Series Three-Phase.......	59	4000•	9600	3.75	36.0
		2000▲	4800	3.75	18.0
Half-Wave Four-Phase....	60	3500•	4500	4.5* 5.0■	20.0* 22.5■
		1700▲	2250	4.5* 5.0■	10.0* 11.2■
Half-Wave Six-Phase.....	61	3500•	4800	4.75* 5.0■	22.8* 24.0■
		1700▲	2400	4.75* 5.0■	11.4* 12.0■

• For maximum peak inverse anode voltage of 10000 volts and maximum average anode current of 1.25 amperes.
▲ For maximum peak inverse anode voltage of 5000 volts and maximum average anode current of 1.25 amperes.
* Resistive load. ■ Inductive load.

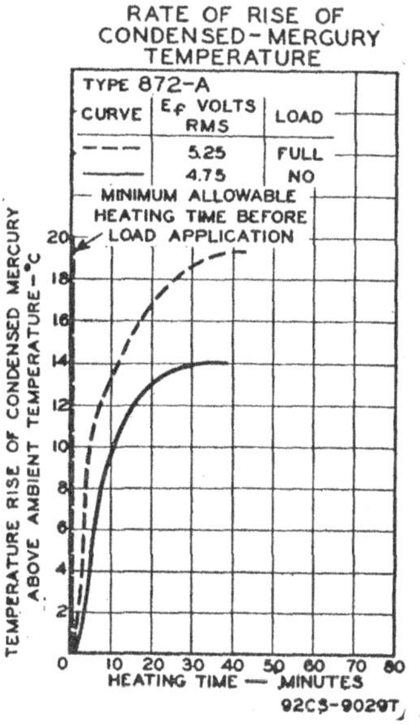

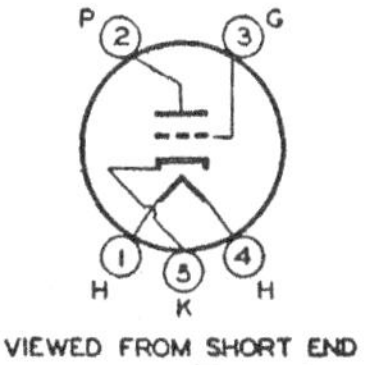

MEDIUM-MU TRIODE

955

Acorn heater-cathode type used as af amplifier and as rf amplifier and oscillator at frequencies up to 600 Mc. Class A_1 Amplifier maximum CCS plate dissipation (design-center value), 1.6 watts. Requires Acorn five-contact socket and may be mounted in any position. OUTLINE 2, *Outlines* Section. Plate shows no color when tube is operated at maximum CCS ratings.

HEATER VOLTAGE (AC/DC)	6.3	volts
HEATER CURRENT	0.15	ampere
DIRECT INTERELECTRODE CAPACITANCES:		
Grid to plate	1.3	μμf
Grid to cathode and heater	1.0	μμf
Plate to cathode and heater	0.4	μμf

AF AMPLIFIER—Class A1

Maximum CCS Ratings, *Design-Center Values:*

DC PLATE VOLTAGE	250 *max*	volts
PLATE DISSIPATION	1.6 *max*	watts
PEAK HEATER-CATHODE VOLTAGE:		
Heater negative with respect to cathode	80 *max*	volts
Heater positive with respect to cathode	80 *max*	volts

Typical Operation and Characteristics:

DC Plate Voltage	90	135	180	250	volts
DC Grid Voltage	-2.5	-3.75	-5	-7	volts
Amplification Factor	25	25	25	25	
Plate Resistance (Approx.)	14700	13200	12500	11400	ohms
Transconductance	1700	1900	2000	2200	μmhos
DC Plate Current	2.5	3.5	4.5	6.3	ma
Load Resistance	—	—	20000	—	ohms
Second-Harmonic Distortion	—	—	5	—	per cent
Useful Power Output	—	—	135	—	mw

Maximum Circuit Values:

Grid-Circuit Resistance:		
For fixed-bias operation	0.1 *max*	megohm
For cathode-bias operation	0.5 *max*	megohm

RF AMPLIFIER AND OSCILLATOR—Class C

Maximum CCS Ratings, *Design-Center Values:*

DC PLATE VOLTAGE	180 *max*	volts
DC PLATE CURRENT	8 *max*	ma
DC GRID CURRENT	2 *max*	ma
PEAK HEATER-CATHODE VOLTAGE:		
Heater negative with respect to cathode	80 *max*	volts
Heater positive with respect to cathode	80 *max*	volts

MEDIUM-MU TRIODE

958-A

VIEWED FROM SHORT END

Acorn coated-filament type used as rf power amplifier and oscillator at frequencies up to 350 Mc. Class C Telegraphy maximum CCS plate dissipation (design-center value), 0.6 watt. Requires Acorn five-contact socket and may be mounted in any position. OUTLINE 2, *Outlines* Section. Plate shows no color when tube is operated at maximum CCS ratings.

FILAMENT VOLTAGE (DC)	1.25	volts
FILAMENT CURRENT	0.10	ampere
DIRECT INTERELECTRODE CAPACITANCES:		
Grid to plate	2.5	μμf
Grid to filament	0.45	μμf
Plate to filament	0.6	μμf

RF POWER AMPLIFIER AND OSCILLATOR—Class C Telegraphy# and RF POWER AMPLIFIER—Class C FM Telephony

Maximum CCS Ratings, *Design-Center Values:*

DC PLATE VOLTAGE	135 *max*	volts
DC GRID VOLTAGE	-30 *max*	volts
DC PLATE CURRENT	7 *max*	ma
DC GRID CURRENT	1 *max*	ma
PLATE INPUT	0.95 *max*	watt
PLATE DISSIPATION	0.6 *max*	watt

Typical Operation:

DC Plate Voltage	135	volts
DC Grid Voltage®	-20	volts
From grid resistor of	20000	ohms
From cathode resistor of	2500	ohms

Peak RF Grid Voltage	40	volts
DC Plate Current	7	ma
DC Grid Current	1	ma
Driving Power (Approx.)	0.085	watt
Power Output	0.600	watt

Maximum Circuit Values:

Grid-Circuit Resistance:		
For fixed-bias operation	0.1 *max*	megohm
For cathode-bias operation	0.5 *max*	megohm

♯ Key-down conditions per tube without amplitude modulation. Amplitude modulation essentially negative may be used if the positive peak of the audio-frequency envelope does not exceed 115 per cent of the carrier conditions.

• Obtained from fixed supply, by grid resistor, by cathode resistor, or by combination methods.

POWER TRIODE

1608

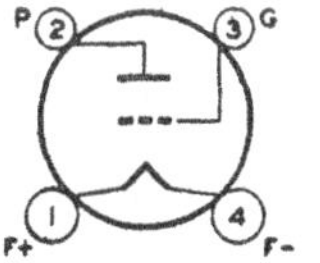

Coated-filament type used as af power amplifier and modulator and as rf power amplifier and oscillator. May be used with full input up to 45 Mc and with reduced input up to 100 Mc. Requires Small four-contact socket and may be mounted in vertical position with base down, or in horizontal position with pins 1 and 4 in vertical plane. OUTLINE 29, *Outlines* Section. Filament volts (ac/dc), 2.5; amperes, 2.5. Direct interelectrode capacitances: grid to plate, 9 μμf; grid to filament, 8.5 μμf; plate to filament, 3 μμf. Maximum CCS ratings as RF POWER AMPLIFIER AND OSCILLATOR: dc plate volts, 425 *max;* dc grid volts, −200 *max;* dc plate milliamperes, 95 *max;* dc grid milliamperes, 25 *max;* plate input, 40 *max* watts; plate dissipation, 20 *max* watts. Plate shows no color when tube is operated at maximum CCS ratings. The 1608 is a DISCONTINUED type listed for reference only.

POWER PENTODE

1610

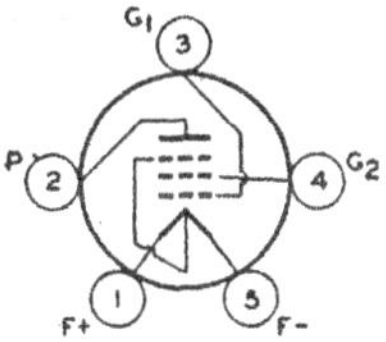

Coated-filament type used as rf power amplifier and oscillator. May be used with full input up to 20 Mc and with reduced input up to 110 Mc. Requires Small five-contact socket and may be mounted in vertical position only, base up or down. OUTLINE 29, *Outlines* Section. Filament volts (ac/dc), 2.5; amperes, 1.75. Direct interelectrode capacitances: grid-No.1 to plate, 1.2 μμf; grid No.1 to filament mid-tap, grid No.3, and grid No.2, 8.6 μμf; plate to filament mid-tap, grid No.3, and grid No.2, 13 μμf. Maximum CCS ratings as RF POWER AMPLIFIER AND OSCILLATOR: dc plate volts, 400 *max;* dc grid-No.2 volts, 200 *max;* dc grid-No.1 volts, −100 *max;* dc plate milliamperes, 30 *max;* dc grid-No.1 milliamperes, 3 *max;* plate input, 9 *max* watts; grid-No.2 input, 2 *max* watts; plate dissipation, 6 *max* watts. Plate shows no color when tube is operated at maximum CCS ratings. The 1610 is a DISCONTINUED type listed for reference only.

POWER PENTODE

1613

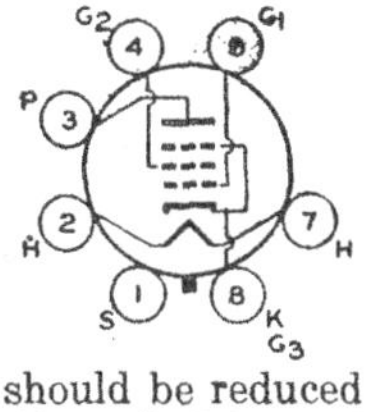

Heater-cathode type having metal shell used as rf power amplifier and oscillator. May be used with full input up to 45 Mc. For operation at 60 Mc, plate voltage and plate input should be reduced to 90 per cent of maximum ratings; at 90 Mc, to 85 per cent. Class C Telegraphy maximum CCS plate dissipation, 10 watts. Requires Octal socket and may be mounted in any position. OUTLINE 11, *Outlines* Section.

HEATER VOLTAGE (AC/DC)	6.3	volts
HEATER CURRENT	0.7	ampere
TRANSCONDUCTANCE (For plate current of 31 milliamperes)	2500	μmhos
DIRECT INTERELECTRODE CAPACITANCES:		
Grid No.1 to plate	0.26	μμf
Grid No.1 to cathode, grid No.3, grid No.2, shell, and heater	6.5	μμf
Plate to cathode, grid No.3, grid No.2, shell, and heater	13.5	μμf

RF POWER AMPLIFIER AND OSCILLATOR—Class C Telegraphy#
and
RF POWER AMPLIFIER—Class C FM Telephony

Maximum CCS Ratings:

DC Plate Voltage	350 *max*	volts
DC Grid-No.2 (screen-grid) Voltage	275 *max*	volts
DC Grid-No.1 (control-grid) Voltage	-100 *max*	volts
DC Plate Current	50 *max*	ma
DC Grid-No.1 Current	5 *max*	ma
Plate Input	17.5 *max*	watts
Grid-No.2 Input	2.5 *max*	watts
Plate Dissipation	10 *max*	watts
Peak Heater-Cathode Voltage:		
Heater negative with respect to cathode	100 *max*	volts
Heater positive with respect to cathode	100 *max*	volts

Key-down conditions per tube without amplitude modulation. Amplitude modulation essentially negative may be used if the positive peak of the audio-frequency envelope does not exceed 115 per cent of the carrier conditions.

BEAM POWER TUBE

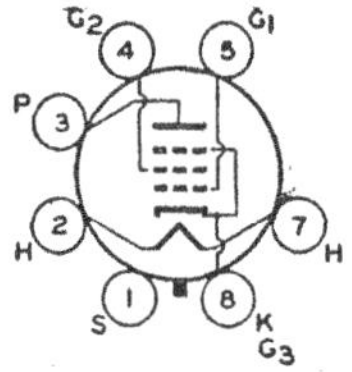

1614

Heater-cathode type having metal shell used as af power amplifier and modulator and as rf power amplifier and oscillator. May be used with full input up to 80 Mc. For operation at 120 Mc, plate voltage and plate input should be reduced to 75 per cent of maximum ratings. Class C Telegraphy maximum plate dissipation, CCS 21 watts, ICAS 25 watts. Requires Octal socket and may be mounted in any position. Outline 21, *Outlines* Section.

Heater Voltage (ac/dc)	6.3	volts
Heater Current	0.9	ampere
Transconductance (For plate current of 72 milliamperes)	6050	μmhos
Direct Interelectrode Capacitances:		
Grid No.1 to plate	0.4 *max*	μμf
Grid No.1 to cathode, grid No.3, grid No.2, shell, and heater	10	μμf
Plate to cathode, grid No.3, grid No.2, shell, and heater	12	μμf

AF POWER AMPLIFIER AND MODULATOR—Class AB1

Maximum Ratings:	*CCS*	*ICAS*	
DC Plate Voltage	375 *max*	550 *max*	volts
DC Grid-No.2 (screen-grid) Voltage	300 *max*	400 *max*	volts
DC Plate Current	110 *max*	110 *max*	ma
Plate Input	40 *max*	60 *max*	watts
Grid-No.2 Input	3.5 *max*	3.5 *max*	watts
Plate Dissipation	21 *max*	25 *max*	watts
Peak Heater-Cathode Voltage:			
Heater negative with respect to cathode	200 *max*	200 *max*	volts
Heater positive with respect to cathode	200 *max*	200 *max*	volts

Typical Operation (*Values are for 2 tubes*):

DC Plate Voltage	360	530	volts
DC Grid-No.2 Voltage	270	340	volts
DC Grid-No.1 (Control-Grid) Voltage	-22.5	-36	volts
Peak AF Grid-No.-1-to-Grid-No.1 Voltage	45	72	volts
Zero-Signal DC Plate Current	88	60	ma
Maximum-Signal DC Plate Current	132	160	ma
Maximum-Signal DC Grid-No.2 Current	15	20	ma
Effective Load Resistance (Plate to plate)	6600	7200	ohms
Total Harmonic Distortion	2	2.5	per cent
Maximum-Signal Power Output	26.5	50	watts

RF POWER AMPLIFIER AND OSCILLATOR—Class C Telegraphy#
and
RF POWER AMPLIFIER—Class C FM Telephony

Maximum Ratings:	*CCS*	*ICAS*	
DC Plate Voltage	375 *max*	450 *max*	volts

DC Grid-No.2 Voltage	300 *max*	300 *max*	volts
DC Grid-No.1 Voltage	-125 *max*	-125 *max*	volts
DC Plate Current	110 *max*	110 *max*	ma
DC Grid-No.1 Current	5 *max*	5 *max*	ma
Plate Input	35 *max*	45 *max*	watts
Grid-No.2 Input	3.5 *max*	3.5 *max*	watts
Plate Dissipation	21 *max*	25 *max*	watts
Peak Heater-Cathode Voltage:			
Heater negative with respect to cathode	200 *max*	200 *max*	volts
Heater positive with respect to cathode	200 *max*	200 *max*	volts

Typical Operation:

DC Plate Voltage	375	450	volts
DC Grid-No.2 Voltage•	250	250	volts
From series resistor of	12500	25000	ohms
DC Grid-No.1 Voltage♦	-40	-45	volts
From grid-No.1 resistor of	20000	22500	ohms
From cathode resistor of	425	410	ohms
Peak RF Grid-No.1 Voltage	51	73	volts
DC Plate Current	80	100	ma
DC Grid-No.2 Current	10	8	ma
DC Grid-No.1 Current (Approx.)	2	2	ma
Driving Power (Approx.)	0.1	0.15	watt
Power Output (Approx.)	21	31	watts

ƒ Key-down conditions per tube without amplitude modulation. Amplitude modulation essentially negative may be used if the positive peak of the audio-frequency envelope does not exceed 115 per cent of the carrier conditions.

• Obtained from separate source, from plate-voltage supply with a voltage divider, or through series resistor of value shown.

♦ Obtained from fixed supply, by grid-No.1 resistor, by cathode resistor, or by combination methods.

HALF-WAVE VACUUM RECTIFIER

1616

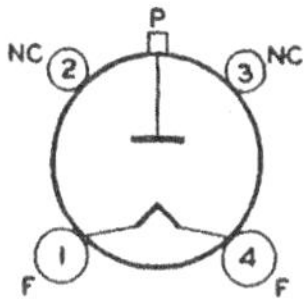

Coated-filament type used in power supply of transmitting and industrial equipment. Maximum peak inverse plate volts, 6000; maximum average plate amperes, 0.13. Requires a Small four-contact socket and may be mounted in vertical position with base down, or in horizontal position with pins 1 and 4 in vertical plane. Outline 45, *Outlines* Section.

Filament Voltage	2.5 ± 10%	volts
Filament Current	5.0	amperes

HALF-WAVE RECTIFIER

Maximum Ratings:

Peak Inverse Plate Voltage	6000 *max*	volts
Plate Current:		
Peak	800 *max*	ma
Average	130 *max*	ma
Fault	2.5 *max*	amperes

BEAM POWER TUBE

1619

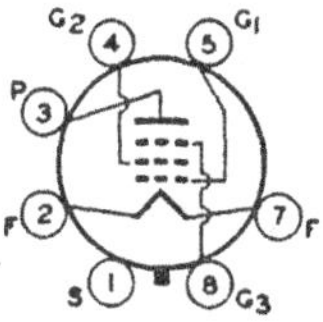

Coated-filament type having metal shell used as af power amplifier and modulator and as rf power amplifier and oscillator. May be used with full input up to 45 Mc. For operation at 60 Mc, plate voltage and plate input should be reduced to 90 per cent of maximum ratings; at 90 Mc, to 77 per cent. Requires Octal socket and may be mounted in vertical position only, base down or up. Outline 21, *Outlines* Section. The 1619 is used principally for renewal purposes.

Filament Voltage (ac/dc)	2.5	volts
Filament Current	2.0	amperes
Transconductance (For plate current of 50 milliamperes)	4500	μmhos
Direct Interelectrode Capacitances:		
Grid No.1 to plate	0.45 *max*	μμf
Grid No.1 to filament, grid No.3, grid No.2, and shell	9.6	μμf
Plate to filament, grid No.3, grid No.2, and shell	12.5	μμf

AF POWER AMPLIFIER AND MODULATOR—Class AB1

Maximum CCS Ratings:

DC Plate Voltage	400 *max*	volts
DC Grid-No.2 (screen-grid) Voltage	300 *max*	volts
Maximum-Signal DC Plate Current■	75 *max*	ma
Maximum-Signal Plate Input■	30 *max*	watts
Grid-No.2 Input■	3.5 *max*	watts
Plate Dissipation■	15 *max*	watts

■ Averaged over any audio-frequency cycle of sine-wave form.

RF POWER AMPLIFIER AND OSCILLATOR—Class C Telegraphy# and RF POWER AMPLIFIER—Class C FM Telephony

Maximum CCS Ratings:

DC Plate Voltage	400 *max*	volts
DC Grid-No.2 Voltage	300 *max*	volts
DC Grid-No.1 (control-grid) Voltage	-125 *max*	volts
DC Plate Current	75 *max*	ma
DC Grid-No.1 Current	5 *max*	ma
Plate Input	30 *max*	watts
Grid-No.2 Input	3.5 *max*	watts
Plate Dissipation	15 *max*	watts

Maximum Circuit Values:

Grid-No.1-Circuit Resistance	25000 *max*	ohms

Key-down conditions per tube without amplitude modulation. Amplitude modulation essentially negative may be used if the positive peak of the audio-frequency envelope does not exceed 115 per cent of the carrier conditions.

POWER TRIODE

1623

Thoriated-tungsten-filament type used as af power amplifier and modulator and as rf power amplifier and oscillator. May be used with full input up to 60 Mc and with reduced input up to 100 Mc. Requires Small four-contact socket and may be mounted in vertical position with base down, or in horizontal position with pins 1 and 4 in vertical plane. Outline 40, *Outlines* Section. Plate does not show color when tube is operated at maximum CCS ratings. The 1623 is used principally for renewal purposes.

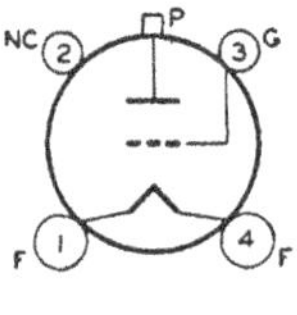

Filament Voltage (ac/dc)	6.3	volts
Filament Current	2.5	amperes
Amplification Factor	20	
Direct Interelectrode Capacitances:		
Grid to plate	6.7	μμf
Grid to filament	5.2	μμf
Plate to filament	0.9	μμf

Maximum CCS Ratings:	Class B Modulator	Class C Telegraphy#	
DC Plate Voltage	750 *max*	750 *max*	volts
DC Grid Voltage	—	-200 *max*	volts
DC Plate Current	100•■ *max*	100 *max*	ma
DC Grid Current	—	25 *max*	ma
Plate Input	75•■ *max*	75 *max*	watts
Plate Dissipation	25■ *max*	25 *max*	watts

Key-down conditions per tube without amplitude modulation. Amplitude modulation essentially neg-

ative may be used if the positive peak of the audio-frequency envelope does not exceed 115 per cent of the carrier conditions.
• For maximum-signal conditions.
▪ Averaged over any audio-frequency cycle of sine-wave form.

BEAM POWER TUBE

1624

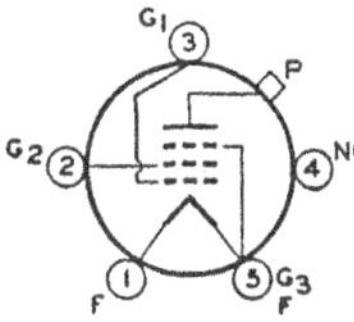

Coated-filament type used as rf power amplifier and oscillator. May be used with full input up to 60 Mc. For operation at 80 Mc, plate voltage and plate input should be reduced to 80 per cent of maximum ratings; at 125 Mc, to 55 per cent. Requires Small five-contact socket and may be mounted in vertical position only, base up or down. OUTLINE 31, *Outlines* Section, except has no bayonet pin. Plate shows no color when tube is operated at maximum CCS ratings. The 1624 is used principally for renewal purposes.

FILAMENT VOLTAGE (AC/DC)	2.5	volts
FILAMENT CURRENT	2.0	amperes
TRANSCONDUCTANCE (For plate current of 50 milliamperes)	4000	μmhos
DIRECT INTERELECTRODE CAPACITANCES:		
Grid No.1 to plate (With external shielding)	0.25 *max*	μμf
Grid No.1 to filament, grid No.3, and grid No.2	11	μμf
Plate to filament, grid No.3, and grid No.2	7.5	μμf

RF POWER AMPLIFIER AND OSCILLATOR—Class C Telegraphy#
and
RF POWER AMPLIFIER—Class C FM Telephony

Maximum CCS Ratings:

DC PLATE VOLTAGE	600 *max*	volts
DC GRID-No.2 (SCREEN-GRID) VOLTAGE	300 *max*	volts
DC GRID-No.1 (CONTROL-GRID) VOLTAGE	-200 *max*	volts
DC PLATE CURRENT	90 *max*	ma
DC GRID-No.1 CURRENT	5 *max*	ma
PLATE INPUT	54 *max*	watts
GRID-No.2 INPUT	3.5 *max*	watts
PLATE DISSIPATION	25 *max*	watts

Maximum Circuit Values:

Grid-No.1-Circuit Resistance	25000 *max*	ohms

Key-down conditions per tube without amplitude modulation. Amplitude modulation essentially negative may be used if the positive peak of the audio-frequency envelope does not exceed 115 per cent of the carrier conditions.

BEAM POWER TUBE

1625

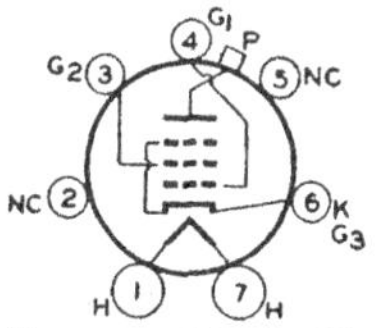

Heater-cathode type used as af power amplifier and modulator and as rf power amplifier and oscillator. Requires Medium seven-contact socket and may be mounted in any position. OUTLINE 31, *Outlines* Section, except has no bayonet pin. Heater volts (ac/dc), 12.6 ± 10%; amperes, 0.45. Except for heater rating and base, this type is identical with type 807.

POWER TRIODE

1626

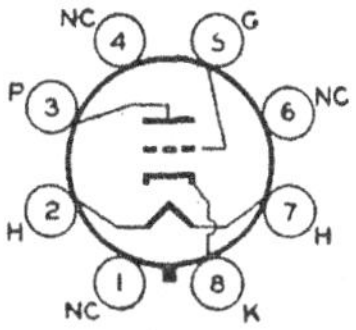

Glass-octal heater-cathode type used as rf power amplifier and oscillator. May be used with full input up to 30 Mc. For operation at 60 Mc, plate voltage and plate input should be reduced to 96 per cent of maximum ratings; at 90 Mc, to 93 per cent. Requires Octal socket and may be mounted in any position. OUTLINE 19, *Outlines* Section. Plate shows no color when tube is operated at maximum CCS ratings. The 1626 is used principally for renewal purposes.

HEATER VOLTAGE (AC/DC)	12.6	volts
HEATER CURRENT	0.25	ampere
AMPLIFICATION FACTOR	5	
DIRECT INTERELECTRODE CAPACITANCES:		
Grid to plate	4.4	μμf
Grid to cathode and heater	3.2	μμf
Plate to cathode and heater	3.0	μμf

RF POWER AMPLIFIER AND OSCILLATOR—Class C Telegraphy# and RF POWER AMPLIFIER—Class C FM Telephony

Maximum CCS Ratings:

DC PLATE VOLTAGE	250 *max*	volts
DC GRID VOLTAGE	-150 *max*	volts
DC PLATE CURRENT	25 *max*	ma
DC GRID CURRENT	8 *max*	ma
PLATE INPUT	6.25 *max*	watts
PLATE DISSIPATION	5 *max*	watts
PEAK HEATER-CATHODE VOLTAGE:		
Heater negative with respect to cathode	100 *max*	volts
Heater positive with respect to cathode	100 *max*	volts

Key-down conditions per tube without amplitude modulation. Amplitude modulation essentially negative may be used if the positive peak of the audio-frequency envelope does not exceed 115 per cent of the carrier conditions.

1635

HIGH-MU TWIN TRIODE

Glass-octal heater-cathode type used as af power amplifier. Class B AF Power Amplifier maximum CCS plate dissipation (design-center value, per plate), 3 watts. Requires Octal socket and may be mounted in any position. OUTLINE 13, *Outlines* Section. Plates show no color when tube is operated at maximum ratings. The 1635 is used principally for renewal purposes.

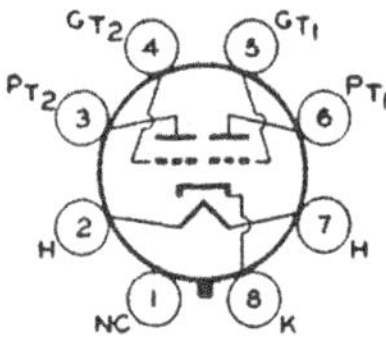

HEATER VOLTAGE (AC/DC)	6.3	volts
HEATER CURRENT	0.6	ampere

AF POWER AMPLIFIER—Class B

Maximum CCS Ratings:

DC PLATE VOLTAGE	300 *max*	volts
PEAK PLATE CURRENT (Per plate)	90 *max*	ma
PLATE DISSIPATION (Per plate)	3 *max*	watts
PEAK HEATER-CATHODE VOLTAGE:		
Heater negative with respect to cathode	90 *max*	volts
Heater positive with respect to cathode	90 *max*	volts

Typical Operation (*Unless otherwise specified, values are for 2 units*):

DC Plate Voltage	300	300	volts
DC Grid Voltage	0	0	volts
Peak AF Grid-to-Grid Voltage	70	108●	volts
Zero-Signal DC Plate Current	6.6	6.6	ma
Maximum-Signal DC Plate Current	54	54	ma
Peak Grid Current (Per unit)	38	39	ma
Plate-Supply Impedance	0	1000■	ohms
Effective Load Resistance (Plate to plate)	12000	12000	ohms
Effective Grid-Circuit Impedance (Per unit)	0	516⊕	ohms
Total Harmonic Distortion	4	5	per cent
Maximum-Signal Power Output	10.4	10.4	watts

● Includes peak voltage drop through the grid-circuit impedance.
■ Practical design value.
⊕ At 400 cycles for class B stage in which the effective resistance per grid circuit is 500 ohms, and the leakage reactance of the coupling transformer is 50 millihenries. The driver stage should be capable of supplying the grids of the class B stage with the specified values at low distortion.

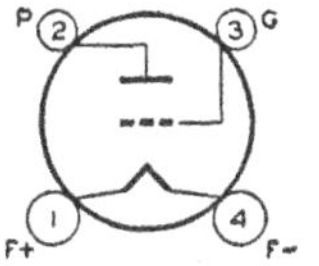

POWER TRIODE

5556

Coated-filament type used as af power amplifier and modulator and as rf power amplifier and oscillator. May be used with full input up to 6 Mc. For operation at 15 Mc, plate voltage and plate input should be reduced to 75 per cent of maximum ratings; at 30 Mc, to 50 per cent. Class C Telegraphy maximum CCS plate dissipation, 10 watts. Requires Small four-contact socket and may be mounted in vertical position with base up or down, or in horizontal position with pins 1 and 4 in vertical plane. OUTLINE 24, *Outlines* Section. Plate shows no color when tube is operated at maximum CCS ratings.

FILAMENT VOLTAGE (AC/DC)	4.5	volts
FILAMENT CURRENT	1.1	amperes
AMPLIFICATION FACTOR*	8.5	
TRANSCONDUCTANCE	1330	μmhos
DIRECT INTERELECTRODE CAPACITANCES:		
Grid to plate	6.7	μμf
Grid to filament	2.3	μμf
Plate to filament	2.2	μμf

* Plate volts, 350; grid volts, -20; plate milliamperes, 19.

AF POWER AMPLIFIER AND MODULATOR—Class A

Maximum CCS Ratings:

DC PLATE VOLTAGE	350 *max*	volts
PLATE DISSIPATION	7.5 *max*	watts

RF POWER AMPLIFIER AND OSCILLATOR—Class C Telegraphy# and RF POWER AMPLIFIER—Class C FM Telephony

Maximum CCS Ratings:

DC PLATE VOLTAGE	350 *max*	volts
DC GRID VOLTAGE	-150 *max*	volts
DC PLATE CURRENT	40 *max*	ma
DC GRID CURRENT (Approx.)	10 *max*	ma
PLATE INPUT	14 *max*	watts
PLATE DISSIPATION	10 *max*	watts

Key-down conditions per tube without amplitude modulation. Amplitude modulation essentially negative may be used if the positive peak of the audio-frequency envelope does not exceed 115 per cent of the carrier conditions.

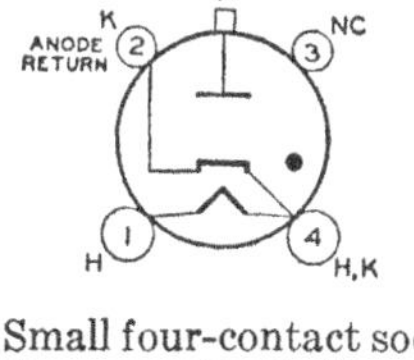

HALF-WAVE MERCURY-VAPOR RECTIFIER

5558

Heater-cathode type used in power supply of transmitting and industrial equipment. Maximum peak inverse anode volts, 5,000; maximum average anode amperes, 2.5. Requires Small four-contact socket and may be mounted in vertical position only, base down, OUTLINE 46, *Outlines* Section.

HEATER VOLTAGE•	5.0	volts
HEATER CURRENT	4.5	amperes
PEAK TUBE VOLTAGE DROP (Approx.)	15	volts

• Heater voltage must be applied at least 5 minutes before application of anode voltage.

HALF-WAVE RECTIFIER

Maximum Ratings *(For power-supply frequency of 60 cps)*:

PEAK INVERSE ANODE VOLTAGE	2000 *max*	5000 *max*	volts
ANODE CURRENT:			
Peak	15 *max*	15 *max*	amperes
Average ♦	2.5 *max*	2.5 *max*	amperes
Fault, for duration of 0.1 second maximum	200 *max*	200 *max*	amperes
CONDENSED-MERCURY-TEMPERATURE RANGE	30 to 80	30 to 60	°C

♦ Averaged over any interval of 15 seconds maximum.

Operating Values:

Circuit (For circuit figures, refer to *Rectifier Considerations* Section)	*Fig.*	*Max. Trans. Sec. Volts (RMS)* E	*Approx. DC Output Volts To Filter* Eav	*Max. DC Output Amperes* Iav	*Max. DC Output KW To Filter* Pdc
In-Phase Operation					
Half-Wave Single-Phase	54	3500•	1600	2.5	1.5
		1400▲	600	2.5	4.0
Full-Wave Single-Phase	55	1700•	1600	5.0	8.0
		700▲	600	5.0	3.0
Series Single-Phase	56	3500•	3200	5.0	16.0
		1400▲	1300	5.0	6.0
Half-Wave Three-Phase	57	2000•	2400	7.5	18.0
		800▲	950	7.5	7.0
Quadrature Operation					
Parallel Three-Phase	58	2000•	2400	15.0	14.0
		800▲	950	15.0	36.0
Series Three-Phase	59	2000•	4800	7.5	36.0
		800▲	1900	7.5	14.0
Half-Wave Four-Phase	60	1700•	2300	13.5* 15.0■	31.0* 34.5■
		700▲	900	13.5* 15.0■	12.0* 13.5■
Half-Wave Six-Phase	61	1700•	2400	14.2* 15.0■	13.5* 14.0■
		700▲	950	14.2* 15.0■	33.1* 34.0■

• For maximum peak inverse anode voltage of 5000 volts and maximum average anode current of 2.5 amperes.

▲ For maximum peak inverse anode voltage of 2000 volts and maximum average anode current of 2.5 amperes.

* Resistive load. ■ Inductive load.

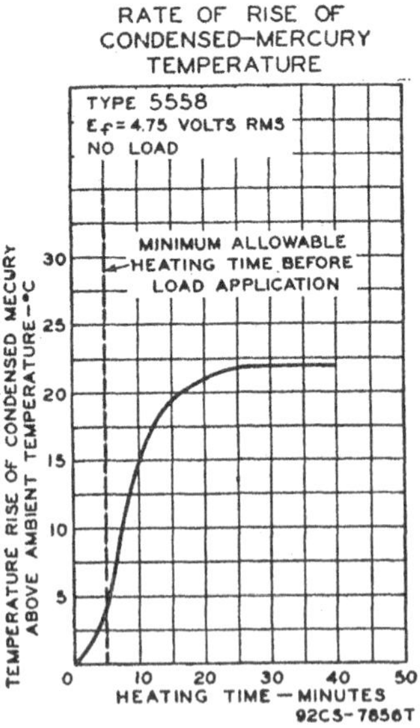

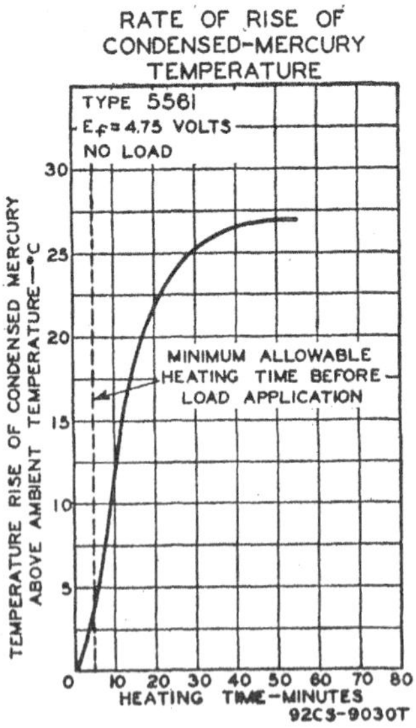

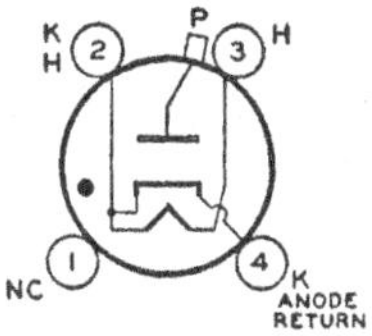

HALF-WAVE MERCURY-VAPOR RECTIFIER

5561

Heater-cathode type used in power supply of transmitting and industrial equipment. Rating I: maximum peak inverse anode volts, 3,000; maximum average anode amperes, 6.4. Rating II: maximum peak inverse anode volts, 10,000; maximum average anode amperes, 4. Requires Super-Jumbo four-contact socket and may be mounted in vertical position only, base down. OUTLINE 61, *Outlines* Section. For curve showing rate of rise of condensed-mercury temperature see preceding page.

HEATER VOLTAGE●	5	volts
HEATER CURRENT	10	amperes
PEAK TUBE VOLTAGE DROP (Approx.)	15	volts

● Heater voltage must be applied at least 5 minutes before application of anode voltage.

HALF-WAVE RECTIFIER

Maximum Ratings *(For power-supply frequency of 60 cps):*

PEAK INVERSE ANODE VOLTAGE	3000 *max*	10000 *max*	volts
ANODE CURRENT:			
Peak	40 *max*	16 *max*	amperes
Average♣	6.4 *max*	4 *max*	amperes
Fault, for duration of 0.1 second maximum	400 *max*	160 *max*	amperes
CONDENSED-MERCURY-TEMPERATURE RANGE	40 to 80	25 to 50	°C

♣ Averaged over any interval of 15 seconds maximum.

Operating Values:

Circuit (For circuit figures, refer to *Rectifier Considerations* Section)	*Fig.*	*Max. Trans. Sec. Volts (RMS)* E	*Approx. DC Output Volts To Filter* Eav	*Max. DC Output Amperes* Iav	*Max. DC Output KW To Filter* Pdc
		In-Phase Operation			
Half-Wave Single-Phase	54	2100•	950	6.4	6.0
Full-Wave Single-Phase	55	1000•	950	12.8	12.0
Series Single-Phase	56	2100•	1900	12.8	24.0
Half-Wave Three-Phase	57	1200•	1450	19.2	27.5
		Quadrature Operation			
Parallel Three-Phase	58	1200•	1450	38.4	55.0
Series Three-Phase	59	1200•	2850	19.2	55.0
Half-Wave Four-Phase	60	1000•	1350	36.0* 40.0■	48.5* 54.0■
Half-Wave Six-Phase	61	1000•	1450	38.0* 40.0■	54.0* 57.0■

•For maximum peak inverse anode voltage of 3000 volts and maximum average anode current of 6.4 amperes.
* Resistive load. ■ Inductive load.

POWER TRIODE

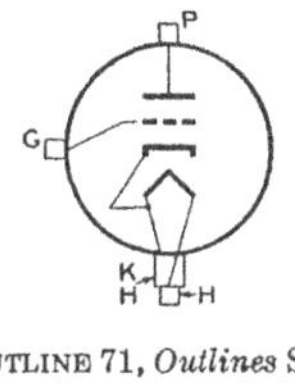

5588

Forced-air-cooled heater-cathode type having integral radiator used in cathode-drive circuits as rf power amplifier and oscillator. May be used with full input up to 1200 Mc. For operation at 1350 Mc, plate voltage and plate input should be reduced to 90 per cent of maximum ratings; at 1500 Mc, to 89 per cent; at 2000 Mc, to 80 per cent. Type 5588 may be mounted in vertical position only, radiator up or down. OUTLINE 71, *Outlines* Section. A minimum air flow of 10 cubic feet per minute should be directed through the radiator toward the bulb and grid terminal when the 5588 is operated at maximum rated dissipation.

Air flow should start before and continue during the application of any voltages to the tube. Maximum temperatures: incoming air, 45°C; radiator, 180°C; and grid terminal, 140°C. The 5588 is used principally for renewal purposes. For new equipment design, refer to type 6161.

HEATER VOLTAGE (AC/DC)°	6.3	volts
HEATER CURRENT	2.5	amperes
AMPLIFICATION FACTOR	16	
DIRECT INTERELECTRODE CAPACITANCES:		
Grid to plate	6.0	μμf
Grid to cathode and heater	13	μμf
Plate to cathode and heater□	0.32 *max*	μμf

° Rated heater voltage must be applied for a minimum time of one minute before voltages are applied to the other electrodes.
□ External shield connected to grid.

RF POWER AMPLIFIER AND OSCILLATOR—Class C Telegraphy# and RF POWER AMPLIFIER—Class C FM Telephony

Maximum CCS Ratings:

DC PLATE VOLTAGE	1000 *max*	volts
DC GRID VOLTAGE	-200 *max*	volts
DC PLATE CURRENT	300 *max*	ma
DC GRID CURRENT	100 *max*	ma
PLATE INPUT	250 *max*	watts
PLATE DISSIPATION	200 *max*	watts

Typical Operation in Cathode-Drive Circuit at 1000 Mc:	*Amplifier*	*Oscillator*	
Heater Voltage°	4.5	3.0	volts
DC Plate Voltage	835	835	volts
DC Grid Voltage♣	-70	-70	volts
From grid resistor of	1750	—	ohms
From cathode resistor of	—	205	ohms
DC Plate Current	300	300	ma
DC Grid Current (Approx.)	40	40	ma
Driving Power (Approx.)●	32	—	watts
Power Output (Approx.)	100	75	watts

Key-down conditions per tube without amplitude modulation. Amplitude modulation essentially negative may be used if the positive peak of the audio-frequency envelope does not exceed 115 per cent of the carrier conditions.
° Rated heater voltage must be applied for a minimum time of one minute before voltages are applied to the other electrodes. Heater voltage may then be reduced to the indicated typical operating value.
♣ Obtained from fixed supply, by grid resistor, by cathode resistor, or by combination methods.
● Required by tube and input circuit. A portion of this power appears in the load circuit.

5618

POWER PENTODE

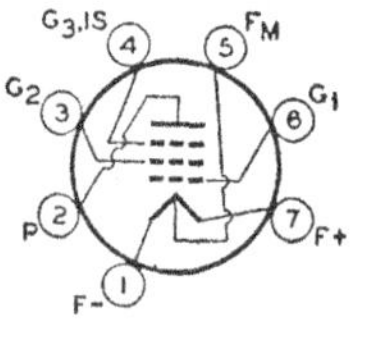

Seven-pin miniature type having quick-heating, mid-tapped, coated filament used as af power amplifier and modulator, rf power amplifier and oscillator, and frequency multiplier in mobile and other communications equipment when compactness and low filament-power consumption are primary requirements. Designed for intermittent operation only. May be used with full input up to 100 Mc and with reduced input up to 165 Mc. Class C Telegraphy maximum ICAS plate dissipation, 5 watts.

FILAMENT ARRANGEMENT	*Series*	*Parallel*	
FILAMENT VOLTAGE (AC/DC)	6.0 ± 10%	3.0 ± 10%	volts
FILAMENT CURRENT	0.23	0.46	ampere
DIRECT INTERELECTRODE CAPACITANCES:			
Grid No.1 to plate	0.24 *max*		μμf
Grid No.1 to filament mid-tap, grid No.3, internal shield, and grid No.2	7.0		μμf
Plate to filament mid-tap, grid No.3, internal shield, and grid No.2	5.0		μμf

AF POWER AMPLIFIER AND MODULATOR—Class A1

Maximum ICAS Ratings:

DC PLATE VOLTAGE	300 *max*	volts
DC GRID-No.2 (SCREEN-GRID) VOLTAGE	125 *max*	volts
GRID-No.2 INPUT	2 *max*	watts
PLATE DISSIPATION	5 *max*	watts

Typical Operation:

	Series	*Parallel*	
DC Plate Voltage	250	250	volts
DC Grid-No.3 (Suppressor-Grid) Voltage	0	0	volts
DC Grid-No.2 Voltage	75	75	volts
DC Grid-No.1 (Control-Grid) Voltage	-8	-8	volts
Peak AF Grid-No.1-to-Grid-No.1 Voltage	8	8	volts
Zero-Signal DC Plate Current	16	19	ma
Maximum-Signal DC Plate Current	17.5	20.5	ma
Zero-Signal DC Grid-No.2 Current	1.5	2	ma
Maximum-Signal DC Grid-No.2 Current	3.5	4.5	ma
Transconductance	3500	3600	μmhos
Effective Load Resistance (Plate to plate)	12000	12000	ohms
Total Harmonic Distortion	10	10	per cent
Maximum-Signal Power Output	1.2	1.4	watts

Circuit Values:

Grid-No.1-Circuit Resistance	5000 *min*	ohms
	100000 *max*	ohms

RF POWER AMPLIFIER AND OSCILLATOR—Class C Telegraphy# and RF POWER AMPLIFIER—Class C FM Telephony

Maximum ICAS Ratings:

DC PLATE VOLTAGE	300 *max*	volts
DC GRID-No.2 VOLTAGE	125 *max*	volts
DC GRID-No.1 VOLTAGE	-125 *max*	volts
DC PLATE CURRENT	30 *max*	ma
DC GRID-No.1 CURRENT	3 *max*	ma
PLATE INPUT	7.5 *max*	watts
GRID-No.2 INPUT	2 *max*	watts
PLATE DISSIPATION	5 *max*	watts

Typical Operation:

	40 Mc	*80 Mc*	
DC Plate Voltage	300	300	volts
DC Grid-No.3 Voltage	0	0	volts
DC Grid-No.2 Voltage⊕	75	75	volts
From series resistor of	32000	32000	ohms
DC Grid-No.1 Voltage♦	-45	-45	volts
From grid-No.1 resistor of	30000	30000	ohms
From cathode resistor of	1400	1400	ohms
Peak RF Grid-No.1 Voltage	65	65	volts
DC Plate Current	25	25	ma
DC Grid-No.2 Current	7	7	ma
DC Grid-No.1 Current (Approx.)	1.5	1.5	ma
Driving Power (Approx.)	0.2	0.3	watt
Power Output (Approx.)	5.4	5.2	watts
Useful Power Output (Approx.)	5.0	4.5	watts

Circuit Values:

Grid-No.1-Circuit Resistance	5000 *min*	ohms
	100000 *max*	ohms

Key-down conditions per tube without amplitude modulation. Amplitude modulation essentially negative may be used if the positive peak of the audio-frequency envelope does not exceed 115 per cent of the carrier conditions.

⊕ Obtained from separate source, from plate-voltage supply with a voltage divider, or from series resistor of value shown.

♦ Obtained from fixed supply, by grid-No.1 resistor, by cathode resistor, or by combination methods.

FREQUENCY MULTIPLIER

Maximum ICAS Ratings:

DC PLATE VOLTAGE	300 *max*	volts
DC GRID-No.2 VOLTAGE	125 *max*	volts

DC GRID-NO.1 VOLTAGE	−125 *max*	volts
DC PLATE CURRENT	30 *max*	ma
DC GRID-NO.1 CURRENT	3 *max*	ma
PLATE INPUT	7.5 *max*	watts
GRID-NO.2 INPUT	2 *max*	watts
PLATE DISSIPATION	5 *max*	watts

Typical Operation at Frequencies up to 80 Mc:

	Doubler	*Tripler*	
DC Plate Voltage	300	300	volts
DC Grid-No.3 Voltage	0	0	volts
DC Grid-No.2 Voltage⊕	75	75	volts
From series resistor of	41000	41000	ohms
DC Grid-No.1 Voltage♣	−125	−125	volts
From grid-No.1 resistor of	68000	68000	ohms
From cathode resistor of	4100	4100	ohms
Peak RF Grid-No.1 Voltage	160	160	volts
DC Plate Current	25	25	ma
DC Grid-No.2 Current	5.5	5.5	ma
DC Grid-No.1 Current (Approx.)	1.85	1.85	ma
Driving Power (Approx.)	0.75	0.75	watt
Power Output (Approx.)	4.2	3.4	watts
Useful Power Output (Approx.)	3.5	2.7	watts

Circuit Values:

Grid-No.1-Circuit Resistance	5000 *min*	ohms
	100000 *max*	ohms

⊕ Obtained from separate source, from plate-voltage supply with a voltage divider, or from series resistor of value shown.

♣ Obtained from fixed supply, by grid-No.1 resistor, by cathode resistor, or by combination methods.

OPERATING CONSIDERATIONS

Type 5618 requires Miniature seven-contact socket and may be mounted in vertical position with base up or down, or in horizontal position with pins 3 and 7 in vertical plane. OUTLINE 8, *Outlines* Section.

For operation at 165 Mc, plate input should be reduced to 90 per cent of maximum rating.

For series-filament arrangement, filament voltage is applied between pins 1 and 7. For parallel-filament arrangement, filament voltage is applied between pin 5 and pins 1 and 7 connected together. In series-filament arrangement, grid-No.1 voltage is referred to pin 1, and pin 4 is connected to pin 1. In parallel-filament arrangement, grid-No.1 voltage is referred to pin 5, and pin 4 is connected to pin 5.

Plate shows no color when tube is operated at maximum ICAS ratings.

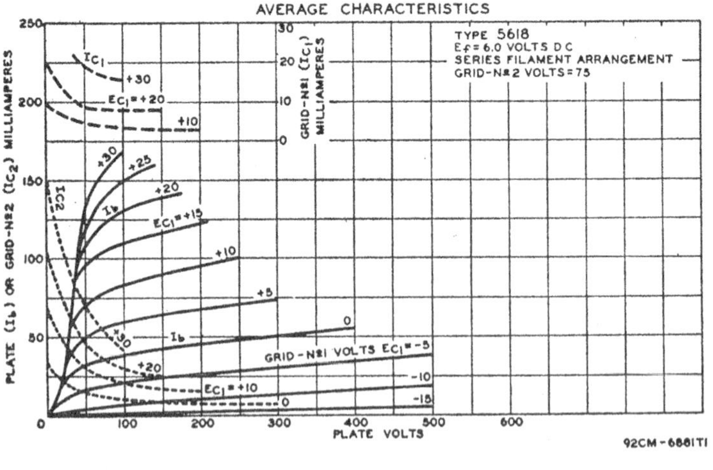

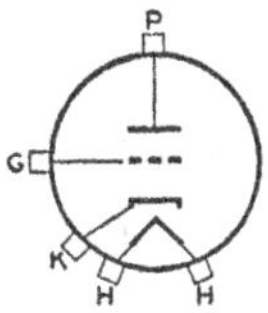

MEDIUM-MU TRIODE

5675

Pencil-type tube used in cathode-drive circuits as rf power amplifier and oscillator. Designed for use in coaxial-cylinder-type circuits, it may also be used in parallel-line or lumped circuits. May be used with full input up to 3000 Mc. Class C maximum CCS plate dissipation, 9 watts. The tube may be mounted in any position. OUTLINE 65, *Outlines* Section.

HEATER VOLTAGE (AC/DC)	6.3 ± 10%	volts
HEATER CURRENT	0.135	ampere
TRANSCONDUCTANCE*	6200	μmhos
AMPLIFICATION FACTOR*	20	
PLATE RESISTANCE (Approx.)*	3225	ohms
DIRECT INTERELECTRODE CAPACITANCES:		
Grid to plate	1.4	μμf
Grid to cathode and heater	2.3	μμf
Plate to cathode and heater	0.09 *max*	μμf

* Plate-supply volts, 135; cathode resistor, 68 ohms; plate milliamperes, 24.

RF POWER AMPLIFIER AND OSCILLATOR—Class C

Maximum CCS Ratings:

DC PLATE VOLTAGE	300 *max*	volts
DC GRID VOLTAGE	-90 *max*	volts
DC PLATE CURRENT	30 *max*	ma
DC GRID CURRENT	8 *max*	ma
PLATE INPUT	9 *max*	watts
PLATE DISSIPATION■	9 *max*	watts
PEAK HEATER-CATHODE VOLTAGE:		
Heater negative with respect to cathode	90 *max*	volts
Heater positive with respect to cathode	90 *max*	volts
PLATE-SEAL TEMPERATURE	175 *max*	°C

Typical Operation as Cathode-Drive Oscillator at 1700 Mc:●

DC Plate Voltage	120	volts
DC Grid Voltage	-8	volts
From a grid resistor of	2000	ohms
DC Plate Current	25	ma
DC Grid Current (Approx.)	4	ma
Power Output (Approx.)	475	mw

■ In applications where the plate dissipation exceeds 2.5 watts, it is important that a large area of contact be provided between the plate cylinder and its lead connector to provide adequate heat conduction.
● At 3000 Mc, and with full ratings, a useful output of approximately 50 milliwatts may be obtained.

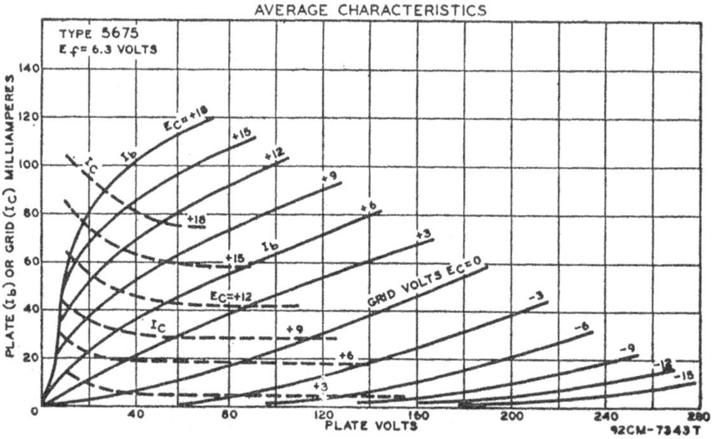

POWER TRIODE

5713

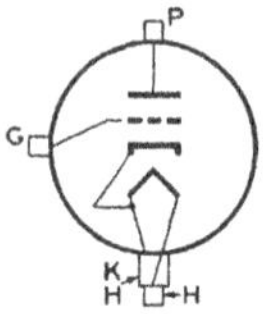

Forced-air-cooled heater-cathode type having integral radiator used in grid-drive circuits and in cathode-drive circuits up to 220 Mc. Class C Telegraphy maximum CCS plate dissipation, 250 watts. This type may be mounted in vertical position only, radiator up or down. OUTLINE 74, *Outlines* Section. A minimum air flow of 18 cubic feet per minute should be directed through the radiator toward the bulb and grid terminal when the tube is operated at maximum rated dissipation. Air flow should start before and continue during the application of any voltages to the tube. Maximum temperatures: incoming air, 45°C; radiator, measured on core at bulb end, 180°C; glass, 180°C; and grid terminal, 140°C.

HEATER VOLTAGE (AC/DC)°	3.3 ± 0.2	volts
HEATER CURRENT	11.5	amperes
AMPLIFICATION FACTOR*	25	
DIRECT INTERELECTRODE CAPACITANCES (Approx.):		
Grid to plate	10.3	μμf
Grid to cathode and heater	26	μμf
Plate to cathode and heater	0.5	μμf

° Heater voltage must be applied for a minimum time of 2 minutes before application of plate voltage.
* Plate volts, 1000; plate milliamperes, 150.

RF POWER AMPLIFIER AND OSCILLATOR—Class C Telegraphy# and RF POWER AMPLIFIER—Class C FM Telephony

Maximum CCS Ratings:

DC PLATE VOLTAGE	1500 *max*	volts
DC GRID VOLTAGE	-250 *max*	volts
DC PLATE CURRENT	300 *max*	ma
DC GRID CURRENT	50 *max*	ma
PLATE INPUT	450 *max*	watts
PLATE DISSIPATION	250 *max*	watts

Typical Operation:	*Grid-Drive*	*Cathode-Drive at 220 Mc*	
DC Plate Voltage	1500	1500	volts
DC Grid Voltage ♁	-175	-175	volts
From cathode resistor of	510	510	ohms
Peak RF Grid Voltage	210	210	volts
DC Plate Current	300	300	ma
DC Grid Current (Approx.)	40	40	ma
Driving Power (Approx.)	8	65•	watts
Power Output (Approx.)	290	325	watts

Key-down conditions per tube without amplitude modulation. Amplitude modulation essentially negative may be used if the positive peak of the audio-frequency envelope does not exceed 115 per cent of the carrier conditions.
♁ Obtained from fixed supply or from cathode resistor of value shown.
• Required by tube and input circuit. A portion of this power appears in the load circuit.

MEDIUM-MU TRIODE

5718

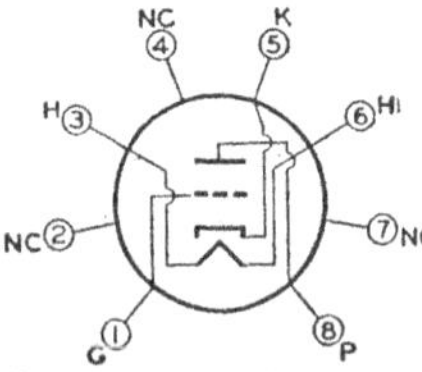

Premium subminiature heater-cathode type used as rf amplifier and oscillator. May be used with full input up to 1000 Mc. Class C maximum CCS plate dissipation, 3.3 watts. Tube may be operated in any position. OUTLINE 3, *Outlines* Section. The flexible leads of the 5718 are usually soldered to the circuit elements. Soldering of the leads may be made close to the glass stem provided care is taken to conduct excessive heat away from the lead seal. Otherwise, the heat of the soldering operation will crack the seals of the leads and damage the tube. Plate shows no color when tube is operated at maximum CCS ratings.

Heater Voltage (ac/dc)	6.3	volts
Heater Current	0.15	ampere
Transconductance*	6500	μmhos
Amplification Factor*	27	
Plate Resistance (Approx.)*	4150	ohms
Direct Interelectrode Capacitances:		
Grid to plate	1.4	μμf
Grid to cathode and heater	2.2	μμf
Plate to cathode and heater	0.7	μμf

* Plate-supply volts, 150; cathode resistor, 180 ohms; plate milliamperes, 13.

RF AMPLIFIER AND OSCILLATOR—Class C

Maximum CCS Ratings:

DC Plate Voltage	165 *max*	volts
DC Grid Voltage	-55 *max*	volts
DC Plate Current	22 *max*	ma
DC Grid Current	5.5 *max*	ma
Plate Dissipation	3.3 *max*	watts
Peak Heater-Cathode Voltage:		
Heater negative with respect to cathode	200 *max*	volts
Heater positive with respect to cathode	200 *max*	volts
Bulb Temperature	250 *max*	°C

Maximum Circuit Values:

Grid-Circuit Resistance:		
For cathode-bias operation	1.2 *max*	megohms
For fixed-bias operation	Not recommended	

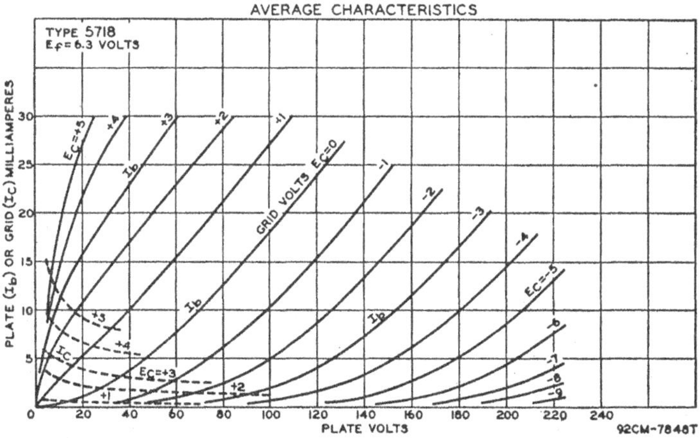

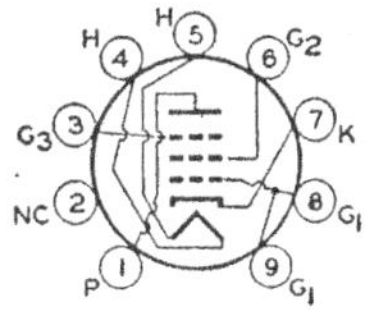

BEAM POWER TUBE

5763

Nine-pin miniature heater-cathode type used as rf power amplifier and oscillator and as frequency multiplier. May be used with full input up to 50 Mc. For operation at 175 Mc, plate input should be reduced to 80 per cent of maximum rating. Class C Telegraphy maximum plate dissipation, CCS 12 watts, ICAS 13.5 watts. Requires

Noval nine-contact socket and may be mounted in any position. OUTLINE 9, *Outlines* Section. Plate shows no color when tube is operated at maximum CCS or ICAS ratings.

HEATER VOLTAGE (AC/DC)	6.0 ± 10%	volts
HEATER CURRENT	0.75	amperes
TRANSCONDUCTANCE*	7000	μmhos
MU-FACTOR, Grid No.2 to Grid No.1*	16	
DIRECT INTERELECTRODE CAPACITANCES:		
Grid No.1 to plate	0.3 *max*	μμf
Grid No.1 to cathode, grid No.3, grid No.2, and heater	9.5	μμf
Plate to cathode, grid No.3, grid No.2, and heater	4.5	μμf

* Plate and grid-No.2 volts, 250; grid-No.1 volts, -7.5; plate milliamperes, 45.

PLATE-MODULATED RF POWER AMPLIFIER—Class C Telephony

Carrier conditions per tube for use with a maximum modulation factor of 1.0

Maximum Ratings:	*CCS*	*ICAS*	
DC PLATE VOLTAGE	250 *max*	300 *max*	volts
DC GRID-NO.3 (SUPPRESSOR-GRID) VOLTAGE	0 *max*	0 *max*	volts
DC GRID-NO.2 (SCREEN-GRID) VOLTAGE	250 *max*	250 *max*	volts
DC GRID-NO.1 (CONTROL-GRID) VOLTAGE	-125 *max*	-125 *max*	volts
DC PLATE CURRENT	40 *max*	50 *max*	ma
DC GRID-NO.2 CURRENT	15 *max*	15 *max*	ma
DC GRID-NO.1 CURRENT	5 *max*	5 *max*	ma
PLATE INPUT	10 *max*	15 *max*	watts
GRID-NO.2 INPUT	1.5 *max*	1.5 *max*	watts
PLATE DISSIPATION	8 *max*	12 *max*	watts
PEAK HEATER-CATHODE VOLTAGE:			
Heater negative with respect to cathode	100 *max*	100 *max*	volts
Heater positive with respect to cathode	100 *max*	100 *max*	volts
BULB TEMPERATURE (At hottest point)	250 *max*	250 *max*	°C

Typical Operation at Frequencies up to 30 Mc:			
DC Plate Voltage	250	300	volts
Grid No.3	Connected to cathode at socket		
DC Grid-No.2 Voltage♦	250	250	volts
DC Grid-No.1 Voltage◊	-39	-42.5	volts
From grid-No.1 resistor of	39000	18000	ohms
Peak RF Grid-No.1 Voltage	46.5	53.5	volts
DC Plate Current	40	50	ma
DC Grid-No.2 Current	5.6	6	ma
DC Grid-No.1 Current (Approx.)	1	2.4	ma
Driving Power (Approx.)	0.05	0.15	watt
Useful Power Output (Approx.)	6.4■	10■	watts

Maximum Circuit Values (CCS or ICAS conditions):

Grid-No.1-Circuit Resistance . . . 0.1 *max* megohm

♦ Obtained preferably from separate source modulated along with the plate supply, or from the modulated plate supply through a series resistor.

◊ Obtained from grid-No.1 resistor of value shown or from a combination of grid-No.1 resistor with either fixed supply or cathode resistor.

■ Measured at load of output circuit.

RF POWER AMPLIFIER AND OSCILLATOR—Class C Telegraphy#
and
RF POWER AMPLIFIER—Class C FM Telephony

Maximum Ratings:	*CCS*	*ICAS*	
DC PLATE VOLTAGE	300 *max*	350 *max*	volts
DC GRID-NO.3 VOLTAGE	0 *max*	0 *max*	volts
DC GRID-NO.2 VOLTAGE	250 *max*	250 *max*	volts
DC GRID-NO.1 VOLTAGE	-125 *max*	-125 *max*	volts
DC PLATE CURRENT	50 *max*	50 *max*	ma
DC GRID-NO.2 CURRENT	15 *max*	15 *max*	ma
DC GRID-NO.1 CURRENT	5 *max*	5 *max*	ma
PLATE INPUT	15 *max*	17 *max*	watts
GRID-NO.2 INPUT	2 *max*	2 *max*	watts
PLATE DISSIPATION	12 *max*	13.5 *max*	watts

PEAK HEATER-CATHODE VOLTAGE:			
Heater negative with respect to cathode	100 *max*	100 *max*	volts
Heater positive with respect to cathode	100 *max*	100 *max*	volts
BULB TEMPERATURE (At hottest point)	250 *max*	250 *max*	°C

Typical Operation:

	30Mc	*50Mc*	*30Mc*	
DC Plate Voltage	300	300	350	volts
Grid No. 3	Connected to cathode at socket			
DC Grid-No.2 Voltage	250	250	250	volts
DC Grid-No.1 Voltageϭ	−28.5	−60	−28.5	volts
From grid-No.1 resistor of	18000	22000	18000	ohms
Peak RF Grid-No.1 Voltage	37.5	80	37	volts
DC Plate Current	50	50	48.5	ma
DC Grid-No.2 Current	6.6	5	6.2	ma
DC Grid-No.1 Current (Approx.)	1.6	3	1.6	ma
Driving Power (Approx.)	0.1	0.35	0.1	watt
Useful Power Output (Approx.)	10.3■	7■	12■	watts

Maximum Circuit Values (CCS or ICAS conditions):

Grid-No.1-Circuit Resistance	0.1 *max*	megohm

‡ Key-down conditions per tube without amplitude modulation. Amplitude modulation essentially negative may be used provided the positive peak of the audio-frequency envelope does not exceed 115 per cent of the carrier conditions.

ϭ Obtained from fixed supply or from grid-No.1 resistor of value shown.

■ Measured at load of output circuit.

FREQUENCY MULTIPLIER

Maximum CCS Ratings:

DC PLATE VOLTAGE	300 *max*	volts
DC GRID-NO.3 VOLTAGE	0 *max*	volts
DC GRID-NO.2 VOLTAGE	250 *max*	volts
DC GRID-NO.1 VOLTAGE	−125 *max*	volts
DC PLATE CURRENT	50 *max*	ma
DC GRID-NO.2 CURRENT	15 *max*	ma
DC GRID-NO.1 CURRENT	5 *max*	ma
PLATE INPUT	15 *max*	watts
GRID-NO.2 INPUT	2 *max*	watts
PLATE DISSIPATION	12 *max*	watts
PEAK HEATER-CATHODE VOLTAGE:		
Heater negative with respect to cathode	100 *max*	volts
Heater positive with respect to cathode	100 *max*	volts
BULB TEMPERATURE (At hottest point)	250 *max*	°C

Typical Operation at Frequencies up to 175 Mc:

	Doubler	*Tripler*	
DC Plate Voltage	300	300	volts
Grid No.3	Connected to cathode at socket		
DC Grid-No.2 Voltage	*	*	volts

AVERAGE CHARACTERISTICS

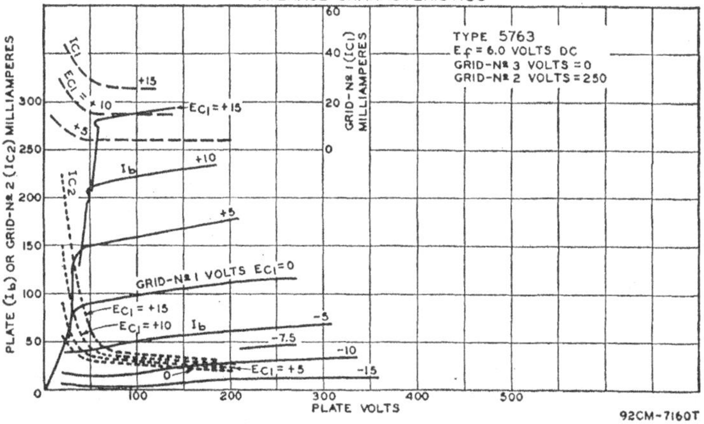

92CM-7160T

DC Grid-No.1 Voltage♁	-75	-100	volts
From grid-No.1 resistor of	75000	100000	ohms
Peak RF Grid-No.1 Voltage	95	120	volts
DC Plate Current	40	35	ma
DC Grid-No.2 Current	4	5	ma
DC Grid-No.1 Current (Approx.)	1	1	ma
Driving Power (Approx.)	0.6	0.6	watt
Useful Power Output (Approx.)	2.1■	1.3■	watts

Maximum Circuit Values:

Grid-No.1-Circuit Resistance	0.1 *max*	megohm

* Obtained from 300-volt supply with series resistor of 12,500 ohms.
♁ Obtained from fixed supply or from grid-No.1 resistor of value shown.
■ Measured at load of output circuit.

POWER TRIODE

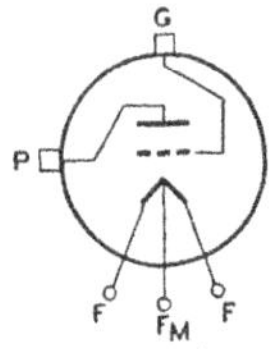

5786

Forced-air-cooled thoriated-tungsten-filament type having integral radiator used as af power amplifier and modulator and as rf power amplifier and oscillator. May be used with full input up to 160 Mc. Class C Telegraphy maximum CCS plate dissipation, 600 watts. May be mounted in vertical position only, filament end up or down. OUTLINE 75, *Outlines* Section. A minimum air flow of 140 cubic feet per minute should be directed by a blower to the radiator and seals when the 5786 is operated at maximum rated dissipation. Air flow should start before and continue during application of any voltages to the tube. Filament power, plate power, and air may be removed simultaneously. Maximum temperatures: incoming air, 45°C; radiator, at core, 180°C; grid and plate seals, 165°C; and filament seals, 220°C.

FILAMENT VOLTAGE (AC/DC)	11 ± 0.6	volts
FILAMENT CURRENT	12.5	amperes
FILAMENT STARTING CURRENT	50 *max*	amperes
AMPLIFICATION FACTOR*	32	
DIRECT INTERELECTRODE CAPACITANCES:		
Grid to plate	5.3	μμf
Grid to filament mid-tap	4.7	μμf
Plate to filament mid-tap	3.8	μμf

* Grid volts, -25; plate milliamperes, 200.

AF POWER AMPLIFIER AND MODULATOR—Class B

Maximum CCS Ratings:

DC PLATE VOLTAGE	4000 *max*	volts
MAXIMUM-SIGNAL DC PLATE CURRENT■	500 *max*	ma
MAXIMUM-SIGNAL PLATE INPUT■	1500 *max*	watts
PLATE DISSIPATION■	600 *max*	watts

Typical Operation *(Values are for 2 tubes):*

DC Plate Voltage	3000	volts
DC Grid Voltage†	-95	volts
Peak AF Grid-to-Grid Voltage	470	volts
Zero-Signal DC Plate Current	75	ma
Maximum-Signal DC Plate Current	800	ma
Effective Load Resistance (Plate to plate)	8600	ohms
Maximum-Signal Driving Power (Approx.)	30	watts
Maximum-Signal Power Output (Approx.)	1640	watts

■ Averaged over any audio-frequency cycle of sine-wave form.
† Grid voltage is given with respect to mid-point of filament operated on ac or dc.

PLATE-MODULATED RF POWER AMPLIFIER—Class C Telephony

Carrier conditions per tube for use with a maximum modulation factor of 1.0

Maximum CCS Ratings:

DC PLATE VOLTAGE	2500 *max*	volts
DC GRID VOLTAGE	-500 *max*	volts

DC PLATE CURRENT	400 *max*	ma
DC GRID CURRENT	150 *max*	ma
PLATE INPUT	1000 *max*	watts
PLATE DISSIPATION	400 *max*	watts

Typical Operation:

DC Plate Voltage	2500	volts
DC Grid Voltage ♁	−350	volts
From grid resistor of	2600	ohms
Peak RF Grid Voltage	620	volts
DC Plate Current	400	ma
DC Grid Current (Approx.)	135	ma
Driving Power (Approx.)	75	watts
Power Output (Approx.)	810	watts

♁ Obtained preferably from grid resistor of value shown or from a combination of grid resistor with either fixed supply or cathode resistor.

RF POWER AMPLIFIER AND OSCILLATOR—Class C Telegraphy#
and
RF POWER AMPLIFIER—Class C FM Telephony

Maximum CCS Ratings:

DC PLATE VOLTAGE	3000 *max*	volts
DC GRID VOLTAGE	−500 *max*	volts
DC PLATE CURRENT	500 *max*	ma
DC GRID CURRENT	150 *max*	ma
PLATE INPUT	1500 *max*	watts
PLATE DISSIPATION	600 *max*	watts

Typical Operation:	*RF Power Amplifier*	*Oscillator at 160 Mc*	
DC Plate Voltage	3000	3000	volts
DC Grid Voltage ♠	−200	−225	volts
From grid resistor of	2200	2000	ohms
From cathode resistor of	330	380	ohms
Peak RF Grid Voltage	450	475	volts
DC Plate Current	500	500	ma
DC Grid Current	90	90	ma
Driving Power (Approx.)	36	—	watts
Power Output (Approx.)	1000	1000	watts
Useful Power Output (Approx.)—85-per-cent circuit efficiency	—	850	watts

Key-down conditions per tube without amplitude modulation. Amplitude modulation essentially negative may be used if the positive peak of the audio-frequency envelope does not exceed 115 per cent of the carrier conditions.

♠ Obtained from fixed supply, by grid resistor, by cathode resistor, or by combination methods.

SELF-RECTIFYING OSCILLATOR OR AMPLIFIER—Class C

Maximum CCS Ratings:

RMS PLATE VOLTAGE	4250 *max*	volts
DC GRID VOLTAGE	−300 *max*	volts
DC PLATE CURRENT	320 *max*	ma
DC GRID CURRENT	85 *max*	ma
PLATE INPUT	1500 *max*	watts
PLATE DISSIPATION	600 *max*	watts

Typical Operation:

RMS Plate Voltage	4250	volts
DC Grid Voltage ♁	−115	volts
From grid resistor of	1500	ohms
DC Plate Current	320	ma
DC Grid Current (Approx.)	77	ma
Driving Power (Approx.)•	46	watts
Power Output (Approx.)	1050	watts

♁ Obtained preferably from grid resistor of value shown or from a combination of grid resistor and fixed supply.

• From a self-rectifying driver.

AMPLIFIER OR OSCILLATOR—Class C

With separate rectified, unfiltered, single-phase, full-wave plate supply

Maximum CCS Ratings:

DC PLATE VOLTAGE	2700 *max*	volts

DC GRID VOLTAGE	-300 *max*	volts
DC PLATE CURRENT	450 *max*	ma
DC GRID CURRENT	120 *max*	ma
PLATE INPUT	1500 *max*	watts
PLATE DISSIPATION	600 *max*	watts

Typical Operation:

DC Plate Voltage	2700	volts
DC Grid Voltage ♁	-180	volts
From a grid resistor of	1530	ohms
DC Plate Current	450	ma
DC Grid Current (Approx.)	118	ma
Driving Power (Approx.)▲	57	watts
Power Output (Approx.)	1150	watts

♁ Obtained preferably from grid resistor of value shown or from a combination of grid resistor and fixed supply.
▲ From a driver having a rectified, unfiltered, single-phase, full-wave plate supply.

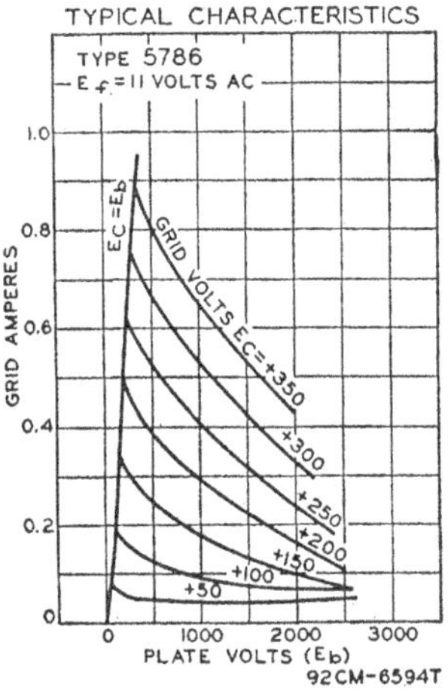

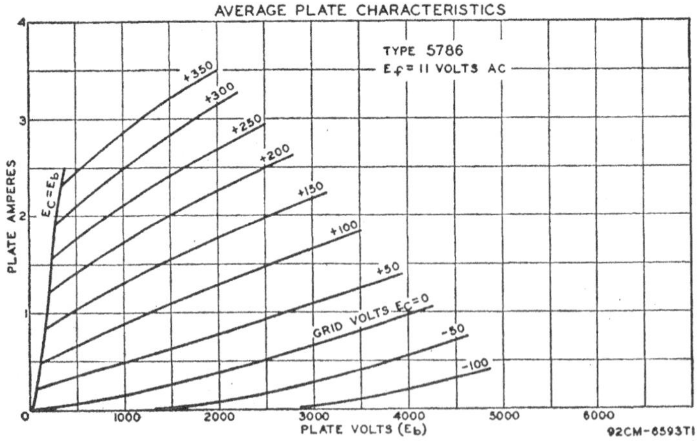

FIXED-TUNED OSCILLATOR TRIODE

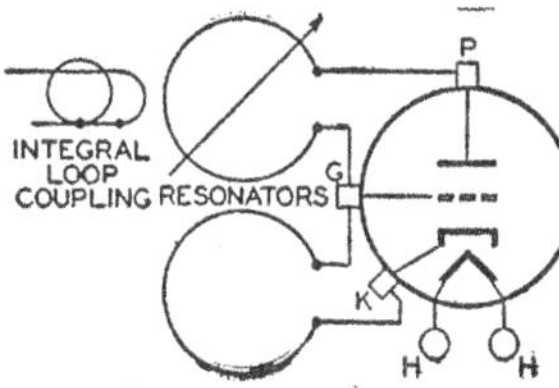

5794

Pencil-type tube having integral resonators used in radiosonde service at 1680 Mc. Fixed-Tuned Oscillator maximum plate dissipation, 3.6 watts. May be mounted in any position. The 5794 is identical with type 6562 except that the 5794 does not have an external connection between the cathode and one side of the heater. OUTLINE 68, *Outlines* Section.

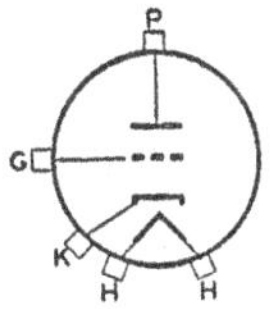

HIGH-MU TRIODE

5876

Pencil-type tube used as rf power amplifier and oscillator at frequencies up to 1700 Mc. Designed for use in coaxial-cylinder-type circuits, it may also be used in parallel-line or lumped circuits. Class C Telegraphy maximum CCS plate dissipation, 6.25 watts. May be mounted in any position. OUTLINE 65, *Outlines* Section.

HEATER VOLTAGE (AC/DC)	6.3 ± 10%	volts
HEATER CURRENT	0.135	ampere
TRANSCONDUCTANCE*	6500	μmhos
AMPLIFICATION FACTOR*	56	
PLATE RESISTANCE (Approx.)*	8625	ohms
DIRECT INTERELECTRODE CAPACITANCES:		
Grid to plate	1.4	μμf
Grid to cathode and heater	2.4	μμf
Plate to cathode and heater	0.035 *max*	μμf

*Plate-supply volts, 250; cathode resistor, 75 ohms; plate milliamperes, 18.

RF POWER AMPLIFIER AND OSCILLATOR—Class C Telegraphy ♯ and RF POWER AMPLIFIER—Class C FM Telephony

Maximum CCS Ratings:

DC PLATE VOLTAGE	360 *max*	volts
DC GRID VOLTAGE	-100 *max*	volts
DC PLATE CURRENT	25 *max*	ma
DC GRID CURRENT	8 *max*	ma
PLATE INPUT	9 *max*	watts
PLATE DISSIPATION♦	6.25 *max*	watts
PEAK HEATER-CATHODE VOLTAGE:		
Heater negative with respect to cathode	90 *max*	volts
Heater positive with respect to cathode	90 *max*	volts
PLATE-SEAL TEMPERATURE	175 *max*	°C

Typical Operation in Cathode-Drive Circuit:	*Amplifier* 500 *Mc*	*Oscillator* 500 *Mc*	*Oscillator* 1700 *Mc*	
DC Plate Voltage	275	250	250	volts
DC Grid Voltage	-51	-12	-2	volts
DC Plate Current	23	23	23	ma
DC Grid Current (Approx.)	7	6	3	ma
Driver Power Output (Approx.)	2	—	—	watts
Useful Power Output (Approx.)	5	3	0.75	watts

Maximum Circuit Values:

Grid-Circuit Resistance	0.1 *max*	megohm

♯ Key-down conditions per tube without amplitude modulation. Amplitude modulation essentially negative may be used if the positive peak of the audio-frequency envelope does not exceed 115 per cent of the carrier conditions.

♦ In applications where the plate dissipation exceeds 2.5 watts, it is important that a large area of contact be provided between the plate cylinder and the connector to provide adequate heat conduction.

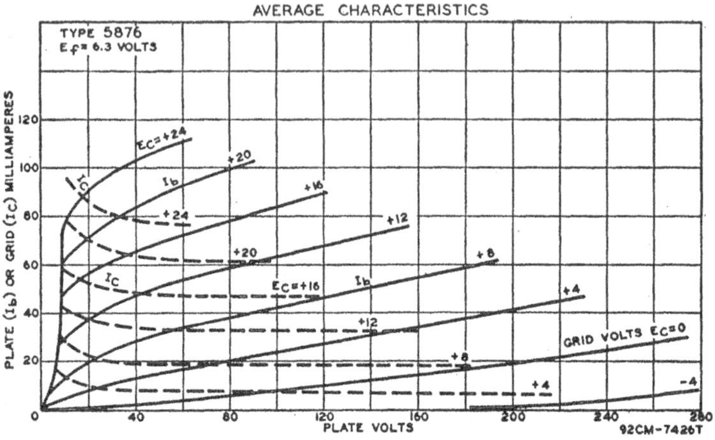

MEDIUM-MU TRIODE

5893

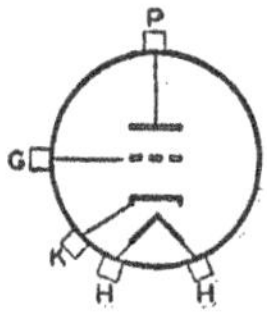

Pencil-type tube used as plate-pulsed oscillator, as rf power amplifier and oscillator, and as frequency doubler. May be used with full input up to 1000 Mc and with reduced input up to 3300 Mc. Designed for use in coaxial-cylinder-type circuits, it may also be used in parallel-line and lumped circuits. Class C Telegraphy maximum plate dissipation, CCS 7 watts, ICAS 8 watts. May be mounted in any position. OUTLINE 66, *Outlines* Section.

HEATER VOLTAGE (AC/DC):	6.0 + 5% −10%	volts
HEATER CURRENT	0.330	ampere
TRANSCONDUCTANCE*	6000	μmhos
AMPLIFICATION FACTOR*	27	
PLATE RESISTANCE (Approx.)*	4500	ohms
DIRECT INTERELECTRODE CAPACITANCES:		
Grid to plate	1.75	μμf
Grid to cathode and heater	2.5	μμf
Plate to cathode and heater	0.07 *max*	μμf

* Plate-supply volts, 200; cathode resistor, 100 ohms; plate milliamperes, 25.

PLATE-PULSED OSCILLATOR▲—Class C

For a maximum "on" time● of 5 microseconds

Maximum Ratings:

PEAK POSITIVE-PULSE PLATE-SUPPLY VOLTAGE♦	1750 *max*	volts
PEAK NEGATIVE-PULSE GRID VOLTAGE	150 *max*	volts
PEAK PLATE CURRENT FROM PULSE SUPPLY	3 *max*	amperes
PEAK RECTIFIED GRID CURRENT	1.3 *max*	amperes
DC PLATE CURRENT	3 *max*	ma
DC GRID CURRENT	1.3 *max*	ma
PLATE DISSIPATION●	6 *max*	watts
DUTY FACTOR■	0.001 *max*	
PULSE DURATION	1.5 *max*	μsec
PLATE-SEAL TEMPERATURE	175 *max*	°C

Typical Operation with Rectangular Wave Shape in Cathode-Drive Circuit at 3300 Mc:

With duty factor■ of 0.001

Peak Positive-Pulse Plate-Supply Voltage♦	1750	volts
Peak Negative-Pulse Grid Voltage♁	110	volts
From grid resistor of	100	ohms
Peak Plate Current from Pulse Supply	3.0	amperes

Peak Rectified Grid Current	1.1	amperes
DC Plate Current	3	ma
DC Grid Current	1.1	ma
Useful Power Output at Peak of Pulse♦ (Approx.)	1200	watts
Pulse Duration	1	μsec
Pulse Repetition Rate	1000	pps

▲ In this class of service, the heater should be allowed to warm up for a minimum of 60 seconds before plate voltage is applied.

• On time for this tube is the sum of the durations of all the individual pulses which occur during any 5000-microsecond interval. Pulse duration is defined as the time interval between the two points on the pulse at which the instantaneous value is 70 per cent of the peak value. The peak value is defined as the maximum value of a smooth curve through the average of the fluctuations over the top portion of the pulse.

♦ The magnitude of any spike on the plate voltage pulse should not exceed a value of 2000 volts with respect to cathode, and its duration should not exceed 0.01 microsecond measured at the peak-pulse-value level.

⊕ In applications where the plate dissipation exceeds 2.5 watts, it is important that a large area of contact be provided between the plate cylinder and the connector in order to provide adequate heat conduction.

■ Duty factor is the product of pulse duration and repetition rate. For variable pulse durations and pulse repetition rates, the duty factor for this tube is defined as the ratio of time "on" to total elapsed time in any 5000-microsecond interval.

ό Obtained from grid resistor of value shown.

♦ This value is determined from the average power output using the duty factor of the peak power-output pulse. This procedure is necessary because the power-output-pulse duty factor may be less than the applied-voltage-pulse duty factor because of a delay in the start of rf power output.

PLATE-MODULATED RF POWER AMPLIFIER—Class C Telephony

Carrier conditions per tube for use with a maximum modulation factor of 1.0

Maximum Ratings:	*CCS*	*ICAS*	
DC Plate Voltage	260 *max*	320 *max*	volts
DC Grid Voltage	-100 *max*	-100 *max*	volts
DC Plate Current	33 *max*	33 *max*	ma
DC Grid Current	15 *max*	15 *max*	ma
Plate Input	8.5 *max*	10.5 *max*	watts
Plate Dissipation⊕	5 *max*	5.5 *max*	watts
Peak Heater-Cathode Voltage:			
Heater negative with respect to cathode	90 *max*	90 *max*	volts
Heater positive with respect to cathode	90 *max*	90 *max*	volts
Plate-Seal Temperature	175 *max*	175 *max*	°C

Typical Operation in Cathode-Drive Circuit at 500 Mc:	*CCS*	*ICAS*	
DC Plate Voltage	250	300	volts
DC Grid Voltage ό	-36	-45	volts
DC Plate Current	30	30	ma
DC Grid Current (Approx.)	11	12	ma
Driver Power Output (Approx.)	1.8	2.0	watts
Useful Power Output (Approx.)	5.5	6.5	watts

Maximum Circuit Values (CCS or ICAS conditions):

Grid-Circuit Resistance	0.1 *max*	megohm

⊕ In applications where the plate dissipation exceeds 2.5 watts, it is important that a large area of contact be provided between the plate cylinder and the connector in order to provide adequate heat conduction.

ό Obtained from grid resistor.

RF POWER AMPLIFIER AND OSCILLATOR—Class C Telegraphy#
and
RF POWER AMPLIFIER—Class C FM Telephony

Maximum Ratings:	*CCS*	*ICAS*	
DC Plate Voltage	320 *max*	400 *max*	volts
DC Grid Voltage	-100 *max*	-100 *max*	volts
DC Plate Current	35 *max*	40 *max*	ma
DC Grid Current	15 *max*	15 *max*	ma
Plate Input	11 *max*	16 *max*	watts
Plate Dissipation⊕	7 *max*	8 *max*	watts
Peak Heater-Cathode Voltage:			
Heater negative with respect to cathode	90 *max*	90 *max*	volts
Heater positive with respect to cathode	90 *max*	90 *max*	volts
Plate-Seal Temperature	175 *max*	175 *max*	°C

Typical Operation as RF Power Amplifier in Cathode-Drive Circuit:

	500 Mc	*1000 Mc*	*500 Mc*	*1000 Mc*	
DC Plate Voltage	300	300	350	350	volts
DC Grid Voltage♁	–47	–30	–51	–33	volts
DC Plate Current	33	33	35	35	ma
DC Grid Current (Approx.)	13	12	13	13	ma
Driver Power Output (Approx.)	2	1.9	2.5	2.4	watts
Useful Power Output (Approx.)	7.5	5.5	8.5	6.5	watts

Typical Operation as Oscillator in Cathode-Drive Circuit at 500 Mc:

DC Plate Voltage	300	350	volts
DC Grid Voltage♁	–47	–51	volts
DC Plate Current	33	35	ma
DC Grid Current (Approx.)	13	13	ma
Useful Power Output (Approx.)	5	6	watts

Maximum Circuit Values (CCS or ICAS conditions):

Grid-Circuit Resistance	0.1 *max*	megohm

♯ Key-down conditions per tube without amplitude modulation. Amplitude modulation essentially negative may be used if the positive peak of the audio-frequency envelope does not exceed 115 per cent of the carrier conditions.

● In applications where the plate dissipation exceeds 2.5 watts, it is important that a large area of contact be provided between the plate cylinder and the connector in order to provide adequate heat conduction.

♁ Obtained from grid resistor.

FREQUENCY DOUBLER

Maximum Ratings:

	CCS	*ICAS*	
DC Plate Voltage	260 *max*	320 *max*	volts
DC Grid Voltage	–100 *max*	–100 *max*	volts
DC Plate Current	33 *max*	33 *max*	ma
DC Grid Current	12 *max*	12 *max*	ma
Plate Input	8.5 *max*	10.5 *max*	watts
Plate Dissipation●	6 *max*	7.5 *max*	watts
Peak Heater-Cathode Voltage:			
Heater negative with respect to cathode	90 *max*	90 *max*	volts
Heater positive with respect to cathode	90 *max*	90 *max*	volts
Plate-Seal Temperature	175 *max*	175 *max*	°C

Typical Operation as Doubler to 1000 Mc in Cathode-Drive Circuit:

DC Plate Voltage	250	300	volts
DC Grid Voltage♁	–40	–50	volts
DC Plate Current	33	33	ma
DC Grid Current (Approx.)	7	8	ma
Driver Power Output (Approx.)	3.2	3.5	watts
Useful Power Output (Approx.)	2.75	3.0	watts

AVERAGE CHARACTERISTICS

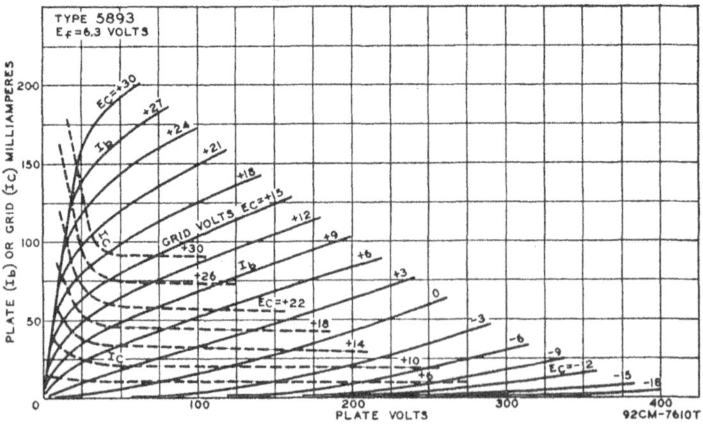

Maximum Circuit Values (CCS or ICAS conditions):

Grid-Circuit Resistance	0.1 *max*	megohm

⊕ In applications where the plate dissipation exceeds 2.5 watts it is important that a large area of contact be provided between the plate cylinder and the connector in order to provide adequate heat conduction.

♁ Obtained from grid resistor.

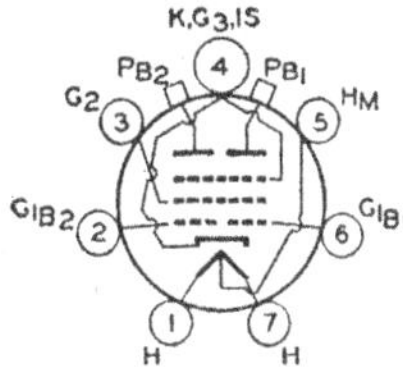

TWIN BEAM POWER TUBE

5894

Small, sturdy, heater-cathode type used as af power amplifier and modulator, as rf power amplifier and oscillator, and as frequency tripler. May be used with full input up to 250 Mc. For operation at 300 Mc, plate voltage and plate input should be reduced to 96 per cent of maximum ratings; at 400 Mc, to 90 per cent; at 500 Mc, to 83 per cent. Class C Telegraphy maximum CCS plate dissipation (per tube), 40 watts. Requires Septar seven-contact socket and may be mounted in vertical position with base up or down, or in horizontal position with plate terminals in horizontal plane. OUTLINE 20, *Outlines* Section. Plates show no color when tube is operated at maximum CCS ratings.

HEATER ARRANGEMENT	*Series*	*Parallel*	
HEATER VOLTAGE (AC/DC)	12.6 ± 10%	6.3 ± 10%	volts
HEATER CURRENT	0.9	1.8	amperes
MU-FACTOR, Grid No.2 to Grid No.1 (Each unit)*	8.2		
DIRECT INTERELECTRODE CAPACITANCES (Each unit):			
Grid No.1 to plate	0.08 *max*		μμf
Grid No.1 to cathode, grid No.3, internal shield, grid No.2, and heater	11		μμf
Plate to cathode, grid No.3, internal shield, grid No.2, and heater	3.4		μμf

* Plate volts, 600; grid-No.2 volts, 250; plate milliamperes, 40.

PUSH-PULL AF POWER AMPLIFIER AND MODULATOR—Class B

Maximum CCS Ratings: *Values are on a per-tube basis*

DC PLATE VOLTAGE	600 *max*	volts
DC GRID-NO.2 (SCREEN-GRID) VOLTAGE	250 *max*	volts
DC GRID-NO.1 (CONTROL-GRID) VOLTAGE	-175 *max*	volts
MAXIMUM-SIGNAL DC PLATE CURRENT■	200 *max*	ma
MAXIMUM-SIGNAL PLATE INPUT■	120 *max*	watts
MAXIMUM-SIGNAL GRID-NO.2 INPUT■	7 *max*	watts
PLATE DISSIPATION■	40 *max*	watts
PEAK HEATER-CATHODE VOLTAGE:		
Heater negative with respect to cathode	100 *max*	volts
Heater positive with respect to cathode	100 *max*	volts

Typical Operation:

DC Plate Voltage	450	600	volts
DC Grid-No.2 Voltage△	250	250	volts
DC Grid-No.1 Voltage	-23	-25	volts
Peak AF Grid-No.1-to-Grid-No.1 Voltage	53	53	volts
Zero-Signal DC Plate Current	67	35	ma
Maximum-Signal DC Plate Current	200	168	ma
Zero-Signal DC Grid-No.2 Current	8	4	ma
Maximum-Signal DC Grid-No.2 Current	26	27	ma
Maximum-Signal DC Grid-No.1 Current	2.3	1.6	ma
Effective Load Resistance (Plate to plate)	4400	8000	ohms
Maximum-Signal Driving Power (Approx.)	0.2	0.2	watt
Maximum-Signal Power Output (Approx.)	60	70	watts

Maximum Circuit Values:

Grid-No.1-Circuit Resistance:		
For fixed-bias operation	50000 *max*	ohms
For cathode-bias operation	Not recommended	

■ Averaged over any audio-frequency cycle of sine-wave form.

△ Obtained preferably from a separate source or from the plate-voltage supply with a voltage divider.

PLATE-MODULATED PUSH-PULL RF POWER AMPLIFIER—Class C Telephony

Carrier conditions per tube with a maximum modulation factor of 1.0

Maximum CCS Ratings:

DC Plate Voltage	450 *max*	volts
DC Grid-No.2 Voltage	250 *max*	volts
DC Grid-No.1 Voltage	-175 *max*	volts
DC Plate Current	160 *max*	ma
DC Grid-No.1 Current	10 *max*	ma
Plate Input	72 *max*	watts
Grid-No.2 Input	4.5 *max*	watts
Plate Dissipation	27 *max*	watts
Peak Heater-Cathode Voltage:		
Heater negative with respect to cathode	100 *max*	volts
Heater positive with respect to cathode	100 *max*	volts

Typical Operation:

	250 Mc	*470 Mc*	
DC Plate Voltage	450	380	volts
DC Grid-No.2 Voltage (Approx.)‡	250	250	volts
From an adjustable series resistor having a maximum value of	20000	30000	ohms
DC Grid-No.1 Voltage♁	-100	-60	volts
From a grid-No.1 resistor of	20000	15000	ohms
Peak RF Grid-No.1-to-Grid-No.1 Voltage	120	—	volts
DC Plate Current	150	160	ma
DC Grid-No.2 Current (Approx.)	16	8	ma
DC Grid-No.1 Current (Approx.)	5	4	ma
Driver Power Output (Approx.)	0.6	13	watts
Useful Power Output (Approx.)▲	50	35	watts

Maximum Circuit Values:

Grid-No.1-Circuit Resistance	50000 *max*	ohms

‡ Obtained preferably from a separate source modulated along with the plate supply, or from the modulated plate supply through a series resistor. It is recommended that this resistor be adjustable to permit obtaining the desired operating plate current after initial tuning adjustments are completed.

♁ Obtained from a grid-No.1 resistor of the value shown or from a combination of grid-No.1 resistor with either fixed supply or cathode resistor.

▲ Measured at load of output circuit.

PUSH-PULL RF POWER AMPLIFIER AND OSCILLATOR—Class C Telegraphy#
and
PUSH-PULL RF POWER AMPLIFIER—Class C FM Telephony

Maximum CCS Ratings:

DC Plate Voltage	600 *max*	volts
DC Grid-No.2 Voltage	250 *max*	volts
DC Grid-No.1 Voltage	-175 *max*	volts
DC Plate Current	220 *max*	ma
DC Grid-No.1 Current	10 *max*	ma
Plate Input	120 *max*	watts
Grid-No.2 Input	7 *max*	watts
Plate Dissipation	40 *max*	watts
Peak Heater-Cathode Voltage:		
Heater negative with respect to cathode	100 *max*	volts
Heater positive with respect to cathode	100 *max*	volts

Typical Operation:

	250 Mc	*470 Mc*		
DC Plate Voltage	600	400	500	volts
DC Grid-No.2 Voltage (Approx.)⊕	250	250	250	volts
From an adjustable series resistor having a maximum value of	33000	22000	47000	ohms
DC Grid-No.1 Voltage♁	-80	-38	-60	volts
From a grid-No.1 resistor of	39000	24000	30000	ohms
From cathode resistor of	360	180	300	ohms
Peak RF Grid-No.1-to-Grid-No.1 Voltage	200	—	—	volts
DC Plate Current	200	220	200	ma
DC Grid-No.2 Current (Approx.)	16	12	8	ma
DC Grid-No.1 Current (Approx.)	2	3	4	ma
Driver Power Output (Approx.)	4	5	13	watts
Useful Power Output (Approx.)▲	85	48	55	watts

Maximum Circuit Values:

Grid-No.1-Circuit Resistance	50000 *max*	ohms

♯ Key-down conditions per tube without amplitude modulation. Amplitude modulation essentially negative may be used if the positive peak of the audio-frequency envelope does not exceed 115 per cent of the carrier conditions.

• Obtained preferably from a separate source, or from the plate-supply voltage with a voltage divider, or through a series resistor. It is recommended that this resistor be adjustable to permit obtaining the desired operating plate current after initial tuning adjustments are completed.

♣ Obtained from a fixed supply, by grid-No.1 resistor, by cathode resistor, or by combination methods.

▲ Measured at load of output circuit.

FREQUENCY TRIPLER—Class C

Values are on a per-tube basis

Maximum CCS Ratings:

DC Plate Voltage	600 *max*	volts
DC Grid-No.2 Voltage	250 *max*	volts
DC Grid-No.1 Voltage	–175 *max*	volts
DC Plate Current	160 *max*	ma
DC Grid-No.1 Current	10 *max*	ma
Plate Input	80 *max*	watts
Grid-No.2 Input	7 *max*	watts
Plate Dissipation	40 *max*	watts
Peak Heater-Cathode Voltage:		
Heater negative with respect to cathode	100 *max*	volts
Heater positive with respect to cathode	100 *max*	volts

Typical Operation as Tripler:

	150 Mc		*225 Mc*	*462 Mc*		
DC Plate Voltage	400	500	400	400	400	volts
DC Grid-No.2 Voltage (Approx.)♦	250	250	250	220	220	volts
From an adjustable series resistor having a maximum value of	16000	39000	20000	56000	56000	ohms
DC Grid-No.1 Voltage♣	–150	–150	–150	–150	–175	volts
From a grid-No.1 resistor of	30000	24000	50000	36000	36000	ohms
Peak RF Grid-No.1-to-Grid-No.1 Voltage	360	360	360	—	—	volts
DC Plate Current	146	120	130	130	140	ma
DC Grid-No.2 Current (Approx.)	16	10	20	5	5	ma
DC Grid-No.1 Current (Approx.)	5	6	3	4	5	ma
Driver Power Output (Approx.)	0.9	1	0.5	4	8	watts
Useful Power Output (Approx.)▲	18	20	12	13	16	watts

Maximum Circuit Values:

Grid-No.1-Circuit Resistance	50000 *max*	ohms

♦ Obtained preferably from a separate source, or from the plate-supply voltage with a voltage divider, or through a series resistor. It is recommended that this resistor be adjustable to permit obtaining the desired operating plate current after initial tuning adjustments are completed.

♣ Obtained from a fixed supply, by grid-No.1 resistor, by cathode resistor, or by combination methods.

▲ Measured at load of output circuit.

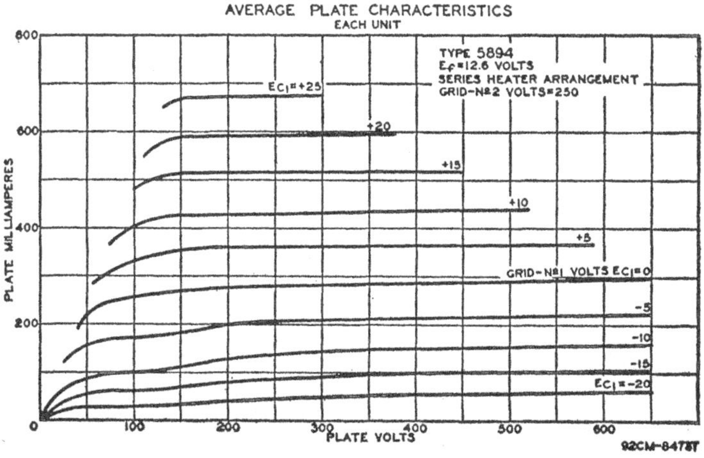

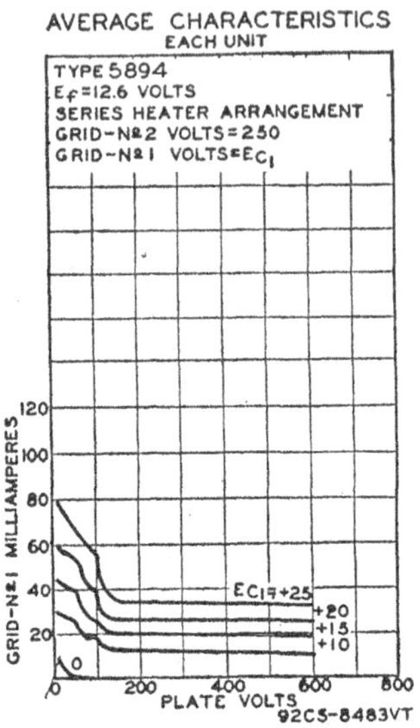

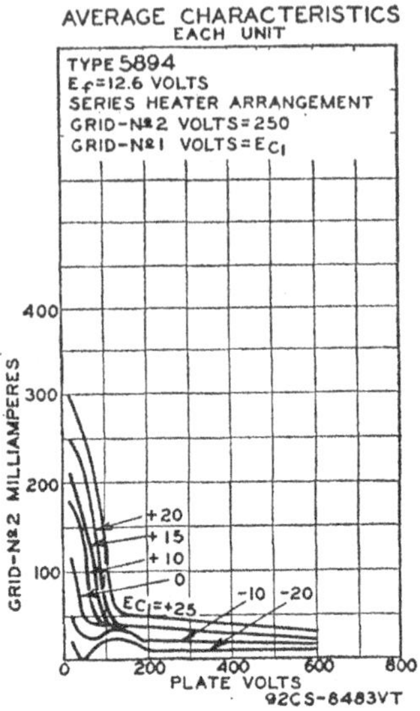

POWER TRIODE

5946

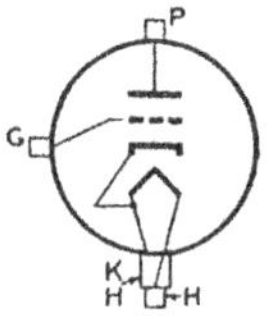

Forced-air-cooled heater-cathode type used as plate-pulsed oscillator and amplifier. May be used with full input up to 1300 Mc. For operation at 2000 Mc, plate voltage and plate input should be reduced to 75 per cent of maximum ratings. Class C maximum plate dissipation, 250 watts. Tube may be mounted in any position. OUTLINE 71, *Outlines* Section. A minimum air flow of 16 cubic feet per minute should be directed through the radiator toward the bulb and grid terminal when the 5946 is operated at maximum rated dissipation. Air flow should start before and continue during application of any voltages to the tube. Heater power, plate power, and air may be removed simultaneously. Maximum temperatures: radiator (measured on core at end adjacent to plate ring), 180°C; grid terminal, 150°C; plate, grid, and cathode seals, 150°C.

HEATER VOLTAGE (AC/DC)°	6.3 ± 10%	volts
HEATER CURRENT	3.4	amperes
AMPLIFICATION FACTOR*	27	
DIRECT INTERELECTRODE CAPACITANCES:		
Grid to plate	6	μμf
Grid to cathode and heater	11	μμf
Plate to cathode and heater□	0.22	μμf

° Heater voltage must be applied for a minimum period of 1 minute before the application of plate voltage.
* Grid volts, −15; plate milliamperes, 250.
□ With external shield connected to grid.

PLATE-PULSED OSCILLATOR AND AMPLIFIER—Class C

Maximum Ratings:

For an "on" time⊕ of	10 *max*	100 *max*	μsec
PEAK POSITIVE-PULSE PLATE-SUPPLY VOLTAGE♦	7500 *max*	7500 *max*	volts
PEAK NEGATIVE-PULSE GRID VOLTAGE	600 *max*	600 *max*	volts
PEAK PLATE CURRENT FROM PULSE SUPPLY	4.5 *max*	3.5 *max*	amperes
PEAK RECTIFIED GRID CURRENT	1.0 *max*	0.75 *max*	amperes
DC PLATE CURRENT	45 *max*	250 *max*	ma
DC GRID CURRENT	10 *max*	70 *max*	ma
PLATE INPUT	340 *max*	340 *max*	watts
PLATE DISSIPATION	250 *max*	250 *max*	watts

Typical Operation with Rectangular Wave Shape in Cathode-Drive Oscillator Circuit at 1250 Mc:

With duty factor■ *of 0.01*

Peak Positive-Pulse Plate-Supply Voltage♦	5500	7500	volts
Peak Negative-Pulse Grid Voltage	375	500	volts
From cathode resistor of△	100	100	ohms
Peak RF Grid Voltage	625	850	volts
Peak Plate Current from Pulse Supply	3.5	4.5	amperes
Peak Rectified Grid Current	0.25	0.50	amperes
DC Plate Current	35	45	ma
DC Grid Current	2.5	5	ma
Useful Power Output at Peak of Pulse♣ (Approx.)	8000	14000	watts

⊕ "On" time for this tube is defined as the sum of the durations of all the individual pulses which occur during any 1000-microsecond interval. Pulse duration is defined as the time interval between the two points on the pulse at which the instantaneous value is 70 per cent of the peak value. The peak value is defined as the maximum value of a smooth curve through the average of the fluctuations over the top portion of the pulse.

♦ The magnitude of any spike on the plate-voltage pulse should not exceed a value of 8.5 kilovolts with respect to cathode, and its duration should not exceed 0.5 microsecond measured at the peak-pulse-value level.

■ Duty factor is the product of pulse duration and repetition rate. For variable pulse durations and pulse repetition rates, the duty factor for this tube is defined as the ratio of "on" to total elapsed time in any 500-microsecond interval.

△ Obtained preferably from cathode resistor of value shown. In certain applications, partial grid-resistor bias may be used.

♣ Determined from the average power output using the duty factor of the peak power output pulse. This procedure is necessary because the power-output-pulse duty factor may be less than the applied-voltage-pulse duty factor because of a delay in the start of rf power output.

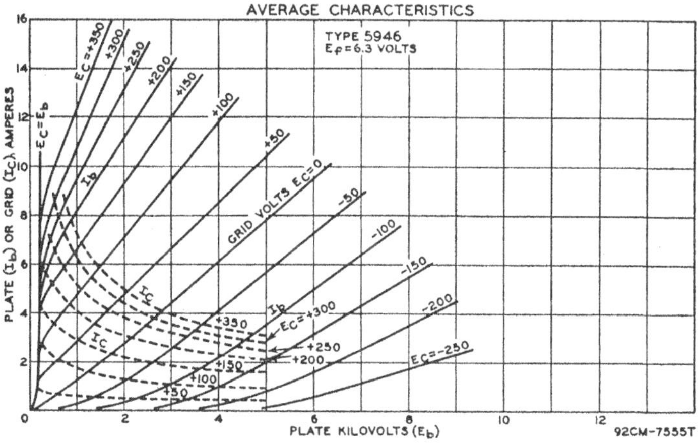

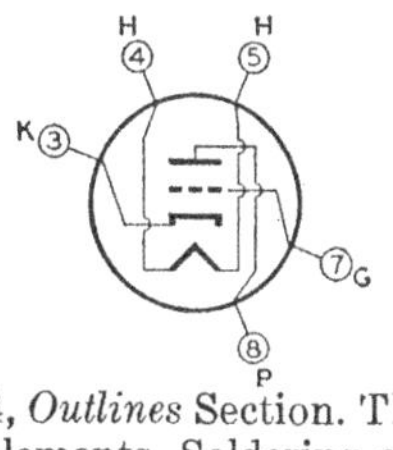

OSCILLATOR TRIODE

6026

Subminiature heater-cathode type used in radiosonde service at 400 Mc. Class C Telegraphy maximum CCS plate dissipation, 3 watts. May be mounted in any position. Outline 4, *Outlines* Section. The flexible leads of the 6026 are usually soldered to the circuit elements. Soldering of the leads may be made close to the glass-button base provided care is taken to conduct excessive heat away from the lead seal. Otherwise, the heat of the soldering operation will crack the seals of the leads and damage the tube. Plate shows no color when tube is operated at maximum CCS ratings.

Heater Voltage Range (ac/dc)°	5.2 to 6.6	volts
Heater Current (At 6.3 volts)	0.2	ampere
Transconductance*	5900	μmhos

Amplification Factor*	24	
Plate Resistance (Approx.)*	4000	ohms
Direct Interelectrode Capacitances:		
Grid to plate	1.3	μμf
Grid to cathode and heater	2.0	μμf
Plate to cathode and heater	0.42	μμf

° For radiosonde applications in which the heater is supplied from batteries and the equipment-design requirements of minimum size, light weight, and high efficiency are the primary considerations even though the average life expectancy of the 6026 in such service is only a few hours.
* Plate-supply volts, 120; cathode resistor, 220 ohms, plate milliamperes 12.

OSCILLATOR—Class C Telegraphy

Maximum CCS Ratings:

DC Plate Voltage	150 *max*	volts
DC Grid Voltage	−50 *max*	volts
Total Cathode Current	40 *max*	ma
DC Grid Current	10 *max*	ma
Plate Input	3.3 *max*	watts
Plate Dissipation	3.0 *max*	watts
Peak Heater-Cathode Voltage	0 *max*	volts

Typical Operation as an Oscillator at 400 Mc:

DC Plate Voltage	135	volts
Grid Resistor	1300	ohms
DC Plate Current	20	ma
DC Grid Current (Approx.)	9.5	ma
Useful Power Output	1.25	watts

BEAM POWER TUBE

6146

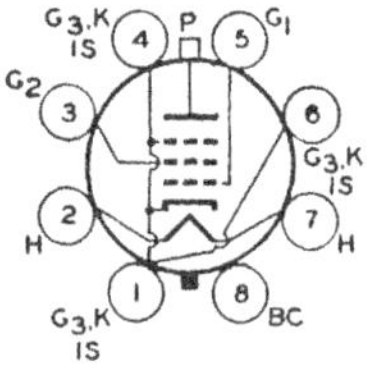

Small, sturdy, glass-octal heater-cathode type used as af power amplifier and modulator and as rf power amplifier and oscillator. May be used with full input up to 60 Mc and with reduced input up to 175 Mc. Class C Telegraphy maximum plate dissipation, CCS 20 watts, ICAS 25 watts.

Heater Voltage (ac/dc)	6.3 ± 10%	volts
Heater Current	1.25	amperes
Transconductance*	7000	μmhos
Mu-Factor, Grid No.2 to Grid No.1*	4.5	
Direct Interelectrode Capacitances:		
Grid No.1 to plate	0.24 *max*	μμf
Grid No.1 to cathode, grid No.3, grid No.2, internal shield, base sleeve, and heater	13.5	μμf
Plate to cathode, grid No.3, grid No.2, internal shield, base sleeve, and heater	8.5	μμf

* Plate and grid-No.2 volts, 200; plate milliamperes, 100.

AF POWER AMPLIFIER AND MODULATOR—CLASS AB_2

Maximum Ratings:	*CCS*	*ICAS*	
DC Plate Voltage	600 *max*	750 *max*	volts
DC Grid-No.2 (screen-grid) Voltage	250 *max*	250 *max*	volts
Maximum-Signal DC Plate Current■	125 *max*	135 *max*	ma
Maximum-Signal Plate Input■	62.5 *max*	90 *max*	watts
Maximum-Signal Grid-No.2 Input■	3 *max*	3 *max*	watts
Plate Dissipation■	20 *max*	25 *max*	watts
Peak Heater-Cathode Voltage:			
Heater negative with respect to cathode	135 *max*	135 *max*	volts
Heater positive with respect to cathode	135 *max*	135 *max*	volts
Bulb Temperature (At hottest point)	220 *max*	220 *max*	°C

Typical Operation *(Values are for 2 tubes)*:

DC Plate Voltage	400	500	600	600	750	volts
DC Grid-No.2 Voltage°	175	175	165	190	165	volts
DC Grid-No.1 (Control-Grid) Voltage	−41	−44	−44	−48	−46	volts
Peak AF Grid-No.1-to-Grid-No.1 Voltage	95	102	97	109	108	volts

Zero-Signal DC Plate Current	33	27	22	28	22	ma
Maximum-Signal DC Plate Current	232	242	207	270	240	ma
Zero-Signal DC Grid-No.2 Current	1.1	0.7	0.6	1.2	0.3	ma
Maximum-Signal DC Grid-No.2 Current	18	18	17	20	20	ma
Maximum-Signal DC Grid-No.1 Current	1.6	1.9	1.1	2	2.6	ma
Effective Load Resistance (Plate to plate)	3700	4600	6800	5000	7400	ohms
Maximum-Signal Driving Power (Approx.)	0.2	0.3	0.2	0.3	0.4	watt
Maximum-Signal Power Output (Approx.)	62	83	90	113	131	watts

Maximum Circuit Values (CCS or ICAS conditions):

Grid-No.1-Circuit Resistance	30000‡ *max*	ohms

■ Averaged over any audio-frequency cycle of sine-wave form.
● Obtained preferably from a separate source or from the plate-voltage supply with a voltage divider.
‡ For operation at less than maximum ratings, this value may be as high as 100000 ohms.

PLATE-MODULATED RF POWER AMPLIFIER—Class C Telephony

Carrier conditions per tube for use with a maximum modulation factor of 1.0

Maximum Ratings:

	CCS	*ICAS*	
DC Plate Voltage	480 *max*	600 *max*	volts
DC Grid-No.2 Voltage	250 *max*	250 *max*	volts
DC Grid-No.1 Voltage	−150 *max*	−150 *max*	volts
DC Plate Current	117 *max*	125 *max*	ma
DC Grid-No.1 Current	3.5 *max*	4.0 *max*	ma
Plate Input	45 *max*	67.5 *max*	watts
Grid-No.2 Input	2 *max*	2 *max*	watts
Plate Dissipation	13.3 *max*	16.7 *max*	watts
Peak Heater-Cathode Voltage:			
Heater negative with respect to cathode	135 *max*	135 *max*	volts
Heater positive with respect to cathode	135 *max*	135 *max*	volts
Bulb Temperature (At hottest point)	220 *max*	220 *max*	°C

Typical Operation:

DC Plate Voltage	400	475	600	volts
DC Grid-No.2 Voltage♠	150	135	150	volts
From series resistor of	33000	51000	56000	ohms
DC Grid-No.1 Voltage♢	−87	−77	−87	volts
From grid-No.1 resistor of	27000	27000	27000	ohms
Peak RF Grid-No.1 Voltage	107	95	107	volts
DC Plate Current	112	94	112	ma
DC Grid-No.2 Current	7.8	6.4	7.8	ma
DC Grid-No.1 Current (Approx.)	3.4	2.8	3.4	ma
Driving Power (Approx.)	0.4	0.3	0.4	watt
Power Output (Approx.)	32	34	52	watts

Maximum Circuit Values (CCS or ICAS conditions):

Grid-No.1-Circuit Resistance	30000‡ *max*	ohms

♠ Obtained preferably from a separate source modulated along with the plate supply, or from the modulated plate supply through a series resistor of value shown.
♢ Obtained from grid-No.1 resistor of value shown or from a combination of grid-No.1 resistor with either fixed supply or cathode resistor.
‡ For operation at less than maximum rated conditions, this value may be as high as 100000 ohms.

RF POWER AMPLIFIER AND OSCILLATOR—Class C Telegraphy#
and
RF POWER AMPLIFIER—Class C FM Telephony

Maximum Ratings:

	CCS	*ICAS*	
DC Plate Voltage	600 *max*	750 *max*	volts
DC Grid-No.2 Voltage	250 *max*	250 *max*	volts
DC Grid-No.1 Voltage	−150 *max*	−150 *max*	volts
DC Plate Current	140 *max*	150 *max*	ma
DC Grid-No.1 Current	3.5 *max*	4.0 *max*	ma
Plate Input	67.5 *max*	90 *max*	watts
Grid-No.2 Input	3 *max*	3 *max*	watts
Plate Dissipation	20 *max*	25 *max*	watts
Peak Heater-Cathode Voltage:			
Heater negative with respect to cathode	135 *max*	135 *max*	volts
Heater positive with respect to cathode	135 *max*	135 *max*	volts
Bulb Temperature (At hottest point)	220 *max*	220 *max*	°C

Typical Operation as Amplifier up to 60 Mc:

DC Plate Voltage	500	600	600	750	volts
DC Grid-No.2 Voltage▲	170	150	180	160	volts
From series resistor of	36000	51000	43000	56000	ohms
DC Grid-No.1 Voltage □‡	−66	−58	−71	−62	volts
From grid-No.1 resistor of	27000	20000	24000	20000	ohms
From cathode resistor of	470	470	430	470	ohms
Peak RF Grid-No.1 Voltage	84	73	91	79	volts
DC Plate Current	135	112	150	120	ma
DC Grid-No.2 Current	9	9	10	11	ma
DC Grid-No.1 Current (Approx.)	2.5	2.8	2.8	3.1	ma
Driving Power (Approx.)	0.2	0.2	0.3	0.2	watt
Power Output (Approx.)	48	52	66	70	watts

Typical Operation as Amplifier at 175 Mc:

DC Plate Voltage	320	400	volts
DC Grid-No.2 Voltage▲	180	190	volts
From series resistor of	13000	20000	ohms
DC Grid-No.1 Voltage □‡	−51	−54	volts
From grid-No.1 resistor of	27000	24000	ohms
From cathode resistor of	330	330	ohms
Peak RF Grid-No.1 Voltage	64	68	volts
DC Plate Current	140	150	ma
DC Grid-No.2 Current	10	10.4	ma
DC Grid-No.1 Current (Approx.)	2	2.2	ma
Driving Power (Approx.)	3	3	watts
Power Output (Approx.)	25	35	watts

Maximum Circuit Values (CCS or ICAS conditions):

Grid-No.1-Circuit Resistance	30000‡ *max*	ohms

Key-down conditions per tube without amplitude modulation. Amplitude modulation essentially negative may be used if the positive peak of the audio-frequency envelope does not exceed 115 per cent of the carrier conditions.

▲ Obtained preferably from separate source, from plate-voltage supply with a voltage divider, or through series resistor of value shown. Grid-No.2 voltage must not exceed 400 volts under key-up conditions.

□ Obtained from fixed supply, by grid-No.1 resistor, by cathode resistor, or by combination methods.

‡ For operation at less than maximum rated conditions, this value may be as high as 100000 ohms.

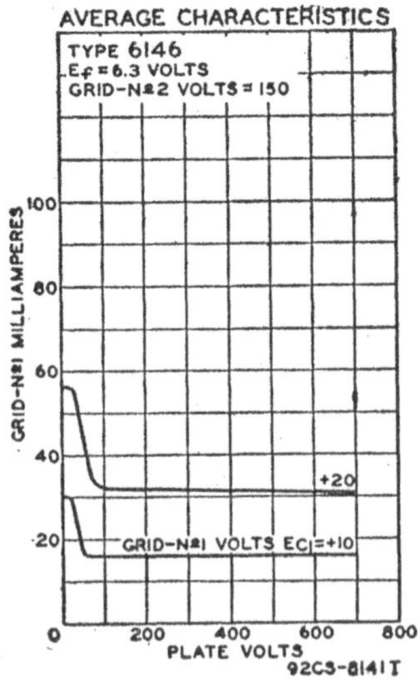

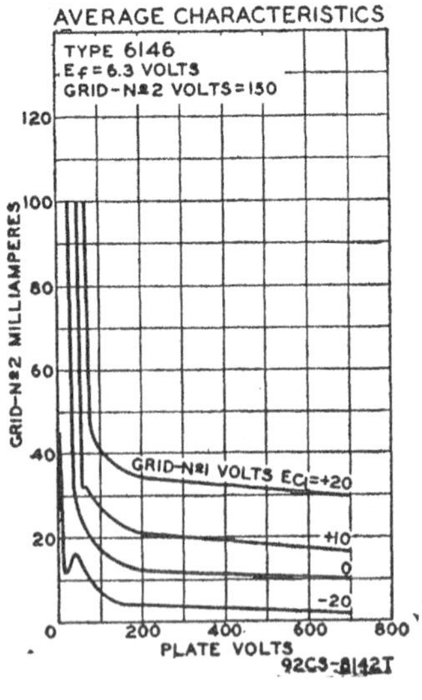

OPERATING CONSIDERATIONS

Type 6146 requires Octal socket and may be mounted in any position. Simplified shielding and good performance are facilitated by the base sleeve with separate base-pin connection and the triple base-pin connection for cathode, grid No.3, and internal shield. OUTLINE 17, *Outlines* Section.

For operation at 120 Mc, plate voltage should be reduced to 67 per cent of maximum rating; plate input to 79 per cent. At 175 Mc, plate voltage should be reduced to 53 per cent of maximum rating; plate input to 66 per cent.

Plate shows no color when tube is operated at maximum CCS or ICAS ratings.

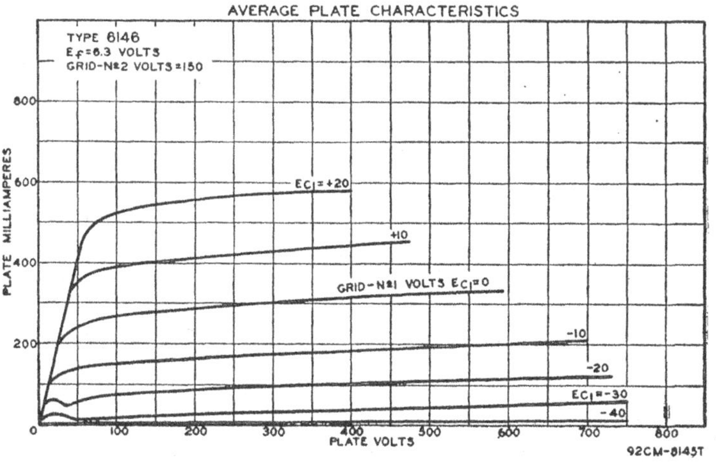

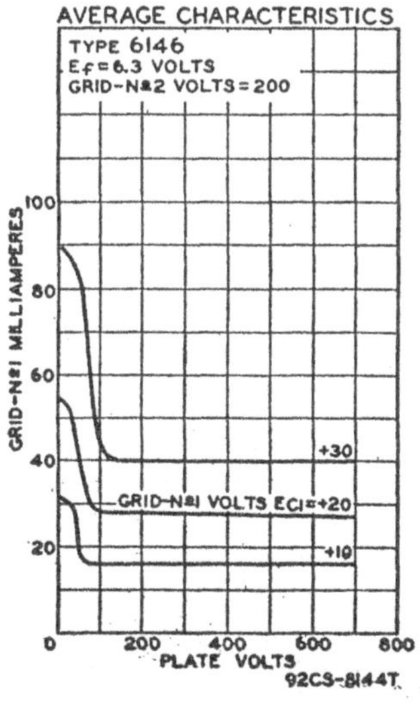

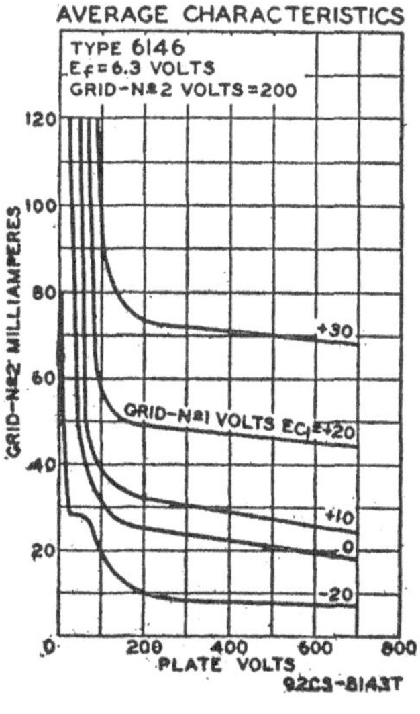

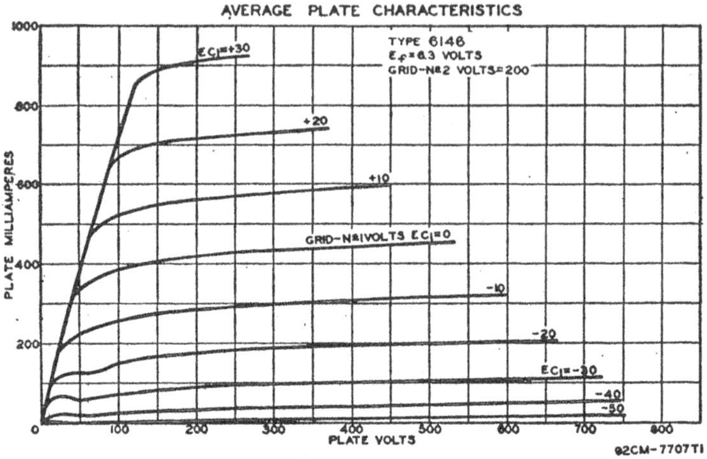

BEAM POWER TUBE

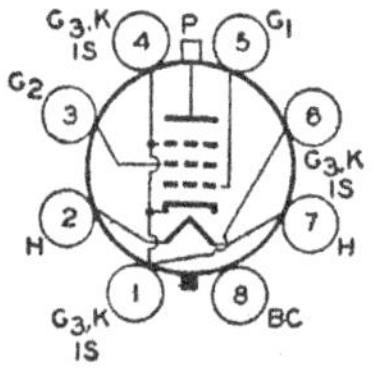

6159

Small, sturdy, glass-octal heater-cathode type used as af power amplifier and modulator and as rf power amplifier and oscillator. May be used with full input up to 60 Mc and with reduced input up to 175 Mc. Class C Telegraphy maximum plate dissipation, CCS 20 watts, ICAS 25 watts. OUTLINE 17, *Outlines* Section. Heater volts, 26.5 ± 10%; amperes, 0.3. Except for heater rating, this type is identical with type 6146.

POWER TRIODE

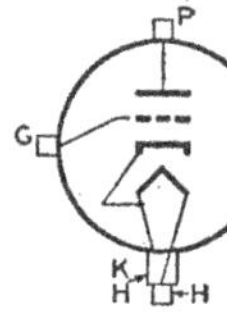

6161

Compact forced-air-cooled heater-cathode type having integral radiator used as rf power amplifier and oscillator and as frequency multiplier. Coaxial terminal arrangement facilitates use in cathode-drive circuits of the coaxial-cylinder type. May be used with full input up to 900 Mc and with reduced input up to 2000 Mc. Class C Telegraphy maximum CCS plate dissipation, 250 watts.

HEATER VOLTAGE (AC/DC):°		
Average	6.3∅	volts
Maximum	6.9	volts
HEATER CURRENT (At 6.3 volts)	3.4	amperes
AMPLIFICATION FACTOR*	27	
DIRECT INTERELECTRODE CAPACITANCES:		
Grid to plate	6	μμf
Grid to cathode and heater	11	μμf
Plate to cathode and heater□	0.22	μμf

° Because the cathode is subjected to considerable back bombardment as the frequency is increased with resultant increase in temperature, the heater voltage should be reduced depending on operating conditions and frequency to prevent overheating the cathode and resultant short life.

∅ Average heater voltage must be applied for a minimum period of one minute before the application of plate voltage.

* Grid volts, −15; plate milliamperes, 250.

□ With external flat shield having minimum diameter of 7½ inches located in plane of grid terminal and perpendicular to axis of tube. Shield is connected to grid terminal.

PLATE-MODULATED RF POWER AMPLIFIER—Class C Telephony

Carrier conditions per tube for use with a maximum modulation factor of 1.0

Maximum CCS Ratings:

DC Plate Voltage	1300 *max*	volts
DC Grid Voltage	-300 *max*	volts
DC Plate Current	210 *max*	ma
DC Grid Current▲	75 *max*	ma
Plate Input	270 *max*	watts
Plate Dissipation	167 *max*	watts

Typical Operation in Cathode-Drive Circuit:

	600 Mc	*900 Mc*	
DC Plate-to-Grid Voltage	1400	1400	volts
DC Cathode-to-Grid Voltage	150	150	volts
Peak RF Cathode-to-Grid Voltage	200	200	volts
DC Plate Current	210	210	ma
DC Grid Current (Approx.)	70	70	ma
Driver Power Output (Approx.)■	70△	75♦	watts
Power Output (Approx.)	180	120	watts

▲ The maximum negative grid current should never exceed 10 milliamperes.
■ In this type of service, the 6161 can be modulated 100 per cent if the rf driver stage is also modulated 100 per cent simultaneously. Care should be taken to insure that the driver-modulation and amplifier-modulation voltages are exactly in phase.
△ This value includes 18 watts of circuit loss and 40 watts added to plate input.
♦ This value includes 23 watts of circuit loss and 40 watts added to plate input.

RF POWER AMPLIFIER AND OSCILLATOR—Class C Telegraphy#
and
RF POWER AMPLIFIER—Class C FM Telephony

Maximum CCS Ratings:

DC Plate Voltage	1600 *max*	volts
DC Grid Voltage	-300 *max*	volts
DC Plate Current	250 *max*	ma
DC Grid Current▲	75 *max*	ma
Plate Input	400 *max*	watts
Plate Dissipation	250 *max*	watts

Typical Operation in Cathode-Drive Circuit:

	600 Mc	*900 Mc*	
DC Plate-to-Grid Voltage	1650	1650	volts
DC Cathode-to-Grid Voltage	150	150	volts
From grid resistor of	3000	15000	ohms
Peak RF Cathode-to-Grid Voltage	200	200	volts
DC Plate Current	250	250	ma
DC Grid Current (Approx.)	50	10	ma
Driver Power Output (Approx.)	75°	80•	watts
Power Output (Approx.)	270	180	watts

Key-down conditions per tube without amplitude modulation. Amplitude modulation essentially negative may be used if the positive peak of the audio-frequency envelope does not exceed 115 per cent of the carrier conditions.
▲ The maximum negative grid current should never exceed 10 milliamperes.
° This value includes 18 watts of circuit loss and 45 watts added to plate input.
• This value includes 23 watts of circuit loss and 45 watts added to plate input.

FREQUENCY MULTIPLIER—Class C

Maximum CCS Ratings:

DC Plate Voltage	1600 *max*	volts
DC Grid Voltage	-300 *max*	volts
DC Plate Current	250 *max*	ma
DC Grid Current▲	75 *max*	ma
Plate Input	400 *max*	watts
Plate Dissipation	250 *max*	watts

Typical Operation as Doubler in Cathode-Drive Circuit:

	600 Mc	*900 Mc*	
DC Plate-to-Grid Voltage	1760	1675	volts
DC Cathode-to-Grid Voltage	260	175	volts
From cathode resistor of	860	645	ohms
Peak RF Cathode-to-Grid Voltage	300	300	volts
DC Plate Current	250	250	ma
DC Grid Current (Approx.)	50	21	ma

Driver Power Output (Approx.) ●	125	100	watts
Power Output (Approx.)	180	140	watts

▲ The maximum negative grid current should never exceed 10 milliamperes.
● Approximate total driving power required. A portion of this power appears in the plate circuit.

OPERATING CONSIDERATIONS

Type 6161 may be mounted in any position. OUTLINE 71, *Outlines* Section.

For operation at 1200 Mc, plate voltage and plate input should be reduced to 80 per cent of maximum ratings; at 1400 Mc, to 71 per cent; at 1650 Mc, to 62.5 per cent; at 2000 Mc, to 62.5 per cent.

A minimum air flow of 16 cubic feet per minute should be directed by a blower through the radiator toward the bulb and the grid terminal when the 6161 is operated at maximum rated dissipation. Air flow should start before and continue during the application of any voltages to the 6161. Maximum temperatures; radiator (measured on core at end adjacent to plate ring), 180°C; grid terminal, 150°C; cathode terminal, 150°C; plate, grid, and cathode seals, 150°C.

The 6161 supersedes the 5588 for new equipment design.

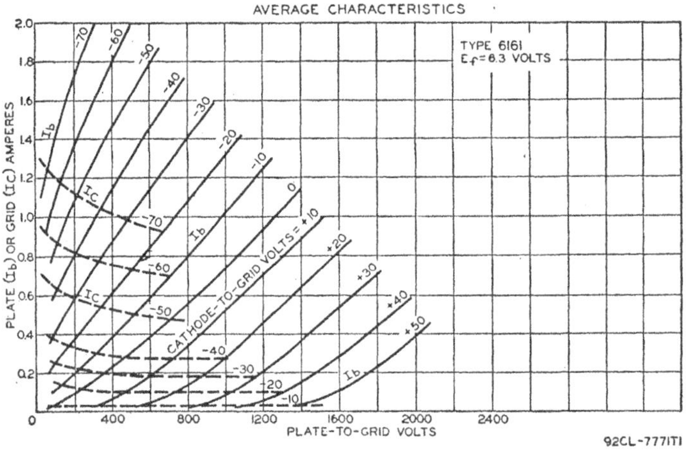

92CL-7771T1

MEDIUM-MU TRIODE

6263

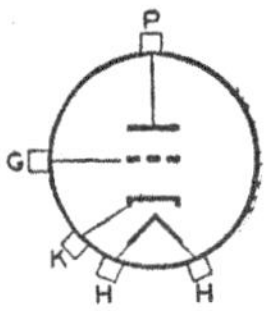

Pencil-type tube having integral radiator used as rf power amplifier and oscillator in mobile equipment and in aircraft transmitters at altitudes up to 60,000 feet without pressurized chambers. May be used with full input up to 500 Mc and with reduced input up to 1700 Mc. Class C Telegraphy maximum plate dissipation, CCS 8 watts, ICAS 13 watts.

HEATER VOLTAGE (AC/DC):		
Under transmitting conditions	6.0 ± 10%	volts
Under stand-by conditions	6.3 *max*	volts
HEATER CURRENT (At 6.0 volts)	0.280	ampere
TRANSCONDUCTANCE*	7000	μmhos
AMPLIFICATION FACTOR	27	
DIRECT INTERELECTRODE CAPACITANCES:		
Grid to plate	1.7	μμf
Grid to cathode and heater	2.9	μμf
Plate to cathode and heater	0.08 *max*	μμf

* Plate volts, 200; plate milliamperes, 27.

PLATE-MODULATED RF POWER AMPLIFIER—Class C Telephony

Carrier conditions per tube for use with a maximum modulation factor of 1.0

Maximum Ratings (*For pressures down to 46 mm of Hg•*):	*CCS*	*ICAS*	
DC Plate Voltage	275 *max*	330 *max*	volts
DC Grid Voltage	−100 *max*	−100 *max*	volts
DC Plate Current	33 *max*	46 *max*	ma
DC Grid Current	25 *max*	25 *max*	ma
DC Cathode Current	50 *max*	60 *max*	ma
Plate Input	9 *max*	15 *max*	watts
Plate Dissipation	5.5 *max*	9 *max*	watts
Peak Heater-Cathode Voltage:			
Heater negative with respect to cathode	90 *max*	90 *max*	volts
Heater positive with respect to cathode	90 *max*	90 *max*	volts

Typical Operation in Cathode-Drive Circuit at 500 Mc:

DC Plate Voltage	275	320	volts
DC Grid Voltage♁	−42	−52	volts
DC Plate Current	35	35	ma
DC Grid Current (Approx.)	13	12	ma
Driver Power Output (Approx.)	2	2.4	watts
Useful Power Output (Approx.) −75-per-cent circuit efficiency	6.7	8	watts

Maximum Circuit Values (CCS or ICAS conditions):

Grid-Circuit Resistance	0.1 *max*	megohm

• Corresponds to altitude of about 60000 feet.

♁ Obtained from grid resistor, or from a combination of grid resistor with either fixed supply or cathode resistor.

RF POWER AMPLIFIER AND OSCILLATOR—Class C Telegraphy#
and
RF POWER AMPLIFIER—Class C FM Telephony

Maximum Ratings (*For pressures down to 46mm of Hg•*):	*CCS*	*ICAS*	
DC Plate Voltage	300 *max*	400 *max*	volts
DC Grid Voltage	−100 *max*	−100 *max*	volts
DC Plate Current	40 *max*	55 *max*	ma
DC Grid Current	25 *max*	25 *max*	ma
DC Cathode Current	55 *max*	70 *max*	ma
Plate Input	13 *max*	22 *max*	watts
Plate Dissipation	8 *max*	13 *max*	watts
Peak Heater-Cathode Voltage:			
Heater negative with respect to cathode	90 *max*	90 *max*	volts
Heater positive with respect to cathode	90 *max*	90 *max*	volts

Typical Operation in Cathode-Drive Circuit at 500 Mc:

	Oscil-lator	*Ampli-fier*	*Oscil-lator*	*Ampli-fier*	
DC Plate Voltage	300	300	350	350	volts
DC Grid Voltage♁	−30	−48	−35	−58	volts
DC Plate Current	35	35	40	40	ma
DC Grid Current (Approx.)	11	13	14	15	ma
Driver Power Output (Approx.)	–	2.2	–	3	watts
Useful Power Output (Approx.) −75-per-cent circuit efficiency	5	7	7	10	watts

Maximum Circuit Values (CCS or ICAS conditions):

Grid-Circuit Resistance	0.1 *max*	megohm

Key-down conditions per tube without amplitude modulation. Amplitude modulation essentially negative may be used if the positive peak of the audio-frequency envelope does not exceed 115 per cent of the carrier conditions.

• Corresponds to altitude of about 60000 feet.

♁ Obtained from grid resistor, or from a combination of grid resistor with either fixed supply or cathode resistor.

OPERATING CONSIDERATIONS

Type 6263 may be mounted in any position. Outline 67, *Outlines* Section.

In many applications, the 6263 does not require forced-air cooling. The radiator in combination with a connector having adequate heat conduction capability will generally provide adequate cooling under conditions of free circulation of air.

The cooling must be sufficient to limit the plate-seal temperature to 175°C. When conditions do not provide adequate circulation of air, provision should be made to direct a blast of air from a small blower through the radiator fins. Maximum temperatures: incoming air, 40°C; radiator, 175°C.

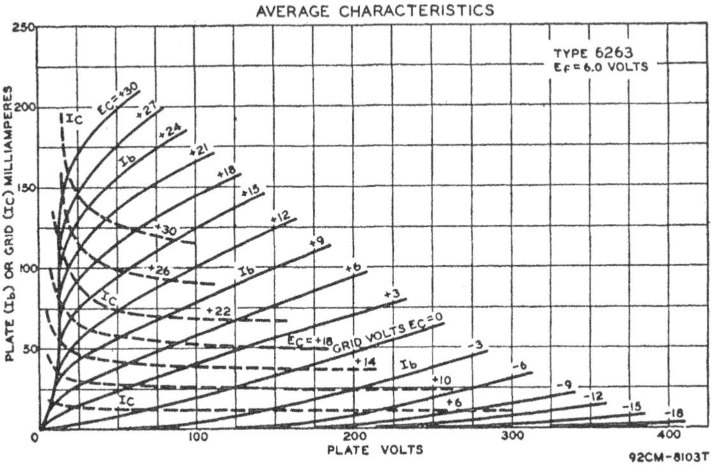

MEDIUM-MU TRIODE

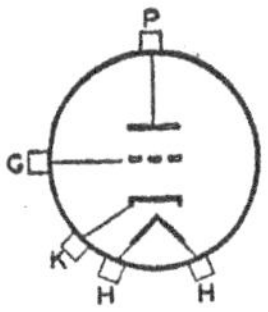

6264

Pencil-type tube having integral radiator used as rf power amplifier and oscillator and as frequency multiplier in mobile equipment and in aircraft transmitters at altitudes up to 60,000 feet without pressurized chambers. May be used with full input up to 500 Mc and with reduced input up to 1700 Mc. Class C Telegraphy maximum plate dissipation, CCS 8 watts, ICAS 13 watts. May be mounted in any position. OUTLINE 67, *Outlines* Section. Cooling requirements for the 6264 are similar to those of type 6263.

HEATER VOLTAGE (AC/DC):		
Under transmitting conditions	6.0 ± 10%	volts
Under stand-by conditions	6.3 *max*	volts
HEATER CURRENT (at 6.0 volts)	0.280	ampere
TRANSCONDUCTANCE*	6800	μmhos
AMPLIFICATION FACTOR	40	
DIRECT INTERELECTRODE CAPACITANCES:		
Grid to plate	1.75	μμf
Grid to cathode and heater	2.95	μμf
Plate to cathode and heater	0.07 *max*	μμf

* Plate volts, 200; plate milliamperes, 18.5.

RF POWER AMPLIFIER AND OSCILLATOR—Class C Telegraphy #
and
RF POWER AMPLIFIER—Class C FM Telephony

Maximum Ratings *(For pressures down to 46mm of Hg•):*	*CCS*	*ICAS*	
DC PLATE VOLTAGE	330 *max*	400 *max*	volts
DC GRID VOLTAGE	-100 *max*	-100 *max*	volts
DC PLATE CURRENT	40 *max*	55 *max*	ma
DC GRID CURRENT	25 *max*	25 *max*	ma
DC CATHODE CURRENT	55 *max*	70 *max*	ma
PLATE INPUT	13 *max*	22 *max*	watts
PLATE DISSIPATION	8 *max*	13 *max*	watts
PEAK HEATER-CATHODE VOLTAGE:			
Heater negative with respect to cathode	90 *max*	90 *max*	volts
Heater positive with respect to cathode	90 *max*	90 *max*	volts

Typical Operation in Cathode-Drive Circuit at 500 Mc:

	Oscil-lator	*Ampli-fier*	*Oscil-lator*	*Ampli-fier*	
DC Plate Voltage	300	300	350	350	volts
DC Grid Voltage♁	-25	-42	-30	-45	volts
DC Plate Current	35	35	35	40	ma
DC Grid Current (Approx.)	11	13	13	15	ma
Driver Power Output (Approx.)	—	2.4	—	3	watts
Useful Power Putput (Approx.)-75-per-cent circuit efficiency	5	7.5	6	10	watts

Maximum Values Circuit (CCS or ICAS conditions):

Grid-Circuit Resistance	0.1 *max*	megohm

Key-down conditions per tube without amplitude modulation. Amplitude modulation essentially negative may be used if the positive peak of the audio-frequency envelope does not exceed 115 per cent of the carrier conditions.

• Corresponds to altitude of about 60000 feet.

♁ Obtained from grid resistor, or from a combination of grid resistor with either fixed supply or cathode resistor.

FREQUENCY MULTIPLIER

Maximum Ratings (*For pressures down to 46mm of Hg•*):	*CCS*	*ICAS*	
DC Plate Voltage	300 *max*	350 *max*	volts
DC Grid Voltage	-125 *max*	-140 *max*	volts
DC Plate Current	33 *max*	45 *max*	ma
DC Grid Current	25 *max*	25 *max*	ma
DC Cathode Current	45 *max*	55 *max*	ma
Plate Input	9.9 *max*	15.8 *max*	watts
Plate Dissipation	6 *max*	9.5 *max*	watts
Peak Heater-Cathode Voltage:			
Heater negative with respect to cathode	90 *max*	90 *max*	volts
Heater positive with respect to cathode	90 *max*	90 *max*	volts

Typical Operation in Cathode-Drive Circuit as Tripler to 510 Mc:

DC Plate Voltage	300	350	volts
DC Grid Voltage♁	-110	-122	volts
DC Plate Current	26	36.5	ma
DC Grid Current (Approx.)	4.1	5.8	ma
Driver Power Output (Approx.)	2.75	4.5	watts
Useful Power Output (Approx.)-75-per-cent circuit efficiency	2.1	3.4	watts

Maximum Circuit Values (CCS or ICAS conditions):

Grid-Circuit Resistance	0.1 *max*	megohm

• Corresponds to altitude of about 60000 feet.

♁ Obtained from grid resistor, or from a combination of grid resistor with either fixed supply or cathode resistor.

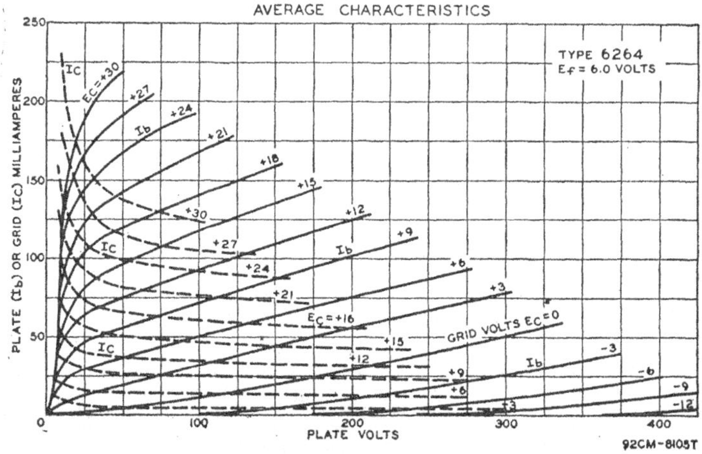

6293

BEAM POWER TUBE

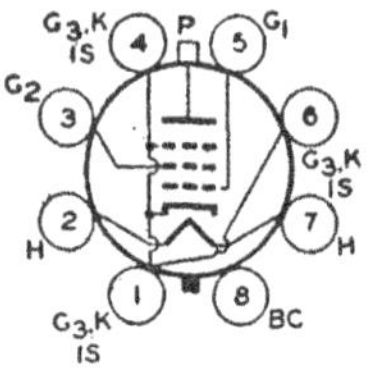

Glass-octal heater-cathode type used as rectangular-wave pulse modulator. Rated for service with duty factors up to 1.0 at a maximum averaging time of 10,000 microseconds. Rectangular-Wave Modulator maximum plate dissipation, 10 watts. Requires Octal socket and may be mounted in any position. OUTLINE 17, *Outlines* Section. Plate shows no color when tube is operated at maximum CCS ratings.

HEATER VOLTAGE (AC/DC)	6.3 ± 10%	volts
HEATER CURRENT	1.25	amperes
TRANSCONDUCTANCE*	7000	μmhos
MU-FACTOR, Grid No.2 to Grid No.1*	4.5	
DIRECT INTERELECTRODE CAPACITANCES:		
Grid No.1 to plate	0.24 *max*	μμf
Grid No.1 to cathode, grid No.3, grid No.2, internal shield, base sleeve, and heater	13.5	μμf
Plate to cathode, grid No.3, grid No.2, internal shield, base sleeve, and heater	8.5	μμf

* Plate and grid-No.2 volts, 200; plate milliamperes, 100.

MODULATOR—Rectangular-Wave Modulation

Maximum and Minimum CCS Ratings:

For Duty Factor■ up to 0.003
and Maximum Averaging Time of 10,000 Microseconds in Any Interval

DC PLATE-SUPPLY VOLTAGE▲	2000 *max*	3500 *max*	volts
INSTANTANEOUS PLATE VOLTAGE●	2300 *max*	4000 *max*	volts
DC GRID-NO.2 (SCREEN-GRID) SUPPLY VOLTAGE▲	500 *max*	200 *max*	volts
DC GRID-NO.1 (CONTROL-GRID) SUPPLY VOLTAGE▲	−300 *max*	−300 *max*	volts
	250 *min*	−130 *min*	volts
GRID-NO.1 VOLTAGE:			
Instantaneous Negative Value	400 *max*	400 *max*	volts
Peak Positive Value	100 *max*	100 *max*	volts
PEAK PLATE CURRENT	3△ *max*	3△ *max*	amperes
PEAK GRID-NO.2 CURRENT	0.75 *max*	0.75 *max*	ampere
PEAK GRID-NO.1 CURRENT	0.5 *max*	0.5 *max*	ampere
PLATE INPUT	80 *max*	80 *max*	watts
GRID-NO.2 INPUT	1.75 *max*	1.75 *max*	watts
GRID-NO.1 INPUT	0.5 *max*	0.5 *max*	watt
PLATE DISSIPATION♦	7 *max*	10 *max*	watts
PEAK HEATER-CATHODE VOLTAGE:			
Heater negative with respect to cathode	135 *max*	135 *max*	volts
Heater positive with respect to cathode	135 *max*	135 *max*	volts
BULB TEMPERATURE (At hottest point)	175 *max*	175 *max*	°C

■ Duty factor is defined as the "on" time in microseconds divided by 10,000 microseconds. "On" time for this tube is defined as the sum of the durations of all the individual pulses which occur during any 10,000-microsecond interval. Pulse duration is defined as the time interval between the two points on the pulse at which the instantaneous value is 70 per cent of the peak value. The peak value is defined as the maximum value of a smooth curve through the average of the fluctuations over the top portion of the pulse.

▲ For tube protection, it is essential that sufficient resistance be used in the plate-supply circuit, the grid-No.-2 supply circuit, and the grid-No.1-supply circuit so that the short-circuit current is limited to 0.5 ampere in each circuit.

● This value is approximately 115 per cent of the maximum dc plate-supply voltage.

△ For higher duty factors, the peak plate current must be reduced. The maximum rated current for a duty factor of 1.0 is 0.2 ampere.

♦ Averaged over any interval not exceeding 10,000 microseconds. Care should be used in determining the plate dissipation. A calculated value based on rectangular pulse can be considerably in error when the actual pulses have a finite rise and fall time. Plate dissipation should preferably be determined by measuring the bulb temperature under actual operating conditions; then, with the tube in the same socket and under the same ambient-temperature conditions, apply to the tube sufficient dc input to obtain the same bulb temperature. This value of dc input is a measure of the plate dissipation.

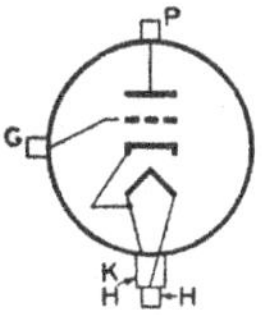

POWER TRIODE

6383

Compact liquid-and-forced-air-cooled type having heater-cathode used as af power amplifier and modulator, as rf power amplifier and oscillator, and as frequency multiplier. Coaxial terminal arrangement facilitates use in cathode-drive circuits of the coaxial-cylinder type. This type is also useful in applications where transmitter design factors of compactness, light weight, and high power output are prime considerations. May be used with full input up to 2000 Mc. Class C Telegraphy maximum CCS plate dissipation, 600 watts.

HEATER VOLTAGE (AC/DC):°		
Average	6.3•	volts
Maximum	6.9	volts
HEATER CURRENT (At 6.3 volts)	3.4	amperes
AMPLIFICATION FACTOR	27	
DIRECT INTERELECTRODE CAPACITANCES:		
Grid to plate	6	μμf
Grid to cathode and heater	11	μμf
Plate to cathode and heater■	0.22	μμf

° Because the cathode is subjected to considerable back bombardment as the frequency is increased with resultant increase in temperature, the heater voltage should be reduced depending on operating conditions and frequency to prevent overheating of the cathode and resultant short life.

• Average heater voltage must be applied for a minimum period of one minute before the application of plate voltage.

■ With external flat shield having a maximum diameter of 7½ inches located in plane of grid terminal and perpendicular to axis of tube. Shield is connected to grid terminal.

AF POWER AMPLIFIER AND MODULATOR—Class A

Maximum CCS Ratings:

DC PLATE VOLTAGE	1500 *max*	volts
DC GRID VOLTAGE	-300 *max*	volts
DC PLATE CURRENT	400 *max*	ma
DC GRID CURRENT	75 *max*	ma
PLATE INPUT	600 *max*	watts
PLATE DISSIPATION	600 *max*	watts

Typical Operation (*Class A_1*):

DC Plate Voltage	1000	1500	volts
DC Grid Voltage	-25	-40	volts
Peak AF Grid Voltage	20	35	volts
DC Plate Current	200	250	ma
Load Resistance	1350	1550	ohms
Power Output♦	20	60	watts

♦ Values are based on maximum power output disregarding distortion.

PLATE-MODULATED RF POWER AMPLIFIER—Class C Telephony

Carrier conditions per tube for use with a maximum modulation factor of 1.0

Maximum CCS Ratings:

DC PLATE VOLTAGE	1200 *max*	volts
DC GRID VOLTAGE	-300 *max*	volts
DC PLATE CURRENT	335 *max*	ma
DC GRID CURRENT	75* *max*	ma
PLATE INPUT	400 *max*	watts
PLATE DISSIPATION	400 *max*	watts

Typical Operation in Cathode-Drive Circuit:	*600 Mc*	*1000 Mc*	*1100 Mc*	*1500 Mc*	
Heater Voltage	5.7	4.5	4.5	4.5	volts
DC Plate-to-Grid Voltage	1340	1315	1290	1280	volts
DC Cathode-to-Grid Voltage	140	115	90	80	volts
From cathode resistor of▲	380	330	260	235	ohms
Peak RF Cathode-to-Grid Voltage	200	175	145	130	volts
DC Plate Current	335	335	335	335	ma
DC Grid Current (Approx.)	35	15	12	4	ma

Driver Power Output (Approx.)△	70	76	80	53	watts
Output-Circuit Efficiency (Approx.)	80	60	55	50	per cent
Useful Power Output (Approx.)	250□	190□	160□	100□	watts

* For frequencies up to 900 Mc. Above 900 Mc, this value must be reduced. At 2000 Mc, rated grid current is 10 milliamperes.

▲ At frequencies below 600 Mc, it is permissible to use a combination of grid resistor and cathode resistor, but the use of a grid resistor alone is not recommended. At frequencies above 600 Mc where the value of grid current may be small, only cathode bias is recommended.

△ In this type of service, the 6383 can be modulated 100 per cent if the rf driver stage is also modulated 100 per cent simultaneously. Care should be taken to insure that the driver-modulation and amplifier-modulation voltages are exactly in phase.

□ Measured at load of output circuit having indicated efficiency.

RF POWER AMPLIFIER AND OSCILLATOR—Class C Telegraphy#
and
RF POWER AMPLIFIER—Class C FM Telephony

Maximum CCS Ratings:

DC Plate Voltage	1500 *max*	volts
DC Grid Voltage	-300 *max*	volts
DC Plate Current	400 *max*	ma
DC Grid Current	75* *max*	ma
Plate Input	600 *max*	watts
Plate Dissipation	600 *max*	watts

Typical Operation as Amplifier in Cathode-Drive Circuit:

	600 Mc	*1000 Mc*	*1100 Mc*	*1500 Mc*	
Heater Voltage	5.7	4.5	4.5	4.5	volts
DC Plate-to-Grid Voltage	1640	1615	1590	1580	volts
DC Cathode-to-Grid Voltage	140	115	90	80	volts
From cathode resistor of▲	315	275	220	200	ohms
Peak RF Cathode-to-Grid Voltage	210	185	155	140	volts
DC Plate Current	400	400	400	400	ma
DC Grid Current (Approx.)	25	20	15	5	ma
Driver Power Output (Approx.)	90	95	80	85	watts
Output-Circuit Efficiency (Approx.)	80	60	55	50	per cent
Useful Power Output (Approx.)	380□	285□	240□	150□	watts

Typical Operation as Oscillator in Cathode-Drive Circuit:

	600 Mc	*1000 Mc*	*1100 Mc*	*1500 Mc*	
Heater Voltage	5.7	4.5	4.5	4.5	volts
DC Plate-to-Grid Voltage	1640	1615	1590	1580	volts
DC Cathode-to-Grid Voltage	140	115	90	80	volts
From cathode resistor of▲	315	275	220	200	ohms
Peak RF Cathode-to-Grid Voltage	175	140	120	110	volts
DC Plate Current	400	400	400	400	ma
DC Grid Current (Approx.)	45	20	15	5	ma
Output-Circuit Efficiency (Approx.)	80	60	55	50	per cent
Useful Power Output (Approx.)	280□	190□	150□	60□	watts

Key-down conditions per tube without amplitude modulation. Modulation essentially negative may be used if the positive peak of the audio-frequency envelope does not exceed 115 per cent of the carrier conditions.

* For frequencies up to 900 Mc. Above 900 Mc, this value must be reduced. At 2000 Mc, rated grid current is 10 milliamperes.

▲ At frequencies below 600 Mc, it is permissible to use a combination of grid resistor and cathode resistor, but the use of a grid resistor alone is not recommended. At frequencies above 600 Mc where the value of grid current may be small, only cathode bias is recommended.

□ Measured at load of output circuit having indicated efficiency.

FREQUENCY MULTIPLIER—Class C

Maximum CCS Ratings:

DC Plate Voltage	1500 *max*	volts
DC Grid Voltage	-300 *max*	volts
DC Plate Current	400 *max*	ma
DC Grid Current	75* *max*	ma
Plate Input	600 *max*	watts
Plate Dissipation	600 *max*	watts

Typical Operation as Doubler in Cathode-Drive Circuit:

	600 Mc	*900 Mc*	
DC Plate-to-Grid Voltage	1760	1675	volts

DC Cathode-to-Grid Voltage	260	175	volts
From cathode resistor of▲	570	415	ohms
Peak RF Cathode-to-Grid Voltage	300	215	volts
DC Plate Current	400	400	ma
DC Grid Current (Approx.)	55	25	ma
Driver Power Output (Approx.)	195	160	watts
Output-Circuit Efficiency (Approx.)	80	60	per cent
Useful Power Output (Approx.)	280	225	watts

* For frequencies up to 900 Mc. Above 900 Mc, this value must be reduced. At 2000 Mc, rated grid current is 10 milliamperes.

▲ At frequencies below 600 Mc, it is permissible to use a combination of grid resistor and cathode resistor, but the use of a grid resistor alone is not recommended. At frequencies above 600 Mc, where the value of grid current may be small, only cathode bias is recommended.

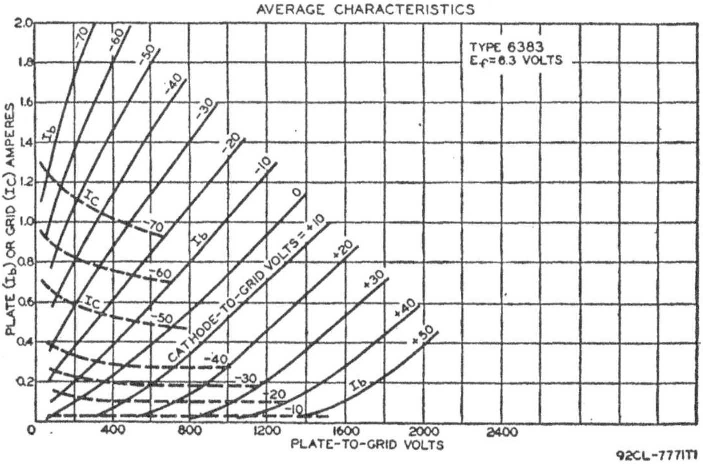

OPERATING CONSIDERATIONS

Type 6383 may be mounted in any position. OUTLINE 72, *Outlines* Section.

Forced-air cooling of the grid terminal, cathode terminal, and glass envelope is required. The air flow must start with the application of any voltages, and be adequate to limit the temperature of the grid terminal, cathode terminal, and glass envelope to their respective maximum values. Maximum temperatures: grid terminal, 200°C; cathode terminal, 200°C; and glass envelope, 175°C. Heater power, plate power, and air flow may be removed simultaneously.

Liquid cooling of the plate is required. The liquid flow must start before the application of any voltages. Interlocking of the liquid flow with all power supplies is recommended to prevent tube damage in case of failure of adequate liquid flow. Suitable coolants are distilled water and a high-temperature hydraulic fluid such as Monsanto 0S45. Maximum plate temperature (measured on side of plate flange opposite the pipes and at junction of flange with tube body), 180°C.

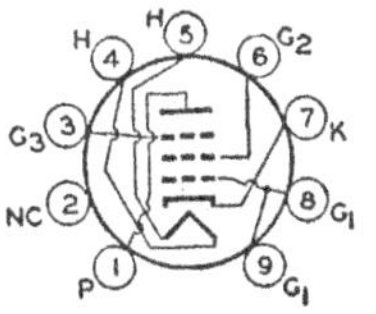

BEAM POWER TUBE

6417

Nine-pin miniature heater-cathode type used as rf power amplifier and oscillator and as frequency multiplier. May be used with full input up to 50 Mc. Class C Telegraphy maximum plate dissipation, CCS 12 watts, ICAS 13.5 watts. Requires Noval nine-contact socket and may be mounted in any position. OUTLINE 9, *Outlines* Section. Heater volts (ac/dc), 12.6 ± 10%; amperes, 0.375. Except for heater ratings, the 6417 is identical with type 5763.

6524

TWIN BEAM POWER TUBE

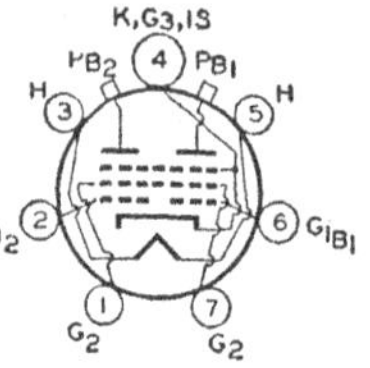

Small, sturdy, heater-cathode type used as af power amplifier and modulator, as push-pull rf power amplifier and oscillator, and as frequency tripler. May be used with full input up to 100 Mc and with reduced input up to 470 Mc. Class C Telegraphy maximum plate dissipation (per tube), CCS 20 watts, ICAS 25 watts.

HEATER VOLTAGE (AC/DC)	6.3 ± 10%	volts
HEATER CURRENT	1.25	amperes
TRANSCONDUCTANCE (Each unit)*	4500	μmhos
MU-FACTOR, Grid No.2 to Grid No.1 (Each unit)*	8.5	
DIRECT INTERELECTRODE CAPACITANCES (Each unit):		
Grid No.1 to plate	0.11 *max*	μμf
Grid No.1 to cathode, grid No.3, internal shield, grid No.2 (pins 1 and 7), and heater	7	μμf
Plate to cathode, grid No.3, internal shield, grid No.2 (pins 1 and 7), and heater	3.4	μμf

* Plate and grid-No.2 volts, 200; plate milliamperes, 50.

PUSH-PULL AF POWER AMPLIFIER AND MODULATOR—Class AB2

Values are on a per-tube basis

Maximum Ratings:	*CCS*	*ICAS*	
DC PLATE VOLTAGE	500 *max*	600 *max*	volts
DC GRID-NO.2 (SCREEN-GRID) VOLTAGE	300 *max*	300 *max*	volts
DC GRID-NO.2 SUPPLY VOLTAGE	400 *max*	400 *max*	volts
MAXIMUM-SIGNAL DC PLATE CURRENT■	150 *max*	150 *max*	ma
MAXIMUM-SIGNAL PLATE INPUT■	70 *max*	85 *max*	watts
MAXIMUM-SIGNAL GRID-NO.2 INPUT■	3 *max*	3 *max*	watts
PLATE DISSIPATION■	20 *max*	25 *max*	watts
PEAK HEATER-CATHODE VOLTAGE:			
Heater negative with respect to cathode	135 *max*	135 *max*	volts
Heater positive with respect to cathode	135 *max*	135 *max*	volts
BULB TEMPERATURE (At hottest point)	210 *max*	210 *max*	°C

Typical Operation:					
DC Plate Voltage	400	500	500	600	volts
DC Grid-No.2 Voltage▲	200	200	200	200	volts
DC Grid-No.1 (Control-Grid) Voltage	−23	−26	−25	−26	volts
Peak AF Grid-No.1-to-Grid-No.1 Voltage	72	70	76	76	volts
Zero-Signal DC Plate Current	25	20	25	21	ma
Maximum-Signal DC Plate Current	145	116	145	135	ma
Zero-Signal DC Grid-No.2 Current	0.1	0.1	0.1	0.1	ma
Maximum-Signal DC Grid-No.2 Current	10	10	10	13	ma
Maximum-Signal DC Grid-No.1 Current	2.4	2.6	2.9	3.3	ma
Effective Load Resistance (Plate to plate)	7100	11100	8900	11400	ohms
Maximum-Signal Driving Power (Approx.)	0.1	0.1	0.1	0.1	watt
Maximum-Signal Power Output (Approx.)	39	40	50	57	watts

Maximum Circuit Values (CCS or ICAS conditions):

Grid-No.1-Circuit Resistance:		
For fixed-bias operation	30000 *max*	ohms
For cathode-bias operation	Not recommended	

■ Averaged over any audio-frequency cycle of sine-wave form.
▲ Obtained preferably from a separate source or from the plate-voltage supply with a voltage divider.

PLATE-MODULATED PUSH-PULL RF POWER AMPLIFIER—Class C Telephony

Carrier conditions per tube for use with a maximum modulation factor of 1.0
Values are on a per-tube basis

Maximum Ratings:	*CCS*	*ICAS*	
DC PLATE VOLTAGE	400 *max*	500 *max*	volts
DC GRID-NO.2 VOLTAGE	300 *max*	300 *max*	volts
DC GRID-NO.2 SUPPLY VOLTAGE	400 *max*	400 *max*	volts
DC GRID-NO.1 VOLTAGE	−200 *max*	−200 *max*	volts
DC PLATE CURRENT	125 *max*	125 *max*	ma

	CCS	ICAS	
DC Grid-No.1 Current	4 *max*	4 *max*	ma
Plate Input	45 *max*	55 *max*	watts
Grid-No.2 Input	2 *max*	2 *max*	watts
Plate Dissipation	13.5 *max*	16.7 *max*	watts
Peak Heater-Cathode Voltage:			
Heater negative with respect to cathode	135 *max*	135 *max*	volts
Heater positive with respect to cathode	135 *max*	135 *max*	volts
Bulb Temperature (At hottest point)	210 *max*	210 *max*	°C

Typical Operation:

	100 Mc	*462 Mc*	*100 Mc*	*462 Mc*	
DC Plate Voltage	400	300	500	300	volts
DC Grid-No.2 Voltage (Approx.)°	200	200	200	240	volts
From an adjustable series resistor having a maximum value of	45000	45000	45000†	25000	ohms
DC Grid-No.1 Voltage•	−61	−60	−61	−60	volts
From combination employing grid-No.1 resistor of	6200	15000	6200	15000	ohms
with fixed bias of	−45	−45	−45	−45	volts
DC Plate Current	100	75	100	95	ma
DC Grid-No.2 Current (Approx.)	7	4	7	5.5	ma
DC Grid-No.1 Current (Approx.)	2.5	1	2.5	1	ma
Driving Power (Approx.)	0.2	–	0.2	–	watt
Driver Power Output (Approx.)	–	7	–	7	watts
Useful Power Output (Approx.)□	29	9	36	12	watts

Maximum Circuit Values (CCS or ICAS conditions):

Grid-No.1-Circuit Resistance	30000 *max*	ohms

° Obtained preferably from a separate source modulated along with the plate supply or from the modulated plate supply through a series resistor. It is recommended that this resistor be adjustable to permit obtaining the desired operating plate current after initial tuning adjustments are completed.

† Connected to a 400-volt tap or suitable voltage divider across the plate-supply voltage.

• Obtained from a combination of grid-No.1 resistor with either fixed supply or cathode resistor. The combination of grid-No.1 resistor and fixed supply has the advantage of not only protecting the tube from damage through loss of excitation but also of minimizing distortion by bias-supply compensation.

□ Measured at load of output circuit.

PUSH-PULL RF POWER AMPLIFIER AND OSCILLATOR—Class C Telegraphy#
and
PUSH-PULL RF POWER AMPLIFIER—Class C FM Telephony

Values are on a per-tube basis

Maximum Ratings:

	CCS	*ICAS*	
DC Plate Voltage	500 *max*	600 *max*	volts
DC Grid-No.2 Voltage	300 *max*	300 *max*	volts
DC Grid-No.2 Supply Voltage	400 *max*	400 *max*	volts
DC Grid-No.1 Voltage	−200 *max*	−200 *max*	volts
DC Plate Current	150 *max*	150 *max*	ma
DC Grid-No.1 Current	4 *max*	4 *max*	ma
Plate Input	70 *max*	85 *max*	watts
Grid-No.2 Input	3 *max*	3 *max*	watts
Plate Dissipation	20 *max*	25 *max*	watts
Peak Heater-Cathode Voltage:			
Heater negative with respect to cathode	135 *max*	135 *max*	volts
Heater positive with respect to cathode	135 *max*	135 *max*	volts
Bulb Temperature (At hottest point)	210 *max*	210 *max*	°C

Typical Operation:

	100 Mc	*462 Mc*	*100 Mc*	*462 Mc*	
DC Plate Voltage	500	300	600	300	volts
DC Grid-No.2 Voltage (Approx.) ♁	200	200	200	250	volts
From an adjustable series resistor having a maximum value of	40000†	60000	40000†	20000	ohms
DC Grid-No.1 Voltage ♦	−44	−31	−44	−38	volts
From grid-No.1 resistor of	12000	12000	12000	12000	ohms
From cathode resistor of	330	240	330	240	ohms
DC Plate Current	120	120	120	150	ma
DC Grid-No.2 Current (Approx.)	8	3	8	6	ma
DC Grid-No.1 Current (Approx.)	3.7	2.6	3.7	3.2	ma
Driving Power (Approx.)	0.2	–	0.2	–	watt
Driver Power Output (Approx.)	–	7	–	7	watts
Useful Power Output (Approx.)□	43	16⊕	52	20•	watts

Maximum Circuit Values: (CCS or ICAS conditions):

Grid-No.1-Circuit Resistance	30000 *max*	ohms

*Key-down conditions per tube without amplitude modulation. Amplitude modulation essentially negative may be used if the positive peak of the audio-frequency envelope does not exceed 115 per cent of the carrier conditions.

♁ Obtained preferably from a separate source, or from the plate-supply voltage with a voltage divider, or through a series resistor. It is recommended that this resistor be adjustable to permit obtaining the desired operating plate current after initial tuning adjustments are completed. Grid-No.2 voltage must not exceed 400 volts under key-up conditions.

†Connected to a 400-volt tap or suitable voltage divider across the plate-supply voltage.

♦ Obtained from fixed supply, by grid-No.1 resistor, by cathode resistor, or by combination methods.

□ Measured at load of output circuit.

● Amplifier power output. For oscillator service, useful power output is approximately 9 watts CCS and 13 watts ICAS at 462 Mc.

FREQUENCY TRIPLER—Class C

Values are on a per-tube basis

Maximum Ratings:	*CCS*	*ICAS*	
DC Plate Voltage	400 *max*	400 *max*	volts
DC Grid-No.2 Voltage	300 *max*	300 *max*	volts
DC Grid-No.2 Supply Voltage	400 *max*	400 *max*	volts
DC Grid-No.1 Voltage	-200 *max*	-200 *max*	volts
DC Plate Current	100 *max*	115 *max*	ma
DC Grid-No.1 Current	4 *max*	4 *max*	ma
Plate Input	36 *max*	45 *max*	watts
Grid-No.2 Input	3 *max*	3 *max*	watts
Plate Dissipation	20 *max*	25 *max*	watts
Peak Heater-Cathode Voltage:			
Heater negative with respect to cathode	135 *max*	135 *max*	volts
Heater positive with respect to cathode	135 *max*	135 *max*	volts
Bulb Temperature (At hottest point)	210 *max*	210 *max*	°C

Typical Operation at Frequencies up to 462 Mc:			
DC Plate Voltage	300	300	volts
DC Grid-No.2 Voltage (Approx.)♁	220	250	volts
From an adjustable series resistor having a maximum value of	30000	20000	ohms
DC Grid-No.1 Voltage♦	-148	-148	volts
From grid-No.1 resistor of	51000	51000	ohms
DC Plate Current	90	110	ma
DC Grid-No.2 Current (Approx.)	5	6.5	ma
DC Grid-No.1 Current (Approx.)	2.9	2.9	ma
Driver Power Output (Approx.)	4	4	watts
Useful Power Output (Approx.)□	7	8.5	watts

Maximum Circuit Values (CCS or ICAS conditions):

Grid-No.1-Circuit Resistance .... 60000 *max* ohms

♁ Obtained preferably from a separate source, or from the plate-supply voltage, with a voltage divider, or through a series resistor. It is recommended that this resistor be adjustable to permit obtaining the

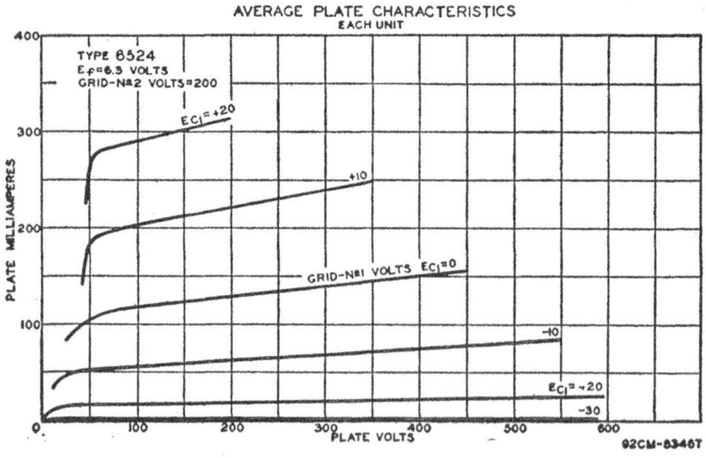

desired operating plate current after initial tuning adjustments are completed. Grid-No.2 voltage must not exceed 400 volts under key-up conditions.
♦ Obtained from fixed supply, by grid-No.1 resistor, by cathode resistor, or by combination methods.
□ Measured at load of output circuit.

OPERATING CONSIDERATIONS

Type 6524 requires Septar seven-contact socket and may be mounted in any position. OUTLINE 14, *Outlines* Section.

For operation in Plate-Modulated Push-Pull RF Power Amplifier Service at 220 Mc, plate voltage should be reduced to 79 per cent of maximum rating, plate input to 80 per cent. At 470 Mc, plate voltage should be reduced to 75 per cent, plate input to 53 per cent.

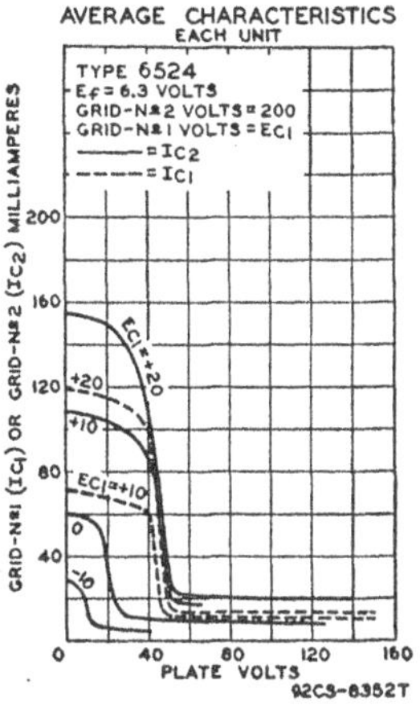

For operation in Class C Telegraphy Service at 220 Mc, plate voltage should be reduced to 79 per cent of maximum rating, plate input to 78 per cent. At 470 Mc, plate voltage should be reduced to 76 per cent, plate input to 51 per cent.

Free circulation of air around the tube is required. In addition, some forced-air cooling will generally be required to prevent exceeding the maximum bulb-temperature rating.

Plates show no color when tube is operated at maximum CCS or ICAS ratings.

FIXED-TUNED OSCILLATOR TRIODE

6562

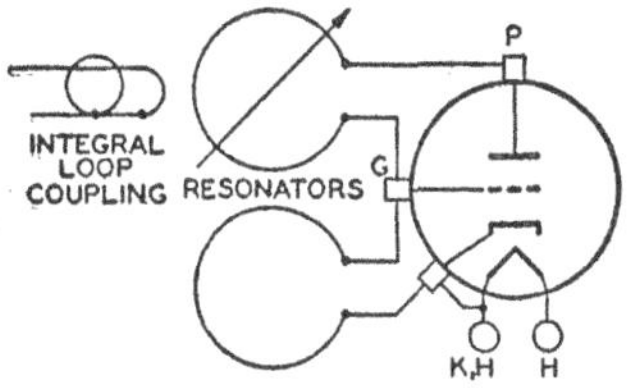

Pencil-type tube having integral resonators used in radiosonde service at a frequency of 1680 Mc. May be used at ambient temperatures ranging from -55°C to +75°C. Fixed-Tuned Oscillator maximum plate dissipation, 3.6 watts.

HEATER VOLTAGE RANGE° (AC/DC)	5.2 to 6.6	volts
HEATER CURRENT (At 6.0 volts)	0.160	ampere
FREQUENCY (Approx.)	1680	Mc
FREQUENCY-ADJUSTMENT RANGE■	±12	Mc

° This range of heater voltage is for radiosonde applications in which the heater is supplied from batteries and in which the equipment design requirements of minimum size, light weight, and high efficiency are the primary considerations even though the average life expectancy of the 6562 in such service is only a few hours.
■ As supplied, tubes are adjusted to 1680 ± 4 megacycles.

FIXED-TUNED OSCILLATOR

Maximum Ratings:

DC PLATE VOLTAGE	120 *max*	volts
DC PLATE CURRENT	34 *max*	ma
DC GRID CURRENT	8 *max*	ma
PLATE INPUT	4 *max*	watts
PLATE DISSIPATION	3.6 *max*	watts
PEAK HEATER-CATHODE VOLTAGE	0 *max*	volts
AMBIENT-TEMPERATURE RANGE	−55 to +75	°C

Operating Frequency Drift:

Maximum Frequency Drift:

For heater-voltage range of 5.2 to 6.6 volts, plate-voltage range of 95 to 117 volts, and ambient-temperature range of +22° to −40°C.....	+4 to −1	Mc

OPERATING CONSIDERATIONS

Type 6562 may be mounted in any position. OUTLINE 68, *Outlines* Section.

The flexible heater leads of the 6562 are usually soldered to the circuit elements. Soldering of these connections should not be made closer than ¾″ from the end of the tube (excluding cathode tab). If this precaution is not followed, the heat of the soldering operation may crack the glass seals of the leads and damage the tube. Under no circumstances should any of the electrodes be soldered to the circuit elements. Connections to the electrodes should be made by spring contact only.

The 6562 should be supported by a suitable clamp around the metal shell either above or below the frequency-adjustment screw. It is essential, however, that the pressure exerted on the shell by the clamp be held to a minimum because excessive pressure can distort the resonators and result in a change of frequency.

The plate connection should have a flexible lead which will accommodate variations in the relative position of the plate terminal in individual tubes.

The 6562 may be mechanically tuned by adjustment of the frequency-adjustment screw located on the metal shell of the tube. A clockwise rotation of the frequency-adjustment screw will decrease the frequency, while a counterclockwise rotation will increase the frequency. The range of adjustment provided by the screw is ± 12 megacycles.

6850

TWIN BEAM POWER TUBE

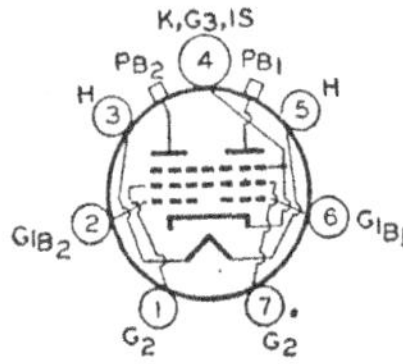

Small, sturdy, heater-cathode type used as af power amplifier and modulator, as push-pull rf power amplifier and oscillator, and as frequency tripler. May be used with full input up to 100 Mc and with reduced input up to 470 Mc. Class C Telegraphy maximum plate dissipation (per tube), CCS 20 watts, ICAS 25 watts. Requires Septar seven-contact socket and may be mounted in any position. OUTLINE 14, *Outlines* Section. Heater volts (ac/dc), 12.6 ± 10%; amperes, 0.625. Except for heater rating, the 6850 is identical with type 6524.

6883

BEAM POWER TUBE

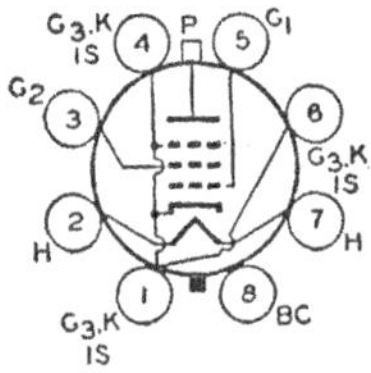

Small, sturdy, glass-octal heater-cathode type used as af power amplifier and modulator and as rf power amplifier and oscillator. May be used with full input up to 60 Mc and with reduced input up to 175 Mc. Class C Telegraphy maximum plate dissipation, CCS 20 watts, ICAS 25 watts. Requires Octal socket and may be mounted in any position. OUTLINE 17, *Outlines* Section. Heater volts (ac/dc), 12.6 ± 10%; amperes, 0.625. Except for heater rating, the 6883 is identical with type 6146.

8000

POWER TRIODE

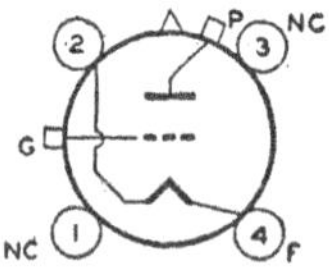

Thoriated-tungsten-filament type used as af power amplifier and modulator and as rf power amplifier and oscillator. May be used with full input up to 30 Mc and with reduced input up to 100 Mc. Class C Telegraphy maximum plate dissipation, CCS 125 watts, ICAS 175 watts.

Filament Voltage (ac/dc)	10	volts
Filament Current	4.5	amperes
Amplification Factor	16.5	
Direct Interelectrode Capacitances:		
Grid to plate	6.4	μμf
Grid to filament	5.0	μμf
Plate to filament	3.3	μμf

AF POWER AMPLIFIER AND MODULATOR—Class B

Maximum Ratings:	*CCS*	*ICAS*	
DC Plate Voltage	2500 *max*	2750 *max*	volts
Maximum-Signal DC Plate Current■	250 *max*	250 *max*	ma
Maximum-Signal Plate Input■	425 *max*	510 *max*	watts
Plate Dissipation■	125 *max*	175 *max*	watts

Typical Operation (*Values are for 2 tubes*):			
DC Plate Voltage	2000	2250	volts
DC Grid Voltage	-120	-130	volts
Peak AF Grid-to-Grid Voltage	520	560	volts
Zero-Signal DC Plate Current	60	65	ma
Maximum-Signal DC Plate Current	425	450	ma
Effective Load Resistance (Plate to plate)	10800	12000	ohms
Maximum-Signal Driving Power (Approx.)	6.5	7.9	watts
Maximum-Signal Power Output (Approx.)	600	725	watts

■ Averaged over any audio-frequency cycle of sine-wave form.

PLATE-MODULATED RF POWER AMPLIFIER—Class C Telephony

Carrier conditions per tube for use with a maximum modulation factor of 1.0

Maximum Ratings:	*CCS*	*ICAS*	
DC Plate Voltage	1600 *max*	2000 *max*	volts
DC Grid Voltage	-500 *max*	-500 *max*	volts
DC Plate Current	210 *max*	250 *max*	ma
DC Grid Current	40 *max*	45 *max*	ma
Plate Input	335 *max*	500 *max*	watts
Plate Dissipation	85 *max*	125 *max*	watts

Typical Operation:			
DC Plate Voltage	1600	2000	volts
DC Grid Voltage♁	-300	-370	volts
From grid resistor of	15000	10000	volts
Peak RF Grid Voltage	470	690	volts
DC Plate Current	210	250	ma
DC Grid Current (Approx.)	20	37	ma
Driving Power (Approx.)	8.5	20	watts
Power Output (Approx.)	250	380	watts

♁ Obtained from grid resistor of value shown or from a combination of grid resistor with either fixed supply or cathode resistor.

RF POWER AMPLIFIER AND OSCILLATOR—Class C Telegraphy# and RF POWER AMPLIFIER—Class C FM Telephony

Maximum Ratings:	*CCS*	*ICAS*	
DC Plate Voltage	2000 *max*	2500 *max*	volts
DC Grid Voltage	-500 *max*	-500 *max*	volts
DC Plate Current	250 *max*	300 *max*	ma
DC Grid Current	40 *max*	45 *max*	ma
Plate Input	500 *max*	750 *max*	watts
Plate Dissipation	125 *max*	175 *max*	watts

Typical Operation:			
DC Plate Voltage	2000	2500	volts
DC Grid Voltage♦	-195	-240	volts
From grid resistor of	8100	6000	ohms
From cathode resistor of	710	700	ohms
Peak RF Grid Voltage	370	480	volts
DC Plate Current	250	300	ma
DC Grid Current (Approx.)	24	40	ma

Driving Power (Approx.)	8	18	watts
Power Output (Approx.)	375	575	watts

♯ Key-down conditions per tube without amplitude modulation. Amplitude modulation essentially negative may be used if the positive peak of the audio-frequency envelope does not exceed 115 per cent of the carrier conditions.

♦ Obtained from fixed supply, by grid resistor, by cathode resistor, or by combination methods.

SELF-RECTIFYING OSCILLATOR—Class C

With separate, rectified, unfiltered, single-phase, full-wave plate supply

Maximum Ratings:

DC PLATE VOLTAGE	1800 *max*	volts
DC GRID VOLTAGE	-300 *max*	volts
DC PLATE CURRENT	225 *max*	ma
DC GRID CURRENT	35 *max*	ma
PLATE INPUT	500 *max*	watts
PLATE DISSIPATION	125 *max*	watts

Typical Push-Pull Operation at 30 Mc (*Values are for 2 tubes*):

DC Plate Voltage	1800	volts
Grid Resistor	5000	ohms
DC Plate Current	450	ma
DC Grid Current▲	35	ma
Power Output (Approx.)	700	watts
Useful Power Output (Approx.)–85-per-cent circuit efficiency	600	watts

▲ For full-load operation. Under no-load operation, grid current and grid voltage should not exceed maximum ratings.

OPERATING CONSIDERATIONS

Type 8000 requires Jumbo four-contact socket and may be mounted in vertical position with base down, or in horizontal position with pins 1 and 2 in vertical plane. OUTLINE 53, *Outlines* Section.

For operation at 60 Mc, plate voltage and plate input should be reduced to 70 per cent of maximum ratings; at 100 Mc, to 50 per cent.

When the 8000 is used in the final amplifier or a preceding stage of a transmitter designed for break-in operation and oscillator keying, a small amount of fixed bias must be used to maintain the plate current at a safe value. With a plate voltage of 2500 volts, a fixed bias of at least -140 volts should be used.

Plate shows a barely perceptible red color when tube is operated at maximum CCS ratings and a cherry-red color at maximum ICAS ratings.

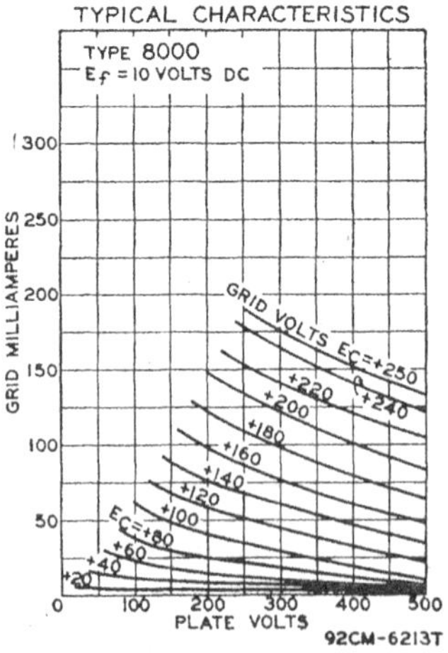

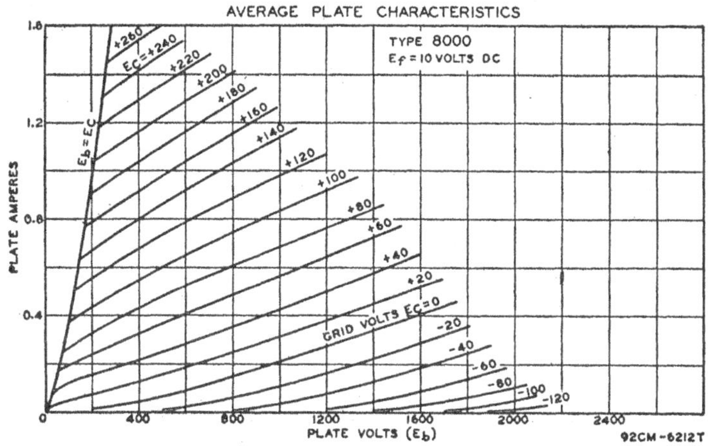

BEAM POWER TUBE

8001

See type 4E27/8001.

POWER TRIODE

8003

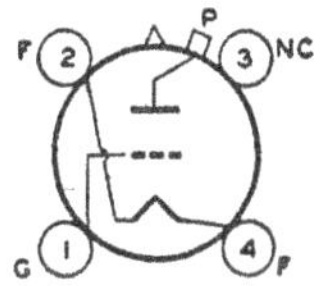

Thoriated-tungsten-filament type used as af power amplifier and modulator and as rf power amplifier and oscillator. May be used with full input up to 30 Mc and with reduced input up to 50 Mc. Requires Jumbo four-contact socket and may be mounted in vertical position with base down, or in horizontal position with pins 1 and 3 in vertical plane. OUTLINE 50, *Outlines* Section. For operation at 50 Mc, plate voltage and plate input should be reduced to 83 per cent of maximum ratings. Filament volts (ac/dc), 10; amperes, 3.25. Direct interelectrode capacitances: grid to plate, 11.7 μμf; grid to filament, 5.8 μμf; plate to filament, 3.4 μμf. Maximum CCS ratings as AF POWER AMPLIFIER AND MODULATOR: dc plate volts, 1350 *max;* maximum-signal dc plate milliamperes, 250 *max;* maximum-signal plate input, 330 *max* watts; plate dissipation, 100 *max* watts. Maximum CCS ratings as RF POWER AMPLIFIER AND OSCILLATOR: dc plate volts, 1350 *max;* dc grid volts, –400 *max;* dc plate milliamperes, 250 *max;* dc grid milliamperes, 50 *max;* plate input, 330 *max* watts; plate dissipation, 100 *max* watts. Plate shows no color when tube is operated at maximum CCS ratings. The 8003 is used principally for renewal purposes.

POWER TRIODE

8005

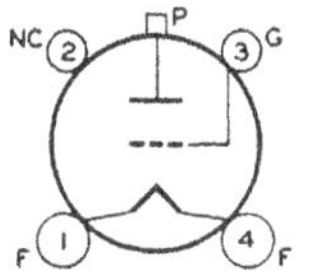

Thoriated-tungsten-filament type used as af power amplifier and modulator and as rf power amplifier and oscillator. May be used with full input up to 60 Mc and with reduced input up to 100 Mc. Class C Telegraphy maximum plate dissipation, CCS 75 watts, ICAS 85 watts.

FILAMENT VOLTAGE (AC/DC)	10	volts
FILAMENT CURRENT	3.25	amperes
AMPLIFICATION FACTOR*	20	
DIRECT INTERELECTRODE CAPACITANCES:		
Grid to plate	5.0	μμf
Grid to filament	6.4	μμf
Plate to filament	1.0	μμf

*Grid volts, 50; plate amperes, 0.5.

AF POWER AMPLIFIER AND MODULATOR—Class B

Maximum Ratings:	*CCS*	*ICAS*	
DC PLATE VOLTAGE	1250 *max*	1500 *max*	volts
MAXIMUM-SIGNAL DC PLATE CURRENT■	200 *max*	200 *max*	ma
MAXIMUM-SIGNAL PLATE INPUT■	225 *max*	250 *max*	watts
PLATE DISSIPATION■	75 *max*	85 *max*	watts

Typical Operation (*Values are for 2 tubes*):			
DC Plate Voltage	1250	1500	volts
DC Grid Voltage†	–55	–67.5	volts
Peak AF Grid-to-Grid Voltage	290	330	volts
Zero-Signal DC Plate Current	40	40	ma
Maximum-Signal DC Plate Current	320	330	ma
Effective Load Resistance (Plate to plate)	8000	9800	ohms
Maximum-Signal Driving Power (Approx.)	4	5.5	watts
Maximum-Signal Power Output (Approx.)	250	330	watts

■ Averaged over any audio-frequency cycle of sine-wave form.
† For ac filament supply.

PLATE-MODULATED RF POWER AMPLIFIER—Class C Telephony

Carrier conditions per tube for use with a maximum modulation factor of 1.0

Maximum Ratings:	*CCS*	*ICAS*	
DC PLATE VOLTAGE	1000 *max*	1250 *max*	volts
DC GRID VOLTAGE	–200 *max*	–200 *max*	volts
DC PLATE CURRENT	160 *max*	200 *max*	ma
DC GRID CURRENT	45 *max*	45 *max*	ma
PLATE INPUT	160 *max*	240 *max*	watts
PLATE DISSIPATION	50 *max*	75 *max*	watts

Typical Operation:			
DC Plate Voltage	1000	1250	volts
DC Grid Voltage♁	–195	–195	volts
From grid resistor of	7000	7000	ohms
Peak RF Grid Voltage	350	350	volts
DC Plate Current	160	190	ma
DC Grid Current (Approx.)	28	28	ma
Driving Power (Approx.)	9	9	watts
Power Output (Approx.)	115	170	watts

♁ Obtained from grid resistor of value shown or from a combination of grid resistor and fixed supply.

RF POWER AMPLIFIER AND OSCILLATOR—Class C Telegraphy#
and
RF POWER AMPLIFIER—Class C FM Telephony

Maximum Ratings:	*CCS*	*ICAS*	
DC PLATE VOLTAGE	1250 *max*	1500 *max*	volts
DC GRID VOLTAGE	–200 *max*	–200 *max*	volts
DC PLATE CURRENT	200 *max*	200 *max*	ma
DC GRID CURRENT	45 *max*	45 *max*	ma
PLATE INPUT	240 *max*	300 *max*	watts
PLATE DISSIPATION	75 *max*	85 *max*	watts

Typical Operation:			
DC Plate Voltage	1250	1500	volts
DC Grid Voltage♠	–115	–130	volts
From grid resistor of	3800	4000	ohms
From cathode resistor of	520	560	ohms
Peak RF Grid Voltage	240	255	volts
DC Plate Current	190	200	ma
DC Grid Current (Approx.)	30	32	ma
Driving Power (Approx.)	6.5	7.5	watts
Power Output (Approx.)	170	220	watts

Key-down conditions per tube without amplitude modulation. Amplitude modulation essentially negative may be used provided the positive peak of the audio-frequency envelope does not exceed 115 per cent of the carrier conditions.
♠ Obtained from fixed supply, by grid resistor, by cathode resistor, or by combination methods.

SELF-RECTIFYING OSCILLATOR OR AMPLIFIER—Class C

Maximum CCS Ratings:

RMS PLATE VOLTAGE	1750 *max*	volts
DC GRID VOLTAGE	-125 *max*	volts
DC PLATE CURRENT	125 *max*	ma
DC GRID CURRENT	25 *max*	ma
PLATE INPUT	240 *max*	watts
PLATE DISSIPATION	75 *max*	watts

Typical Push-Pull Operation at 50 Mc *(Values are for 2 tubes):*

RMS Plate Voltage	1750	volts
Grid Resistor	2000	ohms
DC Plate Current	250	ma
DC Grid Current (At full load)	35	ma
Power Output (Approx.)	330	watts
Useful Power Output (Approx.)-75-per-cent circuit efficiency	250	watts

AMPLIFIER OR OSCILLATOR—Class C

With separate, rectified, unfiltered, single-phase, full-wave plate supply

Maximum CCS Ratings:

DC PLATE VOLTAGE	1125 *max*	volts
DC GRID VOLTAGE	-125 *max*	volts
DC PLATE CURRENT	180 *max*	ma
DC GRID CURRENT	40 *max*	ma
PLATE INPUT	240 *max*	watts
PLATE DISSIPATION	75 *max*	watts

Typical Push-Pull Operation at 27 Mc *(Values are for 2 tubes):*

DC Plate Voltage	1100	volts
Grid Resistor	2000	ohms
DC Plate Current	360	ma
DC Grid Current (At full load)	40	ma
Power Output (Approx.)	330	watts
Useful Power Output (Approx.)-85-per-cent circuit efficiency	280	watts

OPERATING CONSIDERATIONS

Type 8005 requires Small four-contact socket and may be mounted in vertical position with base down, or in horizontal position with pins 2 and 3 in vertical plane. OUTLINE 42, *Outlines* Section.

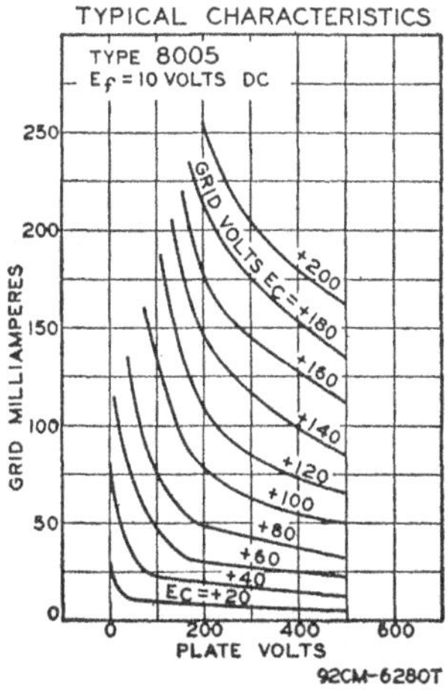

For operation at 80 Mc, plate voltage and plate input should be reduced to 75 per cent of maximum ratings; at 100 Mc, to 60 per cent.

When the 8005 is used in the final amplifier or a preceding stage of a transmitter designed for break-in operation and oscillator keying, a small amount of fixed bias must be used to maintain the plate current at a safe value. With a plate voltage of 1500 volts, a fixed bias of at least -50 volts should be used.

Plate shows a cherry-red color when tube is operated at maximum CCS ratings and an orange-red color at maximum ICAS ratings.

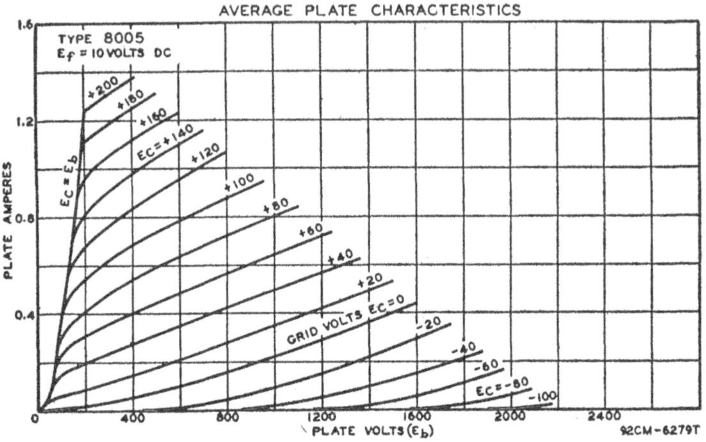

HALF-WAVE MERCURY-VAPOR RECTIFIER

8008

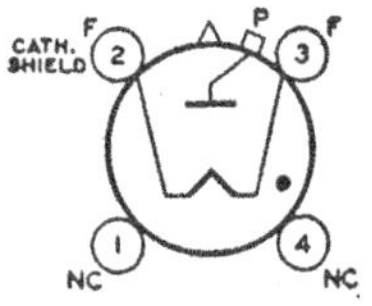

Coated-filament type used in power supply of transmitting and industrial equipment. Maximum peak inverse anode volts, 10,000; maximum average anode amperes, 1.25. Requires Super-Jumbo four-contact socket and may be mounted in vertical position only, base down. OUTLINE 54, *Outlines* Section. Except for physical dimension and base, the 8008 is identical to type 872-A.

POWER TRIODE

8012-A

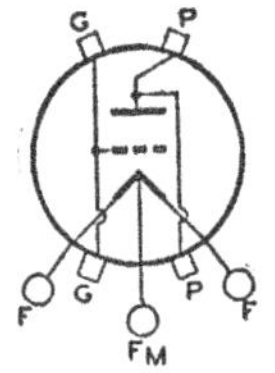

Thoriated-tungsten-filament type having filament mid-tap used as rf power amplifier and oscillator. May be used with full input up to 500 Mc. For operation at 600 Mc, plate voltage should be reduced to 70 per cent of maximum rating. May be mounted in vertical position only, filament end down or up. OUTLINE 18, *Outlines* Section. Forced-air cooling is required when plate dissipation exceeds 75 per cent of the maximum rated value. Plate shows an orange-red color when tube is operated at maximum CCS ratings. The 8012-A is used principally for renewal purposes.

FILAMENT VOLTAGE (AC/DC)	6.3	volts
FILAMENT CURRENT	1.92	amperes
AMPLIFICATION FACTOR	18	
DIRECT INTERELECTRODE CAPACITANCES:		
Grid to plate	2.5	μμf
Grid to filament mid-tap	2.7	μμf
Plate to filament mid-tap	0.4	μμf

Maximum CCS Ratings:	Class C Telephony▲	Class C Telegraphy#	
DC PLATE VOLTAGE	800 *max*	1000 *max*	volts
DC GRID VOLTAGE	-200 *max*	-200 *max*	volts
DC PLATE CURRENT	65 *max*	80 *max*	ma
DC GRID CURRENT	20 *max*	20 *max*	ma

PLATE INPUT	33 *max*	50 *max*	watts
PLATE DISSIPATION	27 *max*	40 *max*	watts

▲ Carrier conditions per tube for use with a maximum modulation factor of 1.0.
Key-down conditions per tube without amplitude modulation. Amplitude modulation essentially negative may be used if the positive peak of the audio-frequency envelope does not exceed 115 per cent of the carrier conditions.

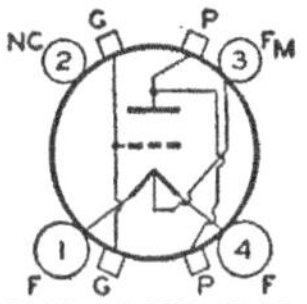

G CAPS NEARER BASE
P CAPS NEARER BULB TIP

POWER TRIODE

8025-A

Thoriated-tungsten-filament type having filament mid-tap used as rf power amplifier and oscillator. May be used with full input up to 500 Mc. For operation at 600 Mc, plate voltage should be reduced to 70 per cent of maximum ratings. Class C Telegraphy maximum plate dissipation, CCS 40 watts with forced-air cooling, ICAS 30 watts with natural cooling. Requires Small four-contact socket and may be mounted in vertical position only, base down or up. OUTLINE 27, *Outlines* Section. When forced-air cooling is required, an air flow from a fan should be directed on the bulb. Plate shows an orange-red color when tube is operated at maximum CCS ratings and a bright orange-red color at maximum ICAS ratings.

FILAMENT VOLTAGE (AC/DC)	6.3	volts
FILAMENT CURRENT	1.92	amperes
AMPLIFICATION FACTOR	18	
DIRECT INTERELECTRODE CAPACITANCES:		
Grid to plate	3.0	μμf
Grid to filament mid-tap	2.7	μμf
Plate to filament mid-tap	0.4	μμf

	Class C Telephony▲		Class C Telegraphy#		
Maximum Ratings:	*Forced-Air Cooling CCS*	*Natural Cooling ICAS*	*Forced-Air Cooling CCS*	*Natural Cooling ICAS*	
DC PLATE VOLTAGE	800 *max*	800 *max*	1000 *max*	1000 *max*	volts
DC GRID VOLTAGE	-200 *max*	-200 *max*	-200 *max*	-200 *max*	volts
DC PLATE CURRENT	65 *max*	65 *max*	80 *max*	80 *max*	ma
DC GRID CURRENT	20 *max*	20 *max*	20 *max*	20 *max*	ma
PLATE INPUT	50 *max*	33 *max*	75 *max*	50 *max*	watts
PLATE DISSIPATION	27 *max*	20 *max*	40 *max*	30 *max*	watts

▲ Carrier conditions per tube for use with a maximum modulation factor of 1.0.
Key-down conditions per tube without amplitude modulation. Amplitude modulation essentially negative may be used if the positive peak of the audio-frequency envelope does not exceed 115 per cent of the carrier conditions.

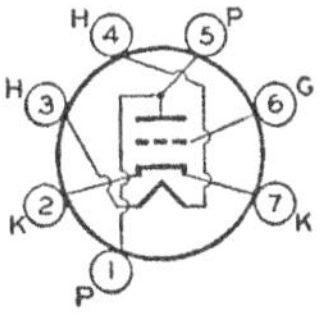

MEDIUM-MU TRIODE

9002

Seven-pin miniature heater-cathode type used as af amplifier and as rf amplifier and oscillator at frequencies up to 500 Mc. Class A_1 Amplifier maximum CCS plate dissipation (design-center value), 1.6 watts. Direct interelectrode capacitances: grid to plate, 1.4 μμf; grid to cathode and heater, 1.2 μμf; plate to cathode and heater, 1.1 μμf. Requires Miniature seven-contact socket and may be mounted in any position. OUTLINE 5, *Outlines* Section. Except for interelectrode capacitances, the 9002 is electrically identical with type 955.

Outlines

OUTLINES 1-10

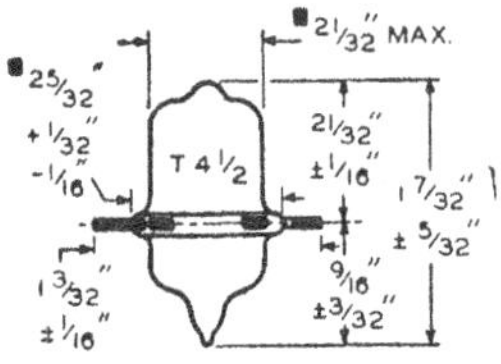

-1-

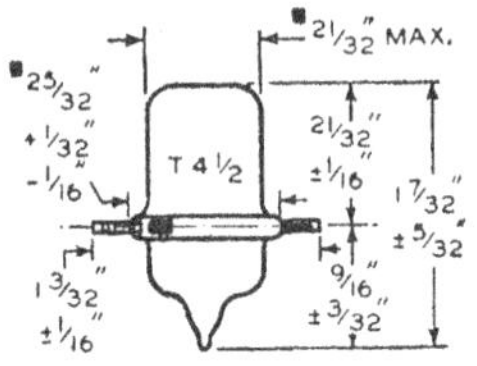

-2-

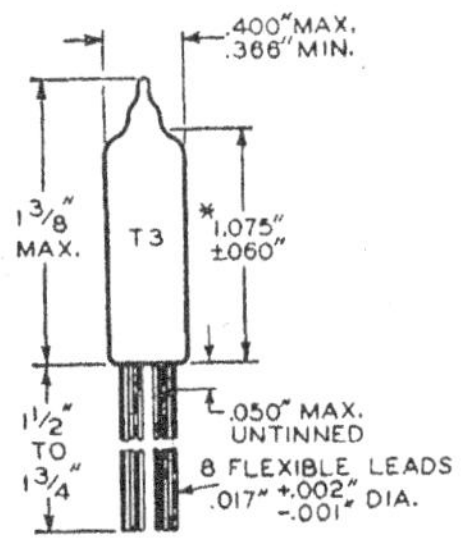

-3-

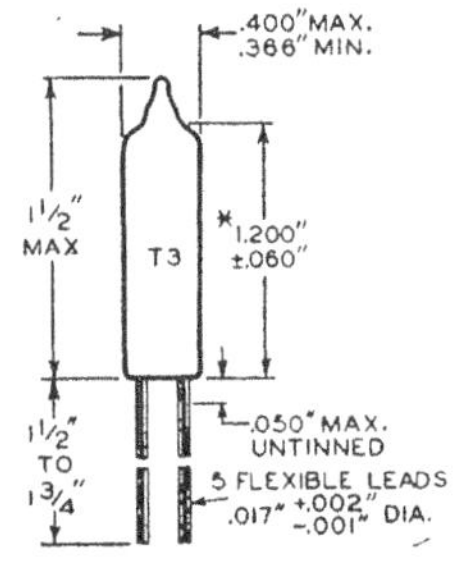

-4-

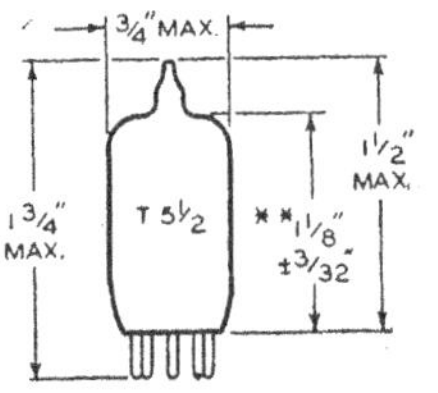

-5-

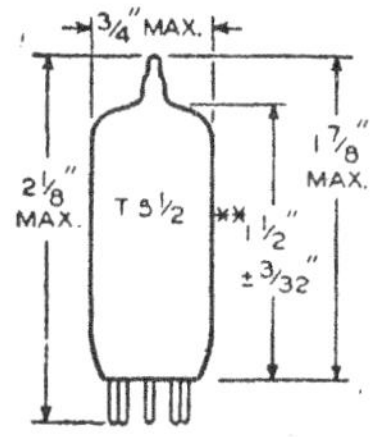

-6-

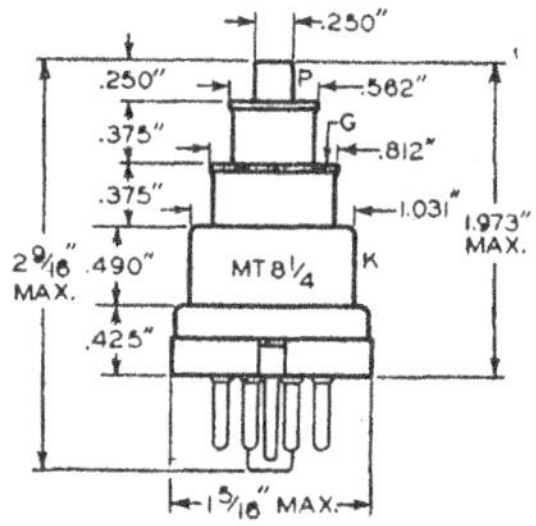

-7-

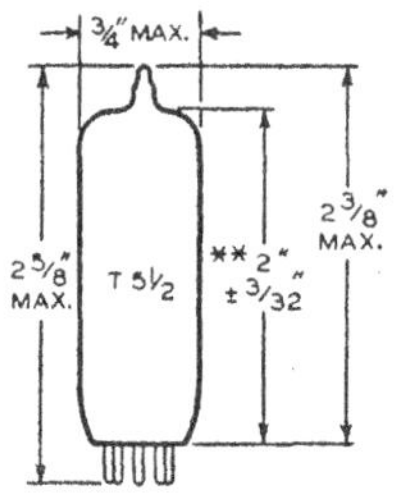

-8-

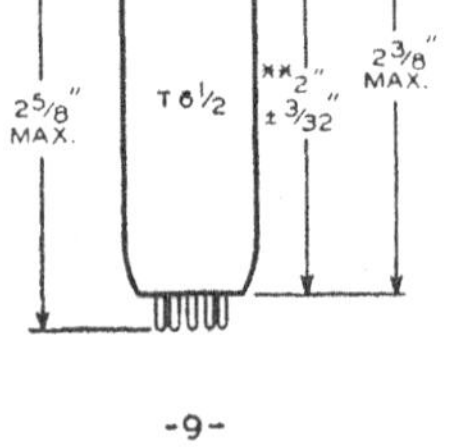

-9-

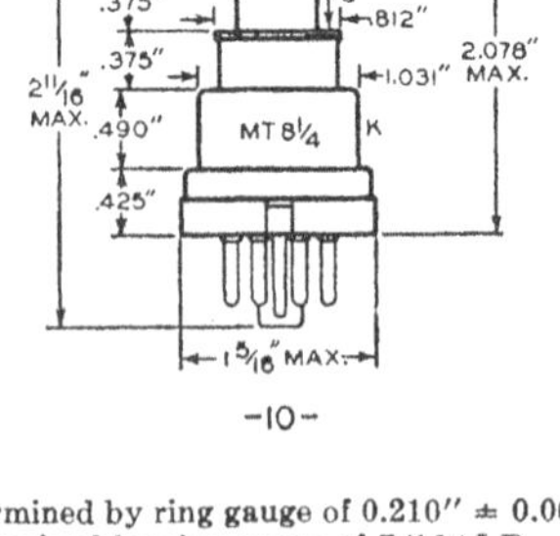

-10-

■ Including eccentricity.
* Measured from bulb seat to bulb-top line as determined by ring gauge of 0.210" ± 0.001" I.D.
** Measured from base seat to bulb-top line as determined by ring gauge of 7/16" I.D.

OUTLINES 11-19

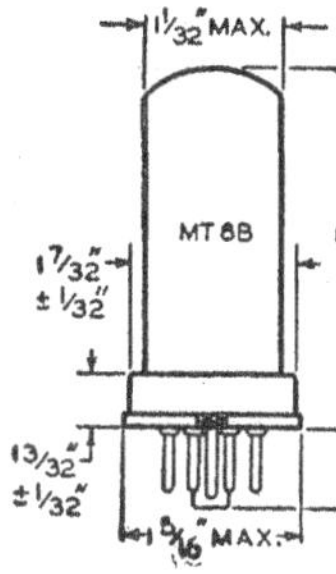

-11-

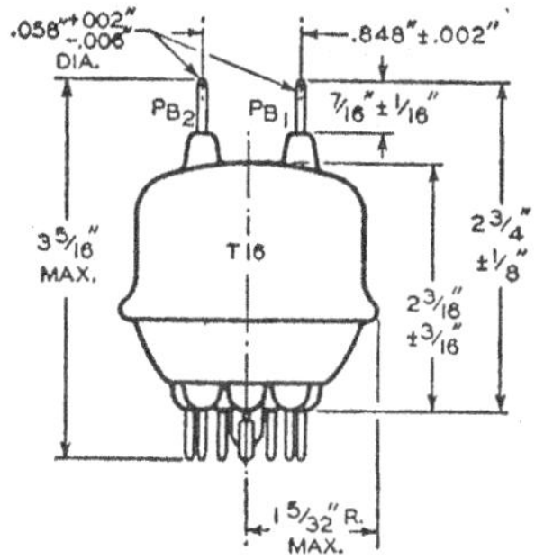

-12-

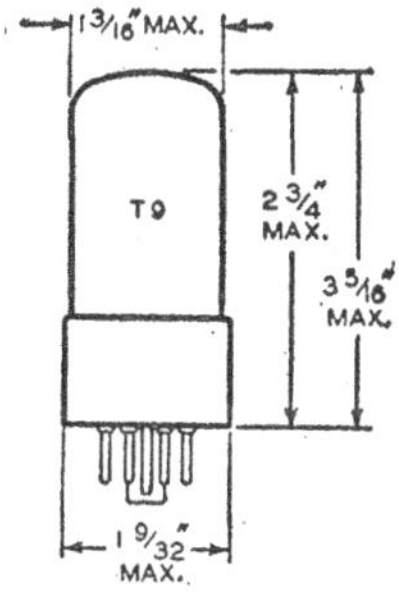

-13-

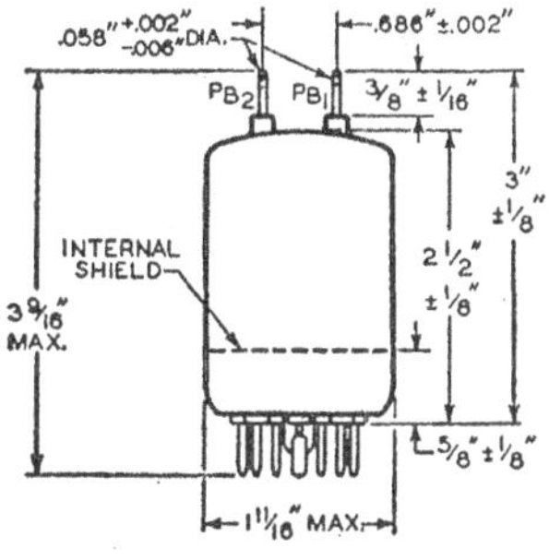

-14-

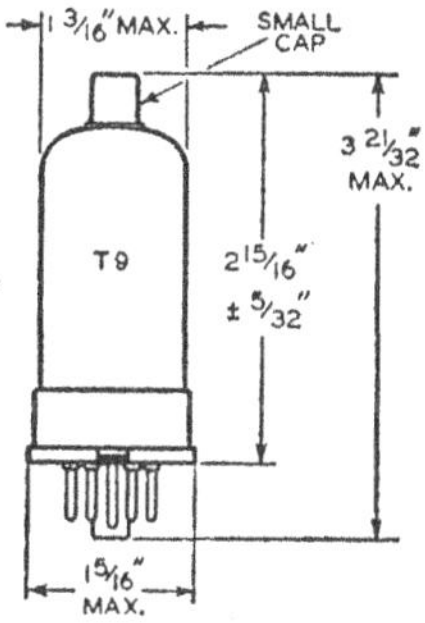

-15-

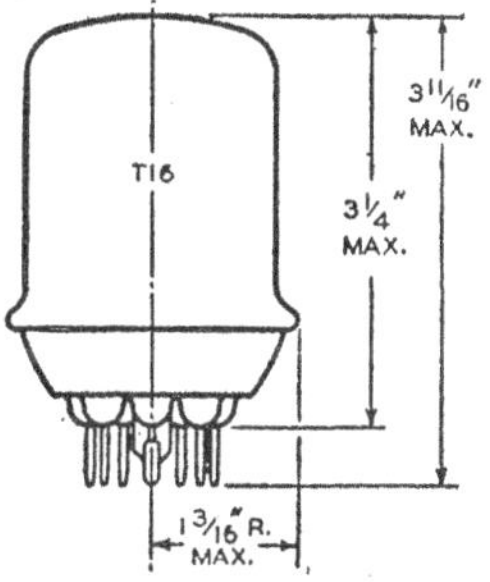

-16-

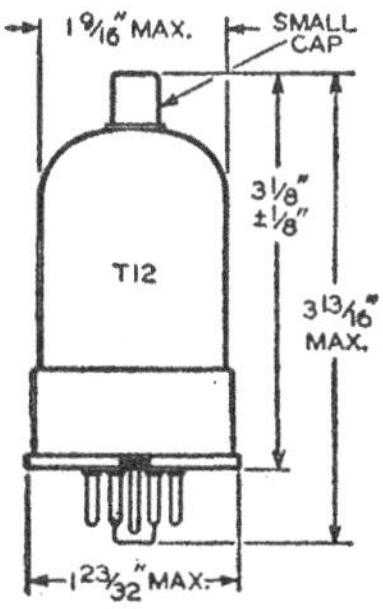

-17-

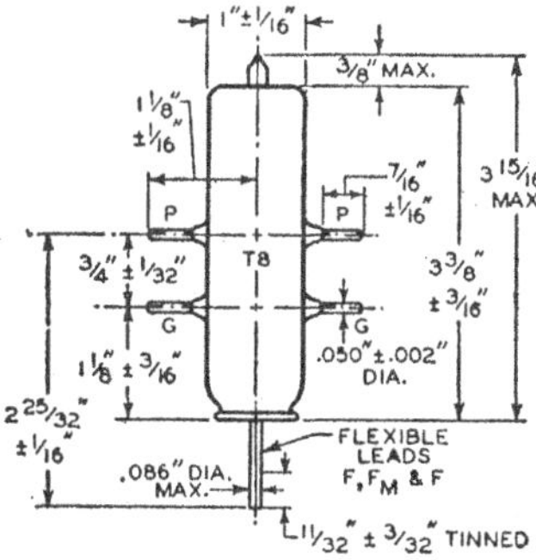

-18-

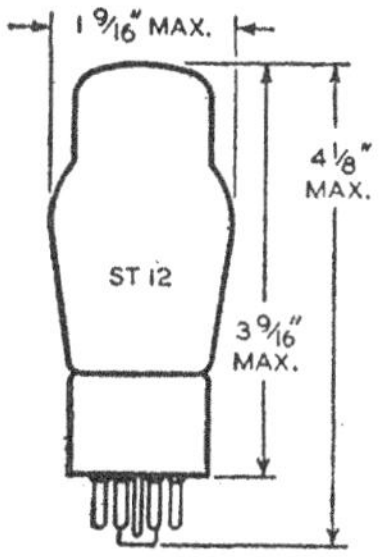

-19-

OUTLINES 20-28

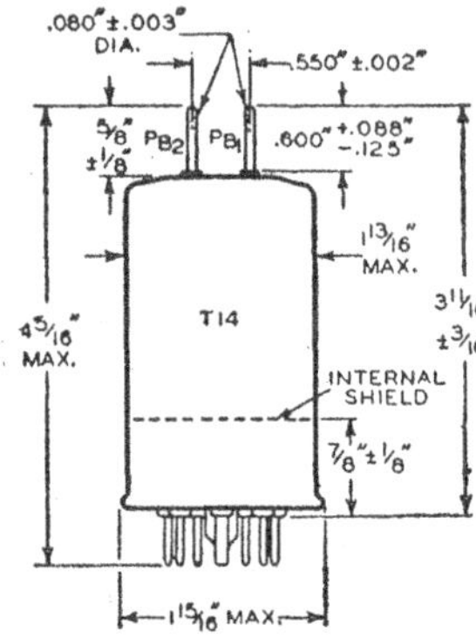

-20-

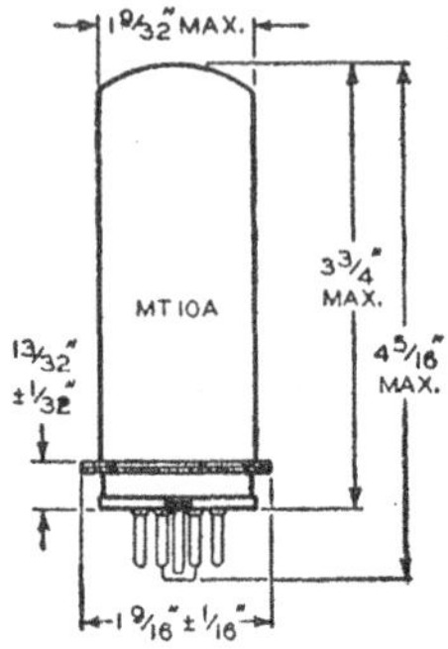

-21-

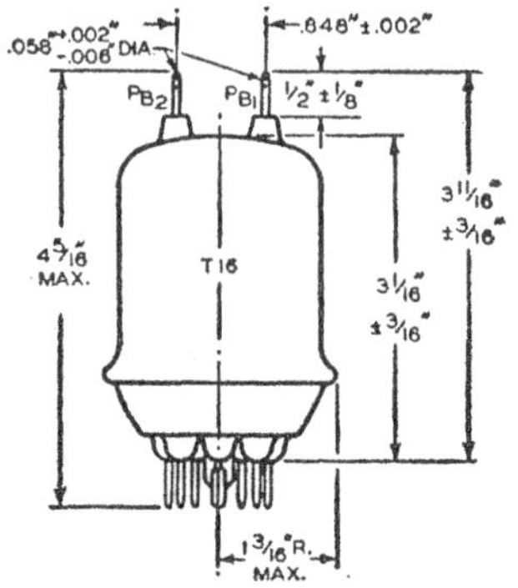

-22-

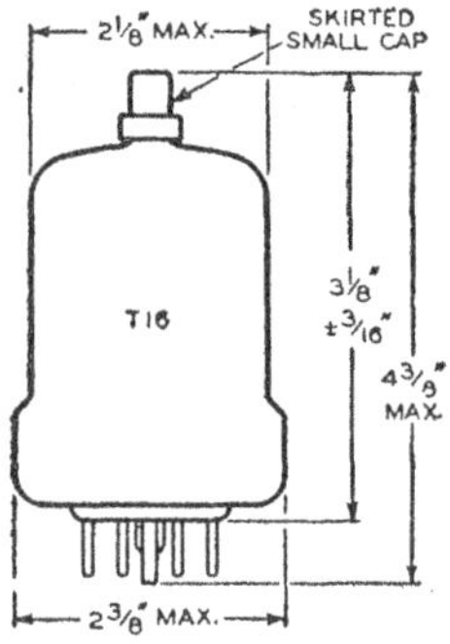

-23-

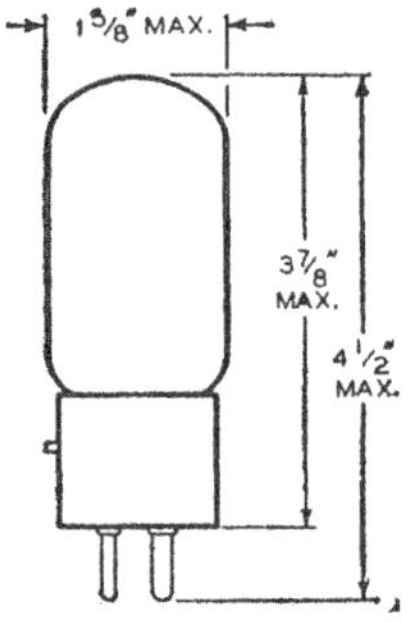

-24-

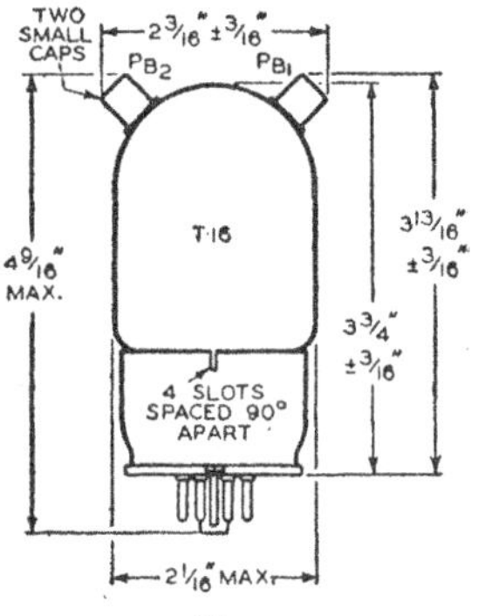

-25-

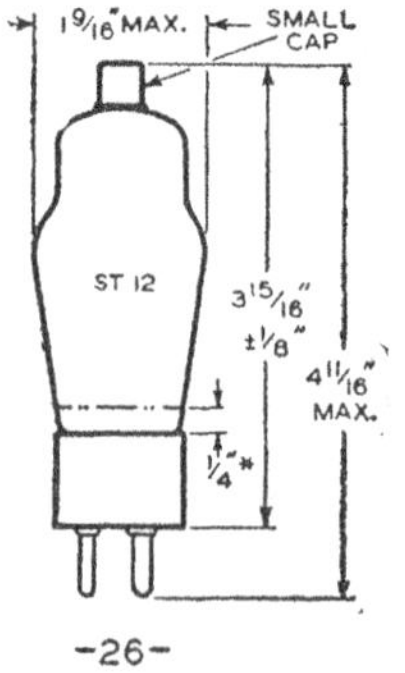

-26-

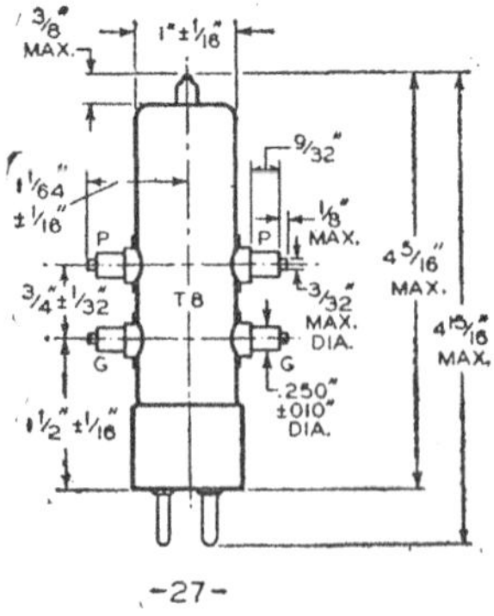

-27-

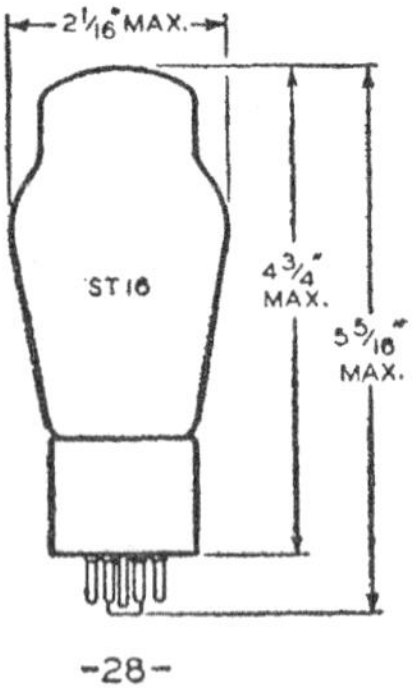

-28-

* Zone where condensed-mercury temperature should be measured.

OUTLINES 29-37

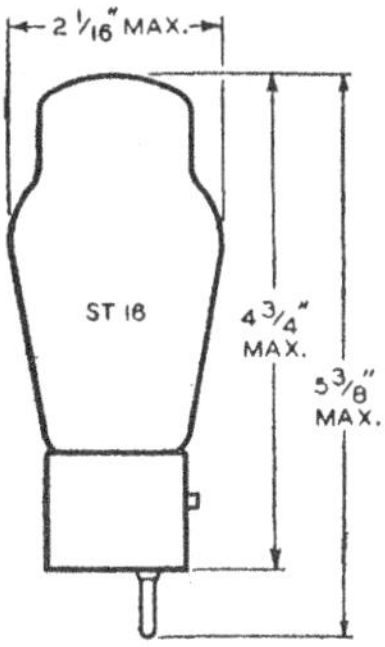

-29-

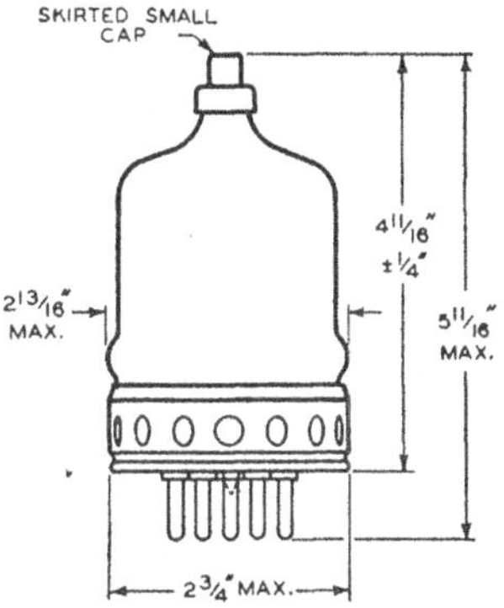

-30-

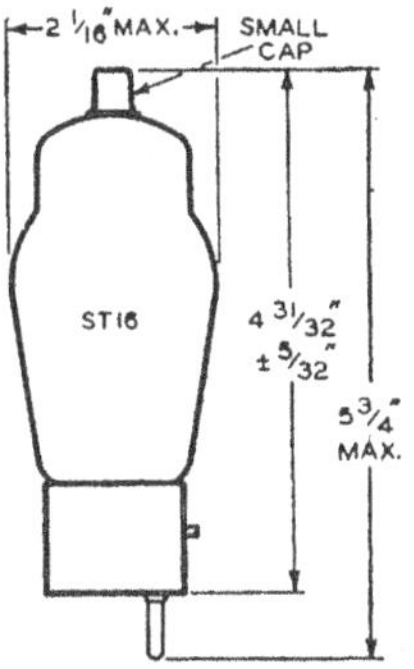

-31-

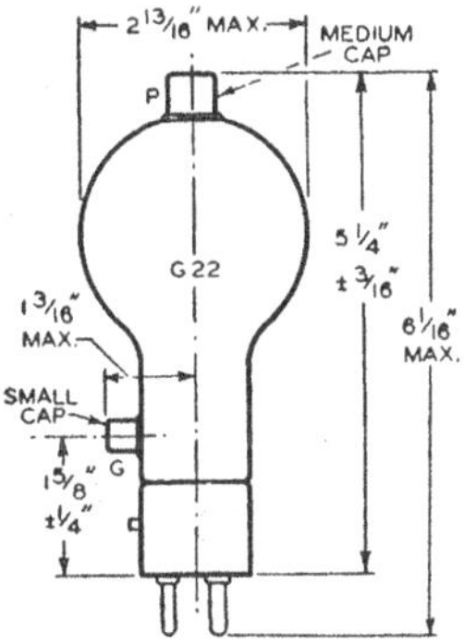

-32-

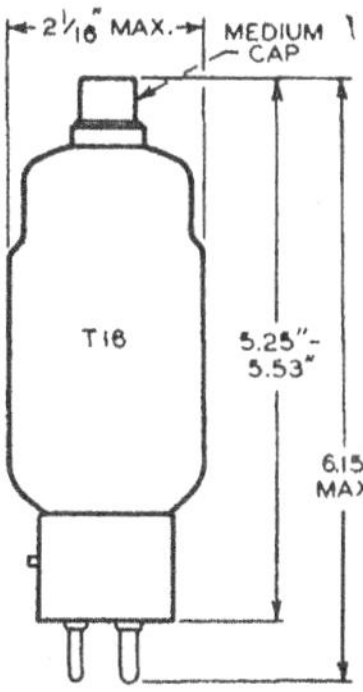

-33-

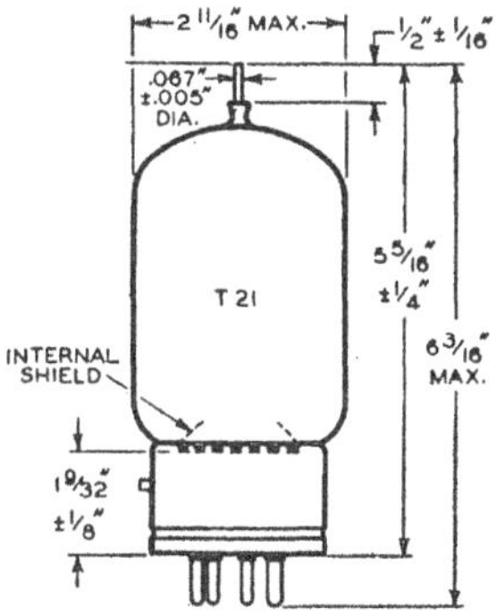

-34-

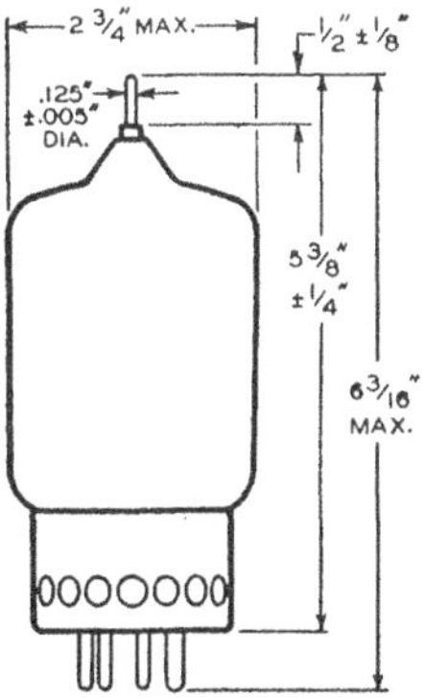

-35-

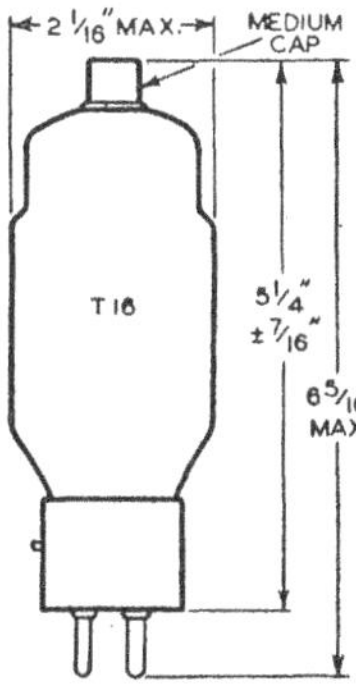

-36-

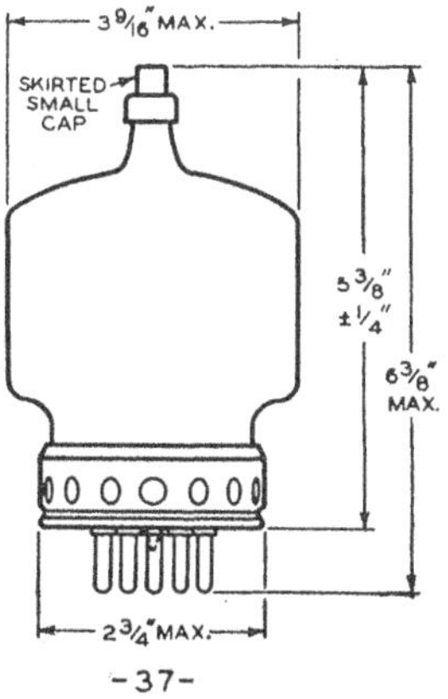

-37-

OUTLINES 38-46

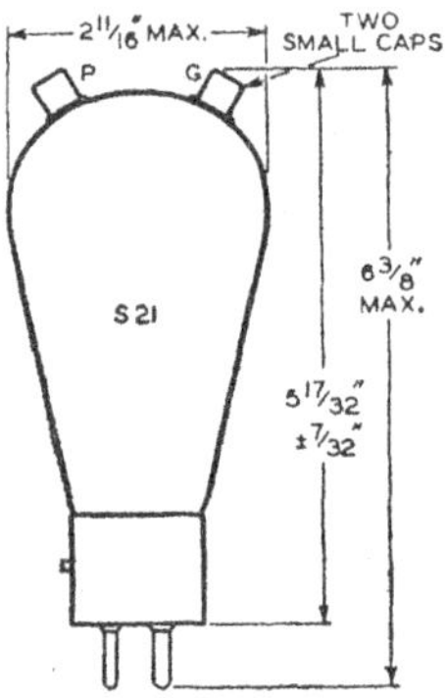

-38-

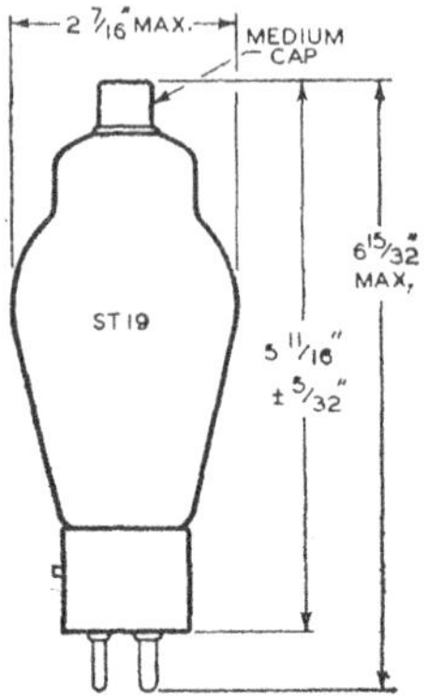

-39-

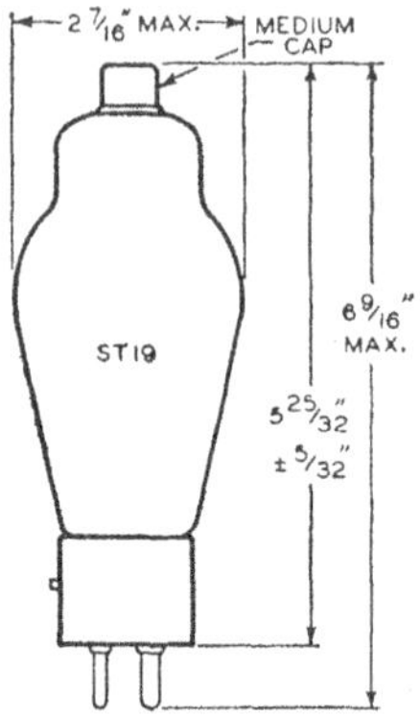

-40-

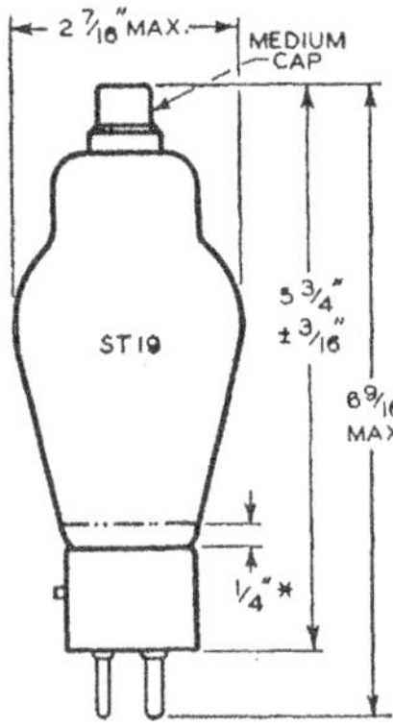

-41-

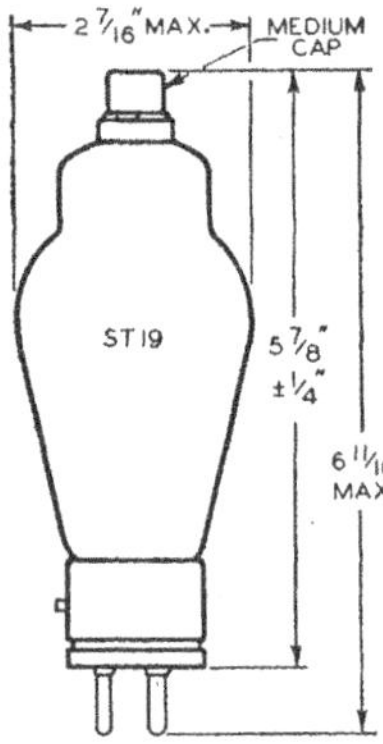

-42-

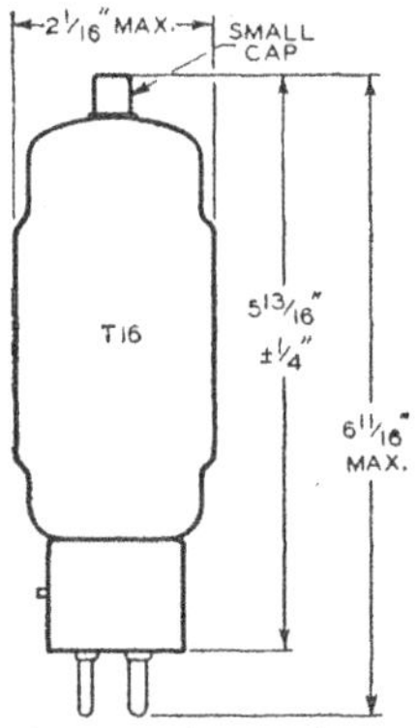

-43-

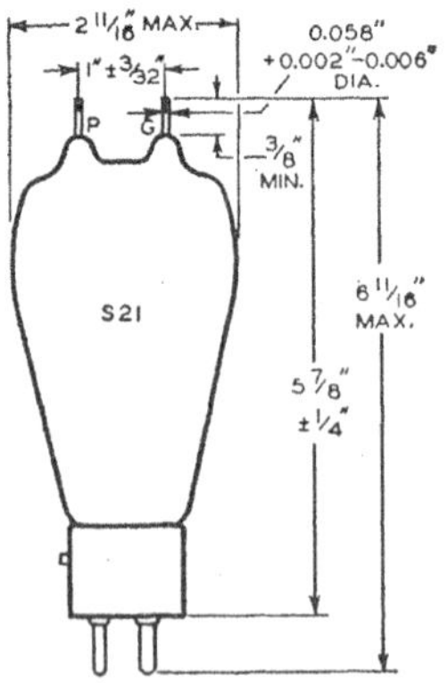

-44-

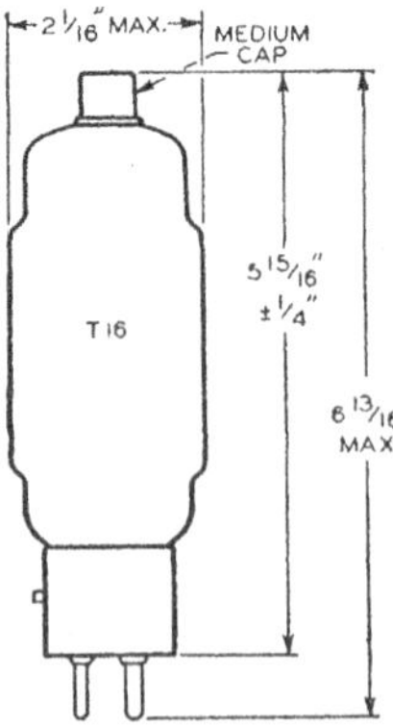

-45-

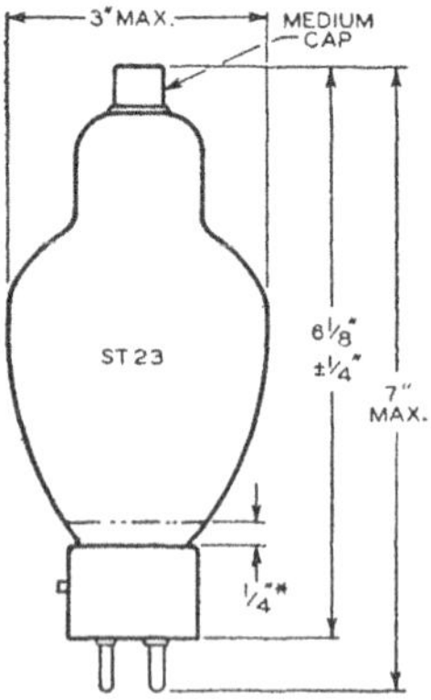

-46-

* Zone where condensed-mercury temperature should be measured.

OUTLINES 47-52

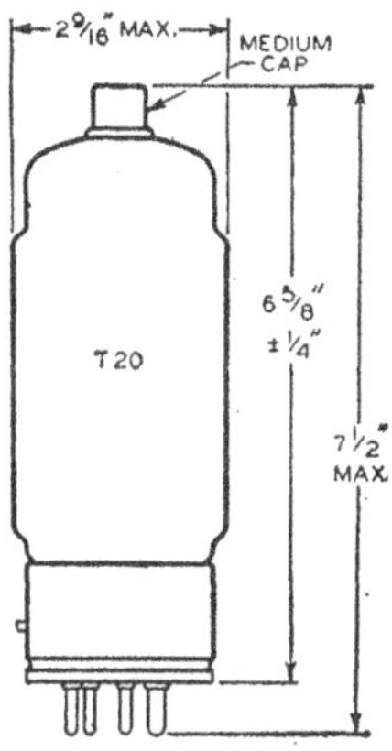

-47-

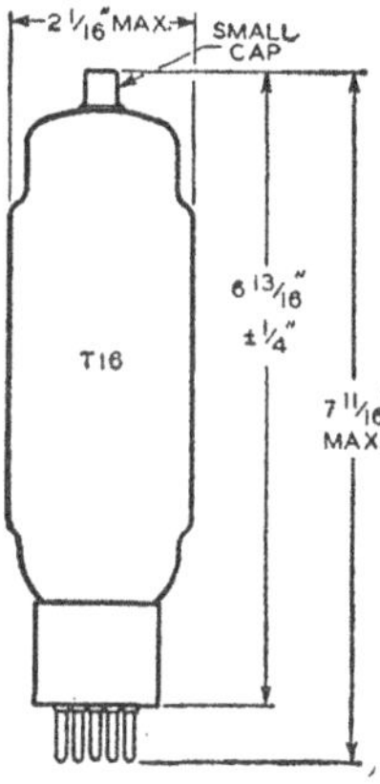

-48-

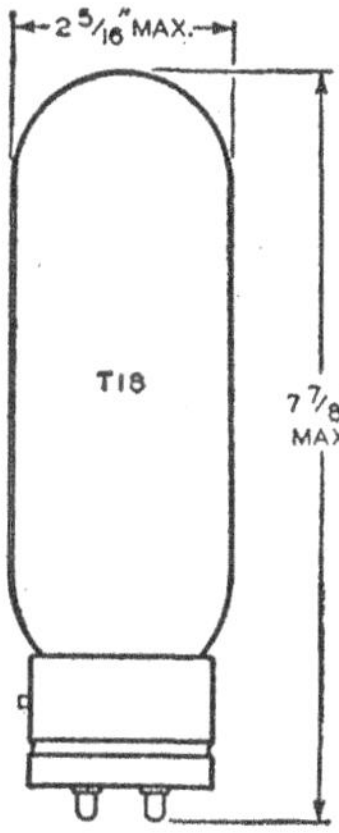

-49-

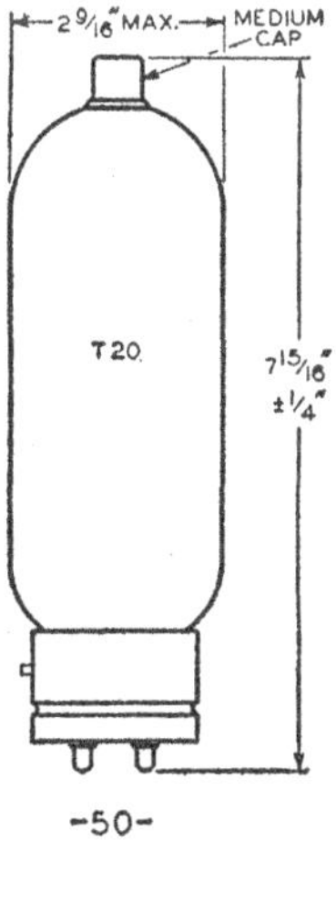

-50-

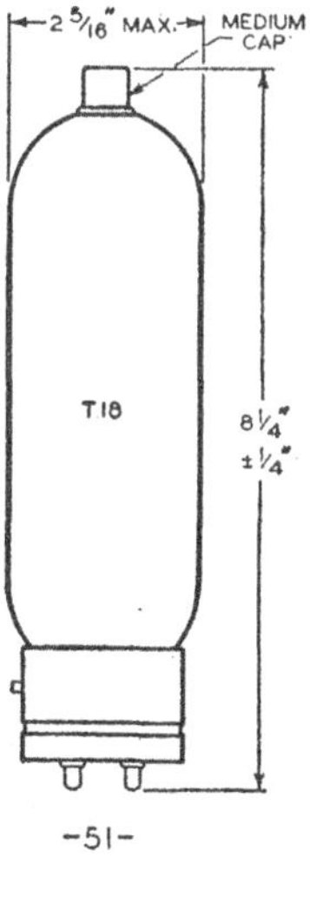

-51-

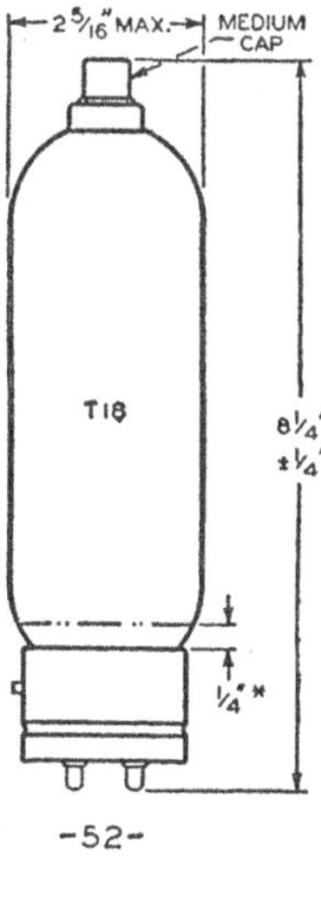

-52-

* Zone where condensed-mercury temperature should be measured.

OUTLINES 53-56

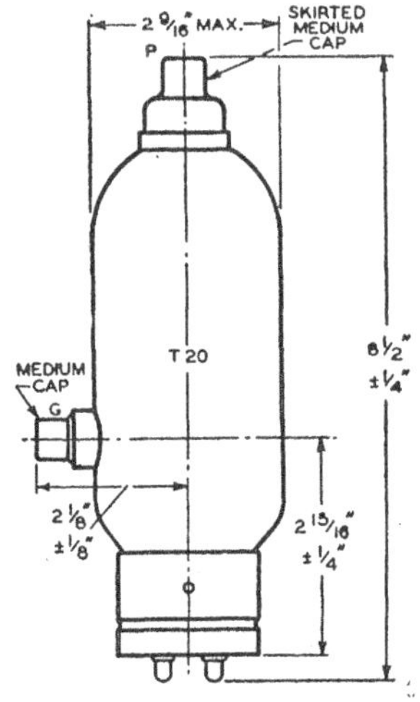

-53-

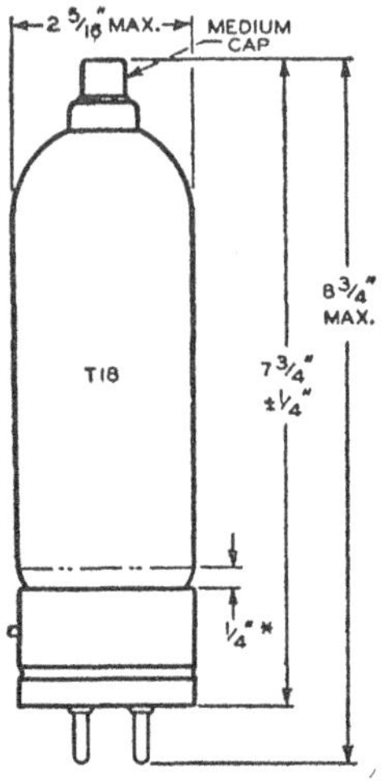

-54-

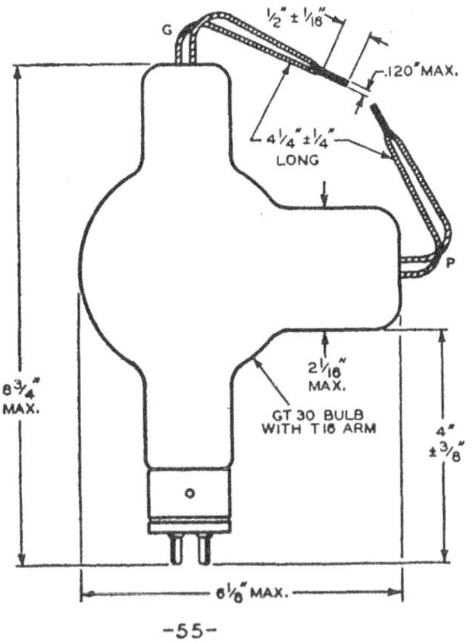

-55-

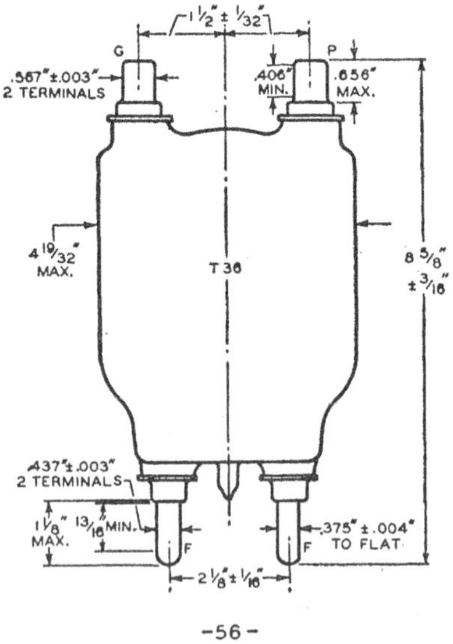

-56-

* Zone where condensed-mercury temperature should be measured.

OUTLINES 57-61

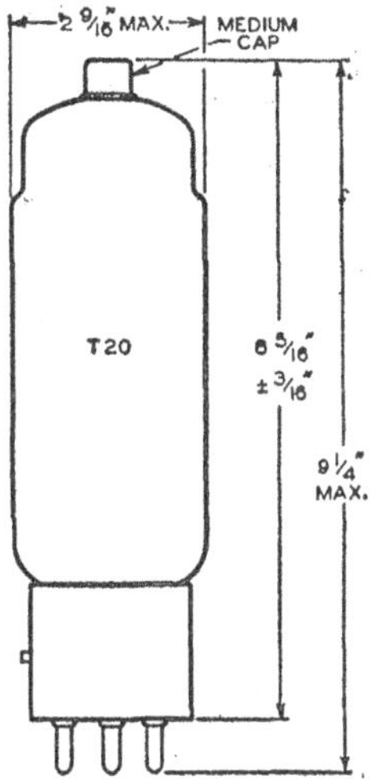

-57-

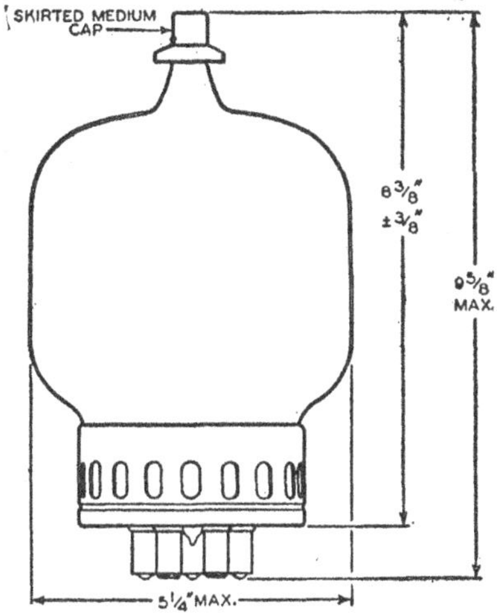

-58-

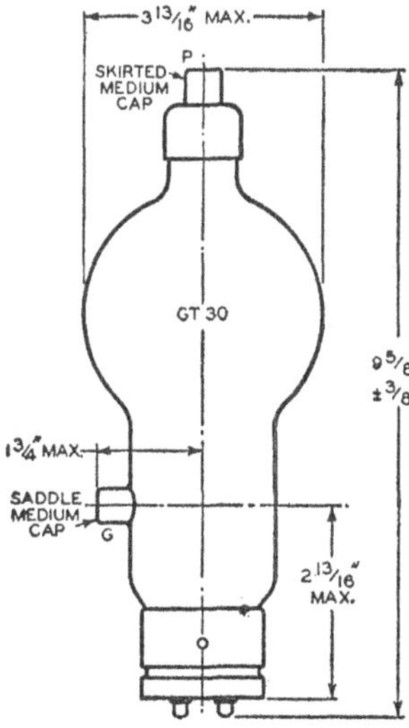

-59-

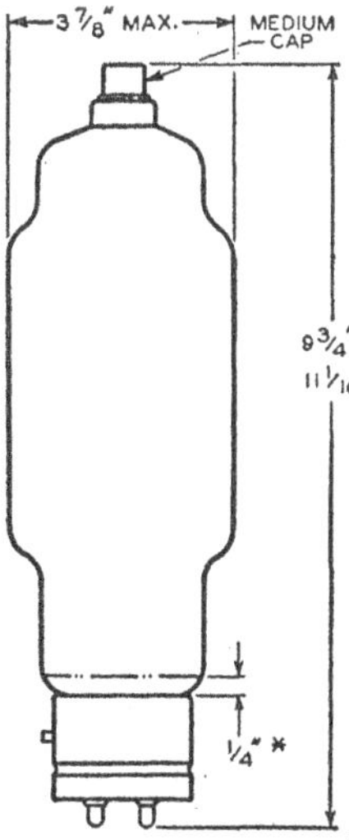

-60-

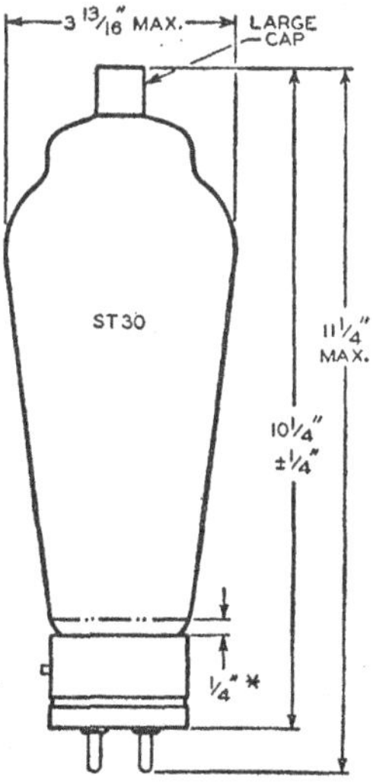

-61-

* Zone where condensed-mercury temperature should be measured.

OUTLINES 62-64

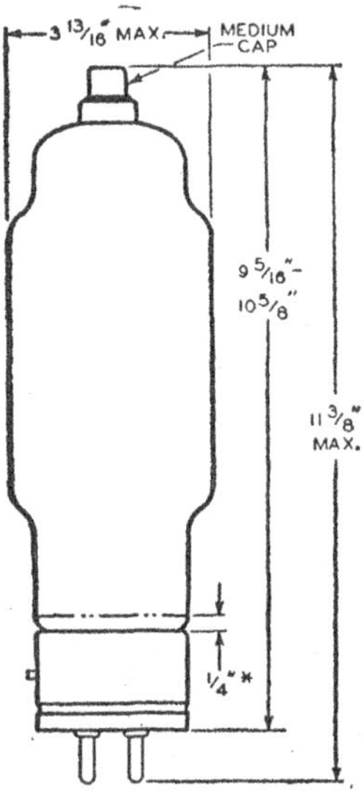

-62-

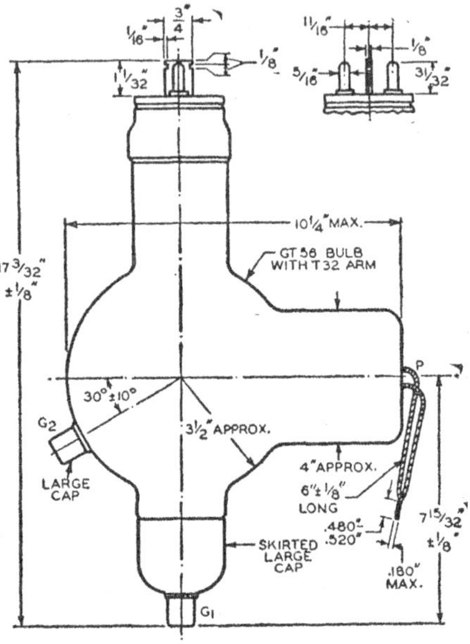

-63-

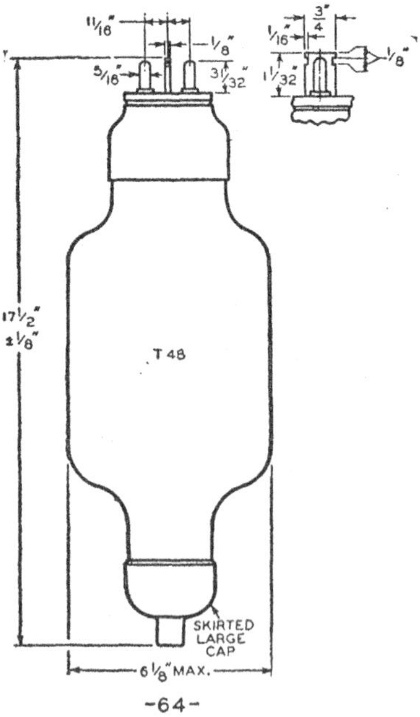

-64-

* Zone where condensed-mercury temperature should be measured.

OUTLINES 65-68

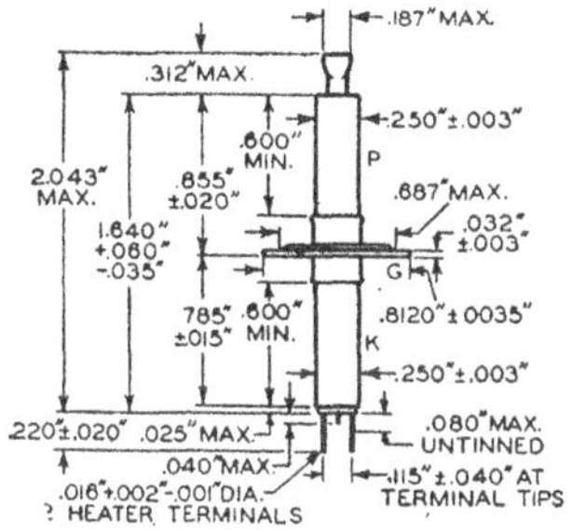

-65-

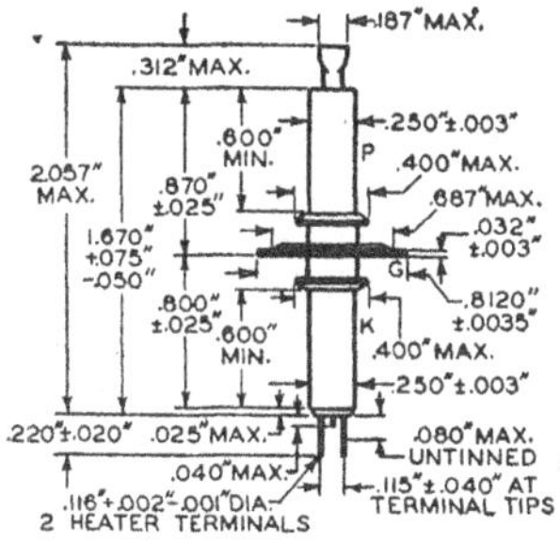

-66-

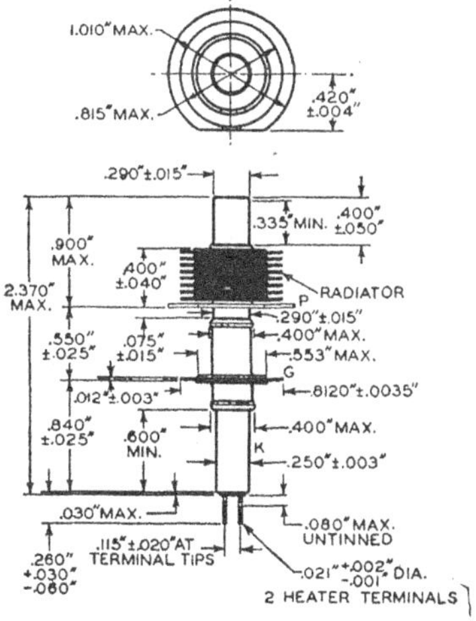

-67-

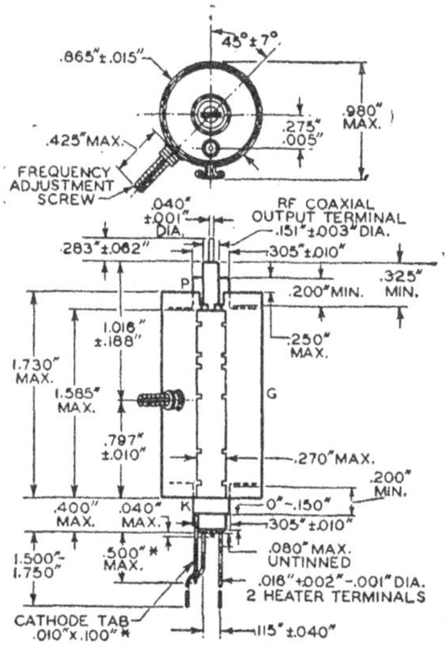

-68-

* Applies to type 6562 only. Type 5794 does not have cathode tab.

OUTLINES 69-72

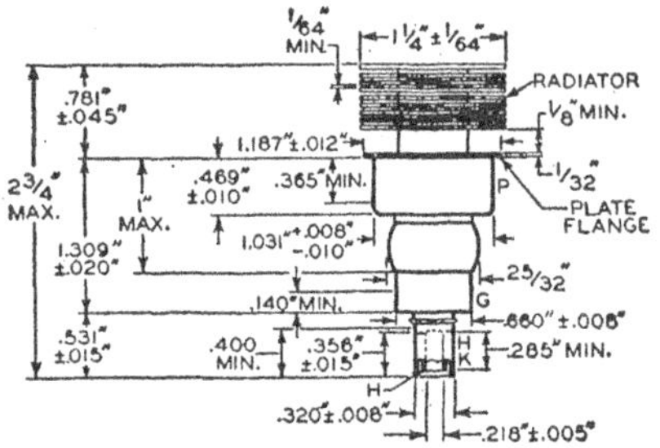

-69-

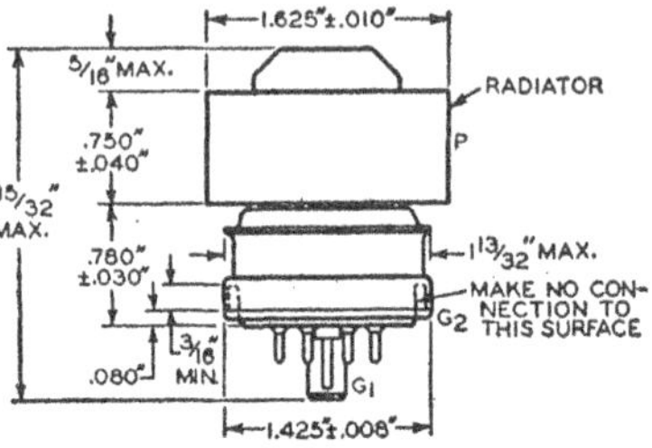

-70-

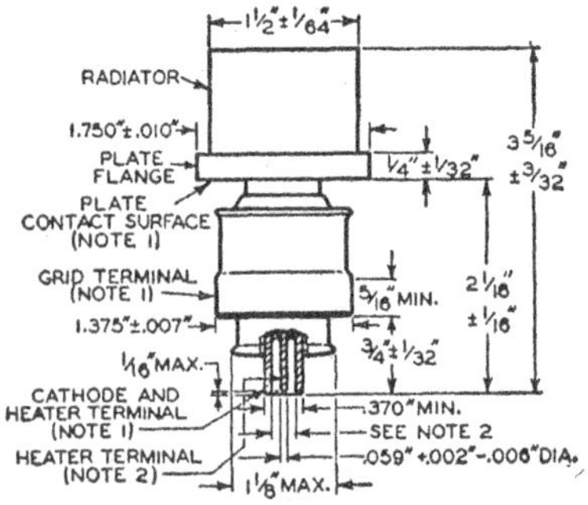

-71-

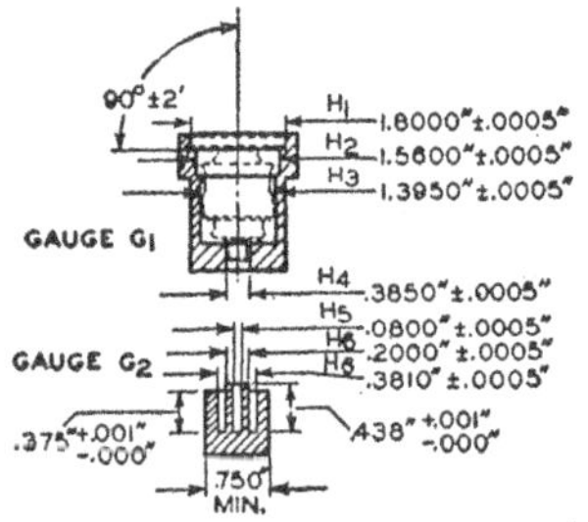

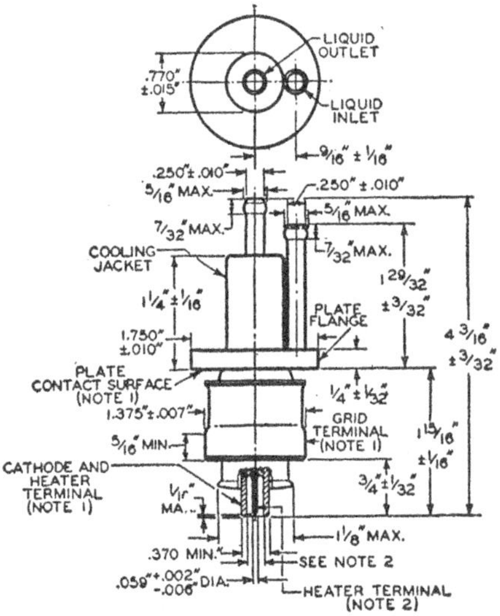

-72-

NOTE 1: With the cylindrical surfaces of its grid and cathode terminals clean, smooth, and free of burrs, the tube will enter a gauge as shown in sketch G_1. The four cylindrical holes H_1, H_2, H_3, and H_4 have axes coincident within 0.0005", lengths determined from the dimensional outline, and successively smaller diameters as shown in the sketch. The plate flange will be entirely engaged by hole H_1, and the contact surface of the plate flange will seat on the shoulder between holes H_1 and H_2. The plane surface of this shoulder is 90° ± 2' to the axes of the holes. Seating is determined by failure of a 0.005" thickness gauge, ⅛" wide, to enter more than 1/16" between the shoulder surface and the plate contact surface. With the tube properly seated as described above, the grid terminal will be entirely engaged by hole H_3, and the cathode terminal will be engaged by hole H_4 to a depth of at least ¼".

NOTE 2: Concentricity of the heater terminal with respect to the cathode terminal is determined by a gauge as shown in sketch G_2. The cylindrical hole H_5 and the annular hole H_6 have axes coincident within 0.0005". The cathode terminal and the heater terminal will enter this gauge to a depth of ⅜".

OUTLINES 73-76

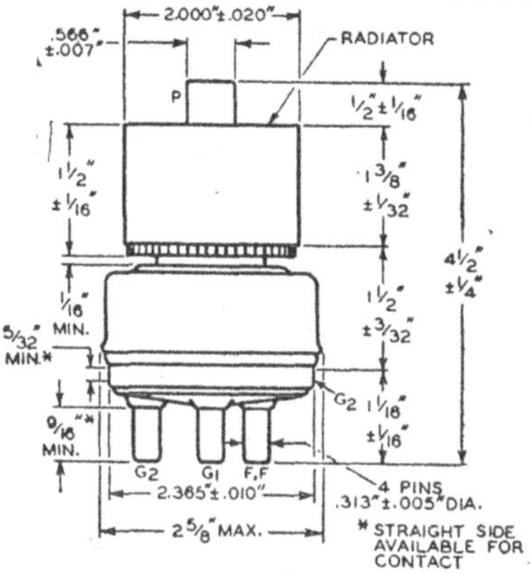

-73-

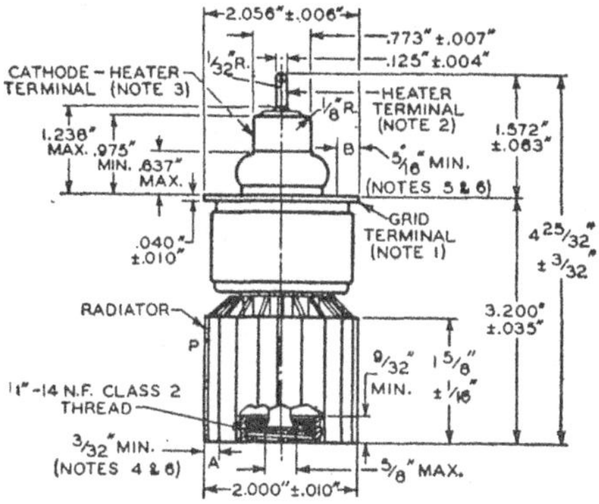

-74-

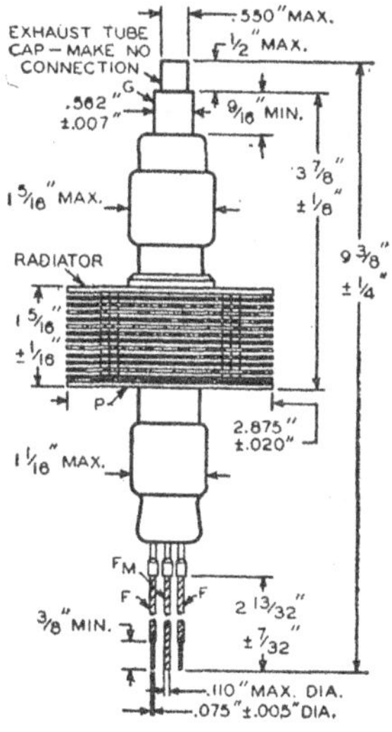

-75-

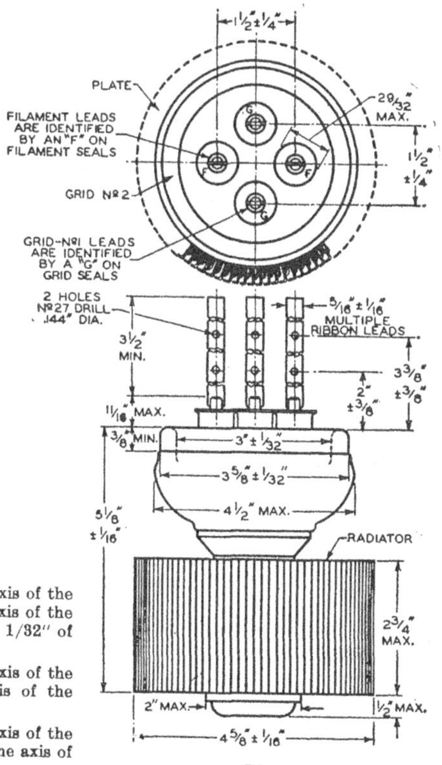

-76-

NOTE 1: Maximum eccentricity of the axis of the grid-terminal flange with respect to the axis of the plate radiator is 0.040", measured within 1/32" of the bottom of the radiator.

NOTE 2: Maximum eccentricity of the axis of the heater terminal with respect to the axis of the cathode-heater terminal is 0.020".

NOTE 3: Maximum eccentricity of the axis of the cathode-heater terminal with respect to the axis of the grid-terminal flange is 0.020".

NOTE 4: Surface of annular area indicated by "A" on bottom of radiator is in the same plane within 0.005", as determined by a gauge 1/16" wide and 0.005" thick. This gauge will not enter more than 1/16" with the bottom of the radiator resting on a flat plate.

NOTE 5: Surface of annular area indicated by "B" on the grid-terminal flange is in the same plane within 0.008", as determined by the gauge method described in Note 4.

NOTE 6: Surface of annular area indicated by "A" on bottom of radiator is parallel within 0.030" to the surface of the annular area indicated by "B" on the grid-terminal flange.

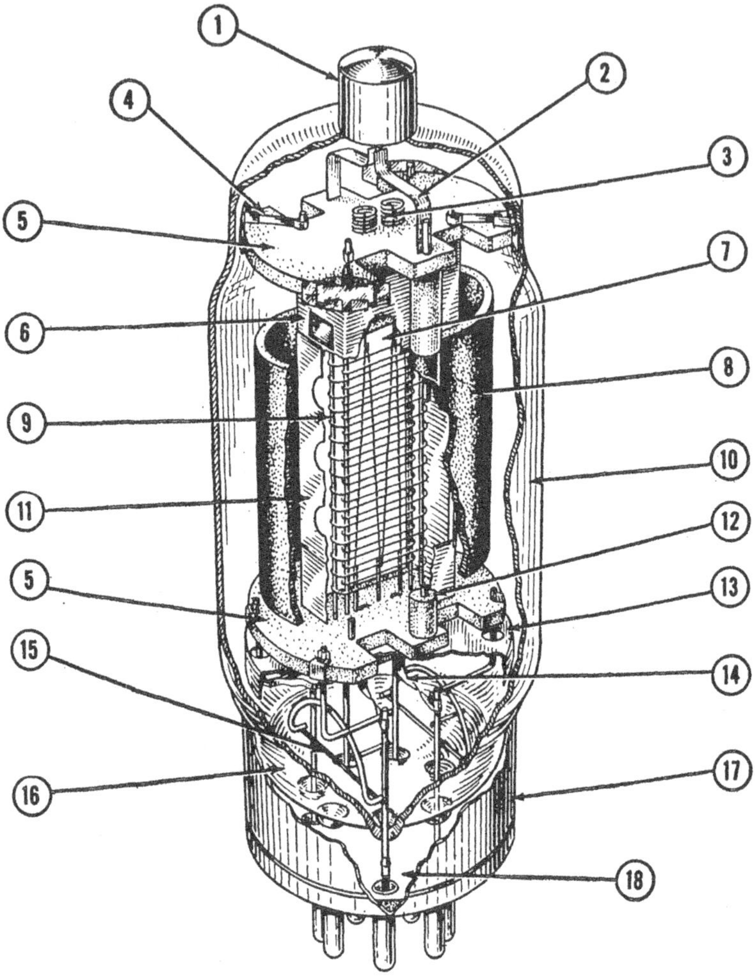

Tube-Part Materials Used in RCA-813 Beam Power Tube

1. *MEDIUM METAL CAP*—nickel-plated brass
2. *SHORT RIBBON PLATE CONNECTOR*—molybdenum
3. *FILAMENT SUPPORT SPRINGS*—tungsten
4. *MOUNT SPACER*—nickel-chromium strip
5. *MOUNT SUPPORT*—ceramic
6. *TOP SHIELD*—nickel
7. *HEAVY-DUTY FILAMENT*—thoriated tungsten
8. *PLATE*—zirconium-coated graphite
9. *ALIGNED-TURN CONTROL GRID (GRID No. 1) AND SCREEN GRID (GRID No. 2)*—molybdenum
10. *BULB OR ENVELOPE*—hard glass
11. *BEAM-FORMING ELECTRODE*—nickel
12. *PLATE-SUPPORT SPACER*—ceramic
13. *BOTTOM SHIELD DISK*—nickel
14. *FILAMENT CONNECTOR*—nickel-plated steel
15. *DIRECTIVE-TYPE GETTER*
16. *MOLDED-FLARE STEM*—hard glass
17. *GIANT BASE*—nickel-plated brass with ceramic insert
18. *TUNGSTEN-TO-GLASS SEAL*

Circuits

The circuits presented in the following pages have been included in this Manual primarily to illustrate the use of generic tube types in diversified transmitting and industrial applications. These circuits have been conservatively designed and are capable of excellent performance. Although relatively few circuits are given, it is often practical to use a portion of one circuit in combination with portions of other circuits to obtain a design meeting specific requirements. In general, almost any circuit shown using a triode, beam power tube, or pentode type is equally suitable for any other tube type in the same generic group, provided the necessary revisions are made to meet the ratings of the tube used.

Electrical specifications are given for the circuit components to assist those interested in home construction. Layouts and mechanical details are omitted because they vary widely with the requirements of individual set builders and with the sizes and shapes of the components employed.

The results that may be expected by those undertaking construction of any of these circuits depend as much on the quality of the components selected and on the care employed in layout, construction, and adjustment as on the circuits themselves.

The voltage ratings specified for capacitors are the minimum dc working voltages required. Where paper, mica, or ceramic capacitors are called for, there is no objection to using capacitors having higher voltage ratings than those specified, except insofar as the physical sizes of such capacitors may affect equipment layout. However, if electrolytic capacitors having substantially higher voltage ratings than those specified are used, they may not "form" completely at the voltages present in these circuits, with the result that the effective capacitances of such units may be below their rated values. The wattage ratings specified for resistors assume methods of construction that provide adequate ventilation; compact installations having poor ventilation may require resistors of higher wattage ratings.

Information on the characteristics and application features of each tube will be found in the *Tube Types* Section of this Manual, or, for the receiving-type tubes, in the *Tube Types* Section of the RCA RECEIVING TUBE MANUAL. This information, as well as the material in the early sections of this Manual on installation, application, and operation of power and rectifier tubes, will prove of assistance in understanding and utilizing the circuits. The following circuits will be found in the subsequent pages:

(4-1)

VARIABLE-FREQUENCY OSCILLATOR

Frequency = 3.5 to 4.0 Mc (80 meters) Output = 3 watts (approx.)

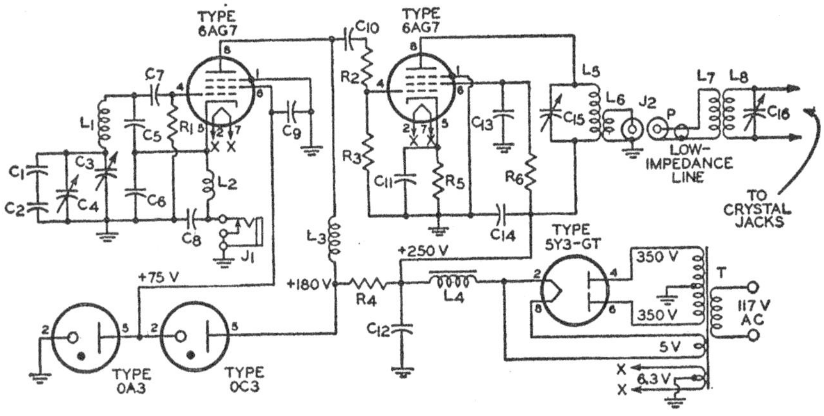

C_1=15 $\mu\mu$f, ceramic, zero temperature coefficient
C_2=100 $\mu\mu$f, ceramic, negative temperature coefficient 750 PPM
C_3=6-75 $\mu\mu$f, trimmer, air gap 0.015 inch, Hammarlund APC-75 or equivalent
C_4=10-75 $\mu\mu$f, trimmer, air gap 0.060 inch, Bud GE-2014 or equivalent
C_5 C_6=0.001 $\mu\mu$f, silver mica, 500 v.
C_7=100 $\mu\mu$f, silver mica, 500 v.
C_8 C_9 C_{11} C_{13} C_{14}=0.01 μf, disk ceramic, 600 v.
C_{10}=15 $\mu\mu$f, silver mica, 500 v.
C_{12}=20 μf, electrolytic 450 v.
C_{15} C_{16}=3-30 $\mu\mu$f, trimmer, mica
J_1=Closed-circuit jack for key
J_2=Coaxial receptacle for P
L_1=28 turns of No. 18 Enam. spaced over 2⅜ inches on 1¾-inch diameter ceramic form, National XR-13 or equivalent
L_2 L_3=2.5 mh, 125 ma, rf choke
L_4=8 henries, 80 ma, choke
L_5=No. 26 Enam., close wound for 13/16 inch on 1-5/16-inch diameter (B & W Miniductor 3016 or equivalent may be used)
L_6=3 turns No. 18 hookup wire wound on L_5 at "cold" end
L_7=56 turns No. 26 Enam. random wound for approx. ¾ inch on 1⅛-inch-diameter coil form
L_8=3 turns No. 18 hookup wire wound over "ground" end of L_7
P=Coaxial plug for J_2
R_1 R_3=100000 ohms, 0.5 watt
R_2=27000 ohms, 0.5 watt
R_4=2000 ohms, 10 watts
R_5=100 ohms, 0.5 watt
R_6=15000 ohms, 1 watt
T=Power transformer; 350-0-350 volts rms, 90 ma; 5 volts rms, 2 amperes; 6.3 volts rms, 3.5 amperes

(4-2)

CRYSTAL OSCILLATOR FOR FUNDAMENTAL OUTPUT

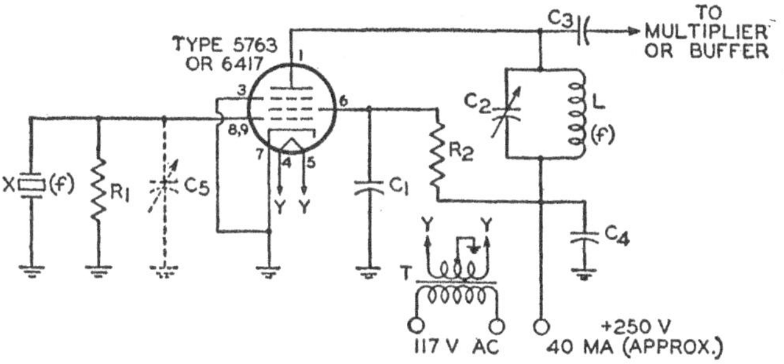

C_1 C_4=0.005 μf, mica, 600 v.
C_2=1.0 μμf per meter (approximate value for resonance at frequency f), variable, air gap 0.015 inch
C_3=50 μμf (approx.), mica (may be in range of 10 to 100 μμf), 600 v.
C_5=3-30 μμf air padder. (Normally omitted. Use only if it is desired to vary operating frequency slightly from crystal frequency)
L=Tune to fundamental frequency f with C_2
R_1=27000 ohms, 0.5 watt
R_2=47000 ohms, 0.5 watt
T=Filament transformer
X=Crystal

(4-3)

CRYSTAL OSCILLATOR FOR HARMONIC OUTPUT

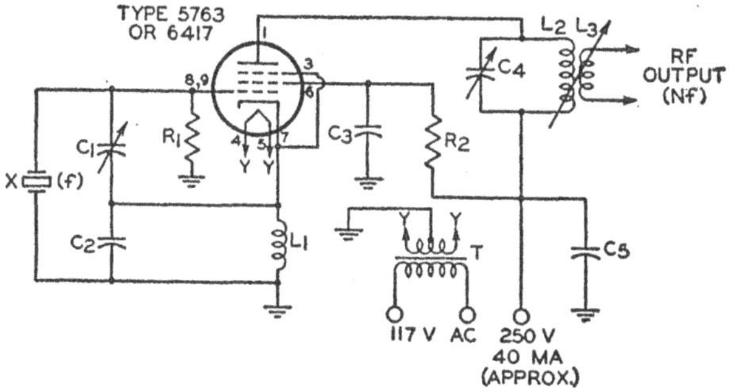

C_1=3-35 μμf, air trimmer
C_2=200 μμf, silver mica, 500 v.
C_3 C_5=0.01 μf, disk ceramic, 600 v.
C_4=1.5 μμf per meter (approximate value for resonance at frequency 2f, 3f, or 4f), variable air gap 0.023 inch
L_1=2.5 mh, rf choke
L_2=Tune to harmonic frequency 2f, 3f, or 4f with C_4 (See note)
L_3=2-turn link at rf ground end of L_2
R_1=100000 ohms, 0.5 watt
R_2=22000 ohms, 0.5 watt
T=Filament transformer
X=Crystal

NOTE: For tank-coil design information, refer to **Parallel-Tuned Tank Circuits** in the *Power-Tube Circuit-Design Considerations* Section

(4-4)

TRIODE AMPLIFIER

Class C Telegraphy Service

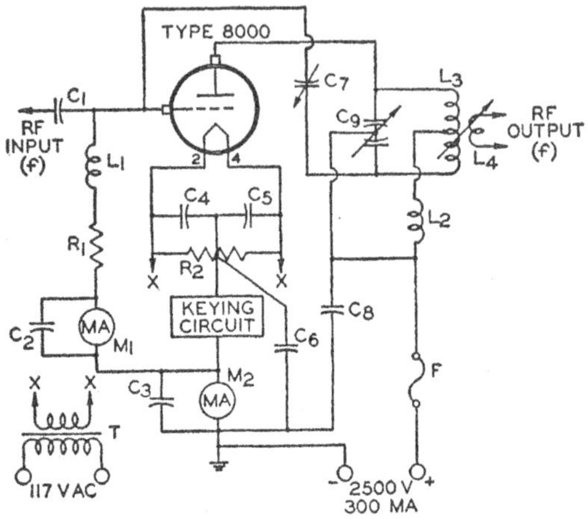

C_1=0.0005 μf, mica, 1500 v.
C_2 C_3 C_4 C_5=0.002 μf, mica, 600 v.
C_6 C_8=0.002 μf, mica, 5000 v.
C_7=5-10 μμf, neutralizing capacitor, air gap 0.3 inch min.
C_9=0.75 μμf per meter per section (approximate value for resonance at frequency f)
F=Fuse, 0.5 amp
L_1=2.5 mh, 100 ma, rf choke
L_2=1 mh, 600 ma, rf choke
L_3=Tune to frequency f with C_9
L_4=2-turn link at center of L_3
M_1=Milliammeter, 0-100 ma, dc
M_2=Milliammeter, 0-500 ma, dc
R_1=6000 ohms, 20 watts
R_2=50 ohms, center-tapped, wire-wound
T=Filament transformer, 10 v., 4.5 amp, insulated for 2500 v.

Keying Circuit: Because this circuit is at a high dc voltage, a relay-type circuit should be used for keying.

(4-5)

BEAM POWER TUBE AMPLIFIER

Class C Telegraphy Service

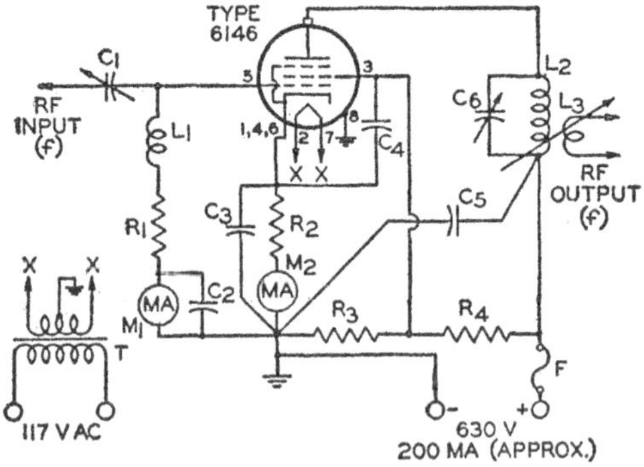

C_1=4-50 μμf, trimmer, air gap 0.015 inch
C_2 C_3 C_4=0.01, disk ceramic, 600 v.
C_5=0.005 μf, mica, 1500 v.
C_6=2 μμf per meter (approximate value, including tube output capacitance, for resonance. For operation above 60 Mc use lowest value which will permit tuning over desired range), air gap 0.075 inch min.
F=Fuse, 0.25 amp
L_1=25 mh, rf choke
L_2=Tune to frequency f with C_6
L_3=2-turn link at rf ground end of L_2
M_1=Milliammeter, 0-10 ma, dc
M_2=Milliammeter, 0-200 ma, dc
R_1=5100 ohms, 1 watt
R_2=390 ohms, 10 watts
R_3=15000 ohms, 10 watts
R_4=25000 ohms, 20 watts
T=Filament transformer, 6.3 v., 1.25 amp

(4-6)

PUSH-PULL TRIODE AMPLIFIER

Class C Plate-Modulated Service

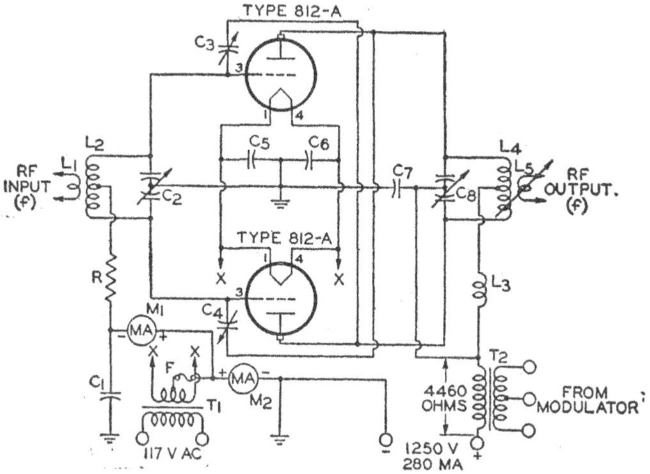

C_1 C_5 C_6=0.005 μf, mica, 600 v.
C_2=2 $\mu\mu$f per meter per section (approximate value for resonance at frequency f), air gap 0.026 inch, min.
C_3 C_4=4-10 $\mu\mu$f neutralizing capacitor, Hammarlund NC-75 or equivalent
C_7=0.002 μf, mica, 5000 v.
C_8=1.5 $\mu\mu$f per meter per section (approximate value for resonance at frequency f), air gap 0.170 inch min.
F=Fuse, 0.5 amp
L_1=3-turn link at center of L_2
L_2=Tune to frequency f with C_2
L_3=2.5 mh, 500 ma, rf choke
L_4=Tune to frequency f with C_8
L_5=3-turn link at center of L_4
M_1=Milliammeter, 0-150 ma, dc
M_2=Milliammeter, 0-500 ma, dc
R=1650 ohms, 20 watts
T_1=Filament transformer, 6.3 v., 8 amp
T_2=Modulation transformer, 125 watts audio level

(4-7)

PUSH-PULL BEAM POWER TUBE AMPLIFIER

Class C Plate-Modulated Service

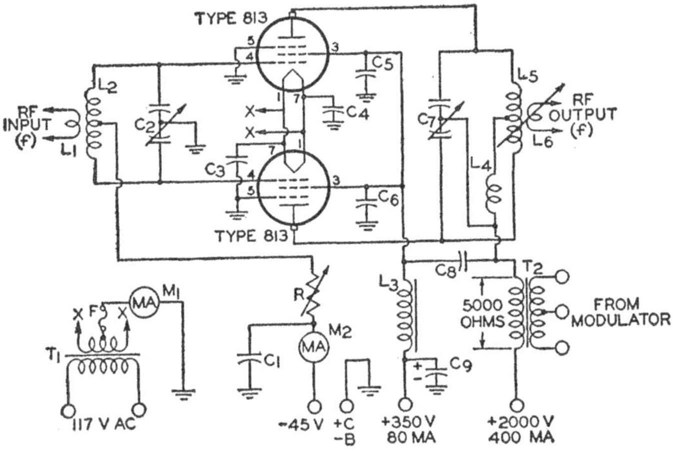

C_1=0.005 μf, mica, 600 v.
C_2=2 $\mu\mu$f per meter per section (approximate value for resonance at frequency f), air gap 0.030 inch min.
C_3 C_4=0.002 μf, mica, 500 v.
C_5 C_6=0.003 μf, mica, 5000 v.
C_7=1.5 $\mu\mu$f per meter per section (approximate value for resonance at frequency f), air gap 0.175 inch min.
C_8=0.002 μf, mica, 6000 v.
C_9=4 μf, electrolytic, 600 v.
F=Fuse, 1 amp
L_1=3-turn link at center of L_2
L_2=Tune to frequency f with C_2
L_3=6 henries, 150 ma, rf choke
L_4=1 mh, 600 ma, rf choke, National R-175 or R-1540, or equivalent
L_5=Tune to frequency f with C_7
L_6=3-turn link at center of L_5
M_1=Milliammeter, 0-800 ma, dc
M_2=Milliammeter, 0-50 ma, dc
R=4000 ohms, adjustable, wire-wound, 25 watts
T_1=Filament transformer, 10 v., 10 amp
T_2=Modulation transformer, 150 watts audio level

(4-8)

CLASS B PUSH-PULL TRIODE MODULATOR

Power Output 590 Watts (Approx.)

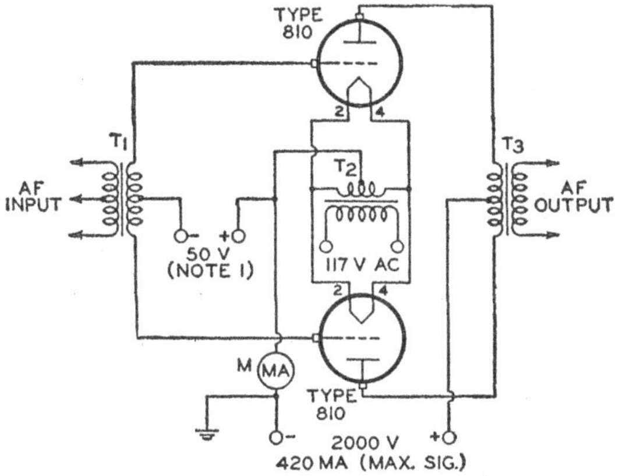

M=Milliammeter, 0-500 ma, dc
T_1=Driver Transformer, plate-to-plate impedance 1500 ohms, turns ratio of total primary to one-half secondary 1.5 to 1 (Note 2)
T_2=Filament transformer, 10 v., 9 amp, center-tapped
T_3=Modulation transformer, load impedance 11000 ohms plate-to-plate; turns ratio depends on modulating impedance of modulated stage

NOTES: 1. This voltage should be obtained from a low-impedance source such as a battery or a power supply having a minimum bleeder current of 100 ma and a minimum filter output capacitance of 150 μf.

2. As the driver for this modulator stage, a circuit having a low output impedance and an output of approximately 25 watts is recommended. For this circuit, four 2A3's in push-pull-parallel Class AB_1, operating with a plate voltage of 300 volts and a fixed bias voltage of -62 volts, with the indicated driver transformer T_1, may be used.

(4-9)

CLASS B MODULATOR WITH TYPE 807 IN SPECIAL TRIODE CONNECTION

Power Output 120 Watts (Approx.)

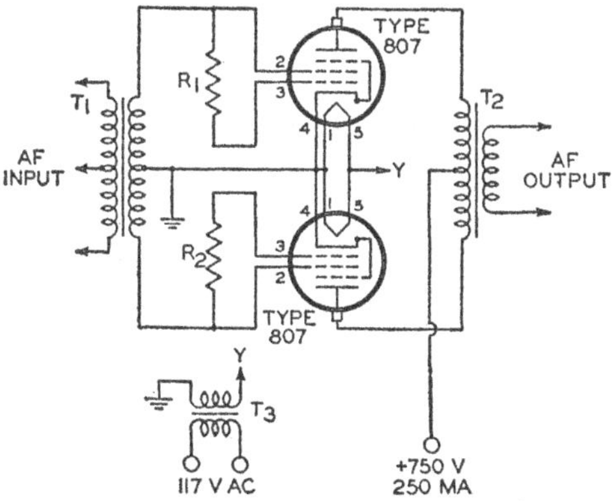

R_1 R_2=20000 ohms, 1 watt, carbon
T_1=Driver transformer, turns ratio of total primary to one-half secondary 1:1.25; Stancor A4761 or equivalent
T_2=Modulation transformer, audio level 120 watts (approx.), primary 6650 ohms (approx.), center-tapped; turns ratio depends on modulating impedance of modulated stage
T_3=Filament transformer, 6.3 volts rms, 1.8 amp

NOTE: As the driver for this modulator stage, a circuit having a low output impedance and an output of approximately 10 watts is recommended. For this circuit, with the indicated driver transformer T_1, two 2A3's in push-pull Class AB_1 operating with a plate voltage of 300 volts and a cathode-bias resistor of 780 ohms may be used.

(4-10)

ELECTRONIC BIAS SUPPLY—30 TO 80 VOLTS

For dc grid-current values to 200 milliamperes

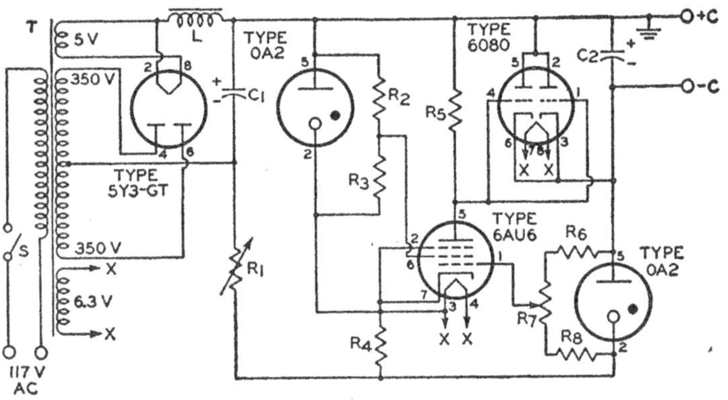

C_1=20 μf, electrolytic, 450 v.
C_2=20 μf, electrolytic, 150 v.
L=8 henries, 50 ma, choke
R_1=Current Balance Control, 5000 ohms, 25 watts, wire-wound (Adjust for 60 volts across R_4)
R_2=24000 ohms, 0.5 watt
R_3=68000 ohms, 0.5 watt
R_4=3000 ohms, 5 watts, wire-wound
R_5=270000 ohms, 0.5 watt
R_6=120000 ohms, 0.5 watt
R_7=Bias control, potentiometer, 100000 ohms
R_8=27000 ohms, 0.5 watt
S=Switch, single-pole, single-throw
T=Power transformer, 350-0-350 volts rms, 50 ma; 5 volts rms, 2 amp; 6.3 volts rms, 3 amp

(4-11)

TWO-METER TRANSMITTER FOR FIXED OR MOBILE OPERATION

Power Output 10 Watts (Approx.)

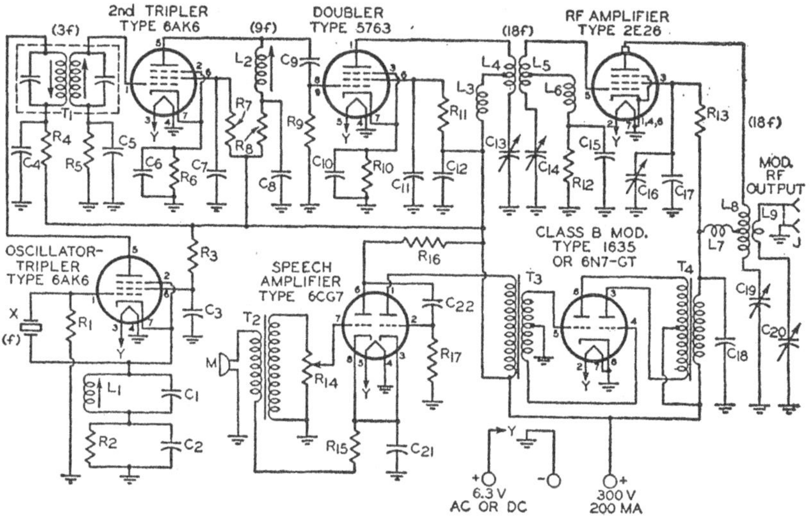

C_1=150 μμf, mica, 600 v.
C_2 C_5 C_{22}=0.005 μf, disk ceramic, 600 v.
C_3 C_4 C_6 C_7 C_{10} C_{11}=0.004 μf, disk ceramic, part of twin capacitor, 600 v.
C_8 C_{12}=0.005 μf, disk ceramic, 1000 v.
C_9=10 μμf, mica, 600 v.
C_{13} C_{14} C_{19} C_{20}=3-25 μμf, trimmer, air gap 0.015 inch
C_{15}=100 μμf, mica, 600 v.
C_{16}=4-30 μμf, trimmer, ceramic
C_{17}=47 μμf, mica, 600 v.
C_{18}=500 μμf, ceramic, feed-through, 500 v.
C_{21}=25 μf, electrolytic, 25 v.
J=Coaxial connector
L_1=15 turns of No. 18 Enam. close wound on ½-inch diameter form, National XR-50 or equivalent, slug tuned
L_2=5 turns of No. 14 Enam. spaced over 11/16 inch on ½-inch diameter form, National XR-50 or equivalent, slug tuned
L_3=40-inch length of No. 32 Enam. close wound on ¼-inch diameter form, rf choke
L_4=5 turns of No. 14 Enam. on ½-inch diameter, space between turns equal to wire diameter
L_5=3 turns of No. 14 Enam. on ½-inch diameter, space between turns equal to wire diameter
L_6 L_7=40-inch length of No. 32 Enam. wire wound on ½-inch diameter form, rf choke
L_8=3 turns of No. 10 Enam. on ¾-inch diameter, winding length 1½ inches
L_9=1 turn of No. 10 Enam. on 1-inch diameter
M=Microphone, single button, carbon
R_1=100000 ohms, 0.5 watt
R_2 R_4 R_6 R_8=1000 ohms, 0.5 watt
R_3 R_7 R_{16} R_{17}=47000 ohms, 0.5 watt
R_5=3300 ohms, 0.5 watt
R_9=82000 ohms, 1 watt
R_{10}=68 ohms, 0.5 watt
R_{11}=22000 ohms, 0.5 watt
R_{12}=33000 ohms, 1 watt
R_{13}=20000 ohms, 1 watt
R_{14}=Volume control, potentiometer, 1 megohm
R_{15}=560 ohms, 0.5 watt
T_1=21.25 Mc TV sound if transformer, RCA-206K1 or equivalent
T_2=Microphone-to-grid transformer, primary 200 or 70 ohms, secondary 80000 ohms, Stancor A4705 or equivalent
T_3=Driver transformer, turns ratio primary to one-half secondary 5.2:1, Thordarson T20D76 or equivalent
T_4=Modulation transformer, audio level 10 watts, primary 10000 ohms, center-tapped, secondary 4500 ohms, Thordarson T21M52 or equivalent
X=Crystal, 8 Mc

(4-12)

TEN-METER TRANSMITTER FOR MOBILE OPERATION

Power Output 10 Watts (Approx.)

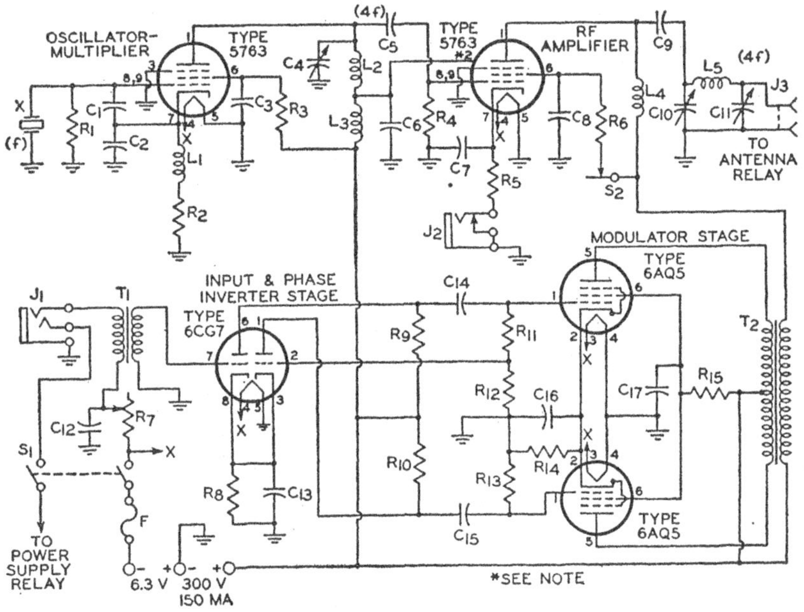

C_1=15 μμf, mica, 500 v.
C_2=50 μμf, mica, 500 v.
C_3 C_7 C_8 C_9=0.001 μf, mica, 500 v.
C_4=4-25 μμf, variable, air gap 0.030 inch
C_5=50 μμf, ceramic
C_6=100 μμf, mica, 600 v.
C_{10}=5-50 μμf, variable, air gap 0.071 inch
C_{11}=5-100 μμf, variable, air gap 0.015 inch
C_{12}=50 μf, electrolytic, 6 v.
C_{13}=10 μf, electrolytic, 25 v.
C_{14} C_{15}=0.01 μf, paper, 400 v.
C_{16}=20 μf, electrolytic, 25 v.
C_{17}=4 μf, electrolytic, 300 v.
F=Fuse, 3 amp

J_1=3-circuit microphone jack
J_2=Closed-circuit jack
J_3=Coaxial connector
L_1 L_3=2.5 mh, rf choke
L_2 L_5=10 turns on ¾-inch diameter, winding length 1½ inches, made from B & W Miniductor 3010
L_4=21 μh, choke, Ohmite Z28 or equivalent
R_1 R_9 R_{10}=100000 ohms, 0.5 watt
R_2=500 ohms, 1 watt
R_3=66000 ohms, 2 watts
R_4=20000 ohms, 1 watt
R_5=68 ohms, 0.5 watt
R_6 R_{15}=10000 ohms, 2 watts
R_7=Potentiometer, 1000 ohms, wire-wound, 2 watts

R_8=3300 ohms, 0.5 watt
R_{11} R_{13}=200000 ohms, 0.5 watt
R_{12}=15000 ohms, 0.5 watt
R_{14}=250 ohms, 2 watts
S_1=Switch, double-pole single-throw
S_2=Momentary push-switch, normally closed
T_1=Microphone-to-grid transformer, primary 100 ohms, secondary 60000 ohms, Stancor A-4706 or equivalent
T_2=Modulation transformer, audio level 10 watts, primary 10000 ohms center-tapped, secondary 4500 ohms, Thordarson T21M52 or equivalent
X=Crystal 7 Mc (approx.)

NOTE: Neutralizing connection is made to pin 2 of socket. Base pin 2 of 5763 has no internal connection.

(4-13)

462-MEGACYCLE TRANSMITTER FOR FIXED OR MOBILE OPERATION

Power Output 20 Watts (Approx.)

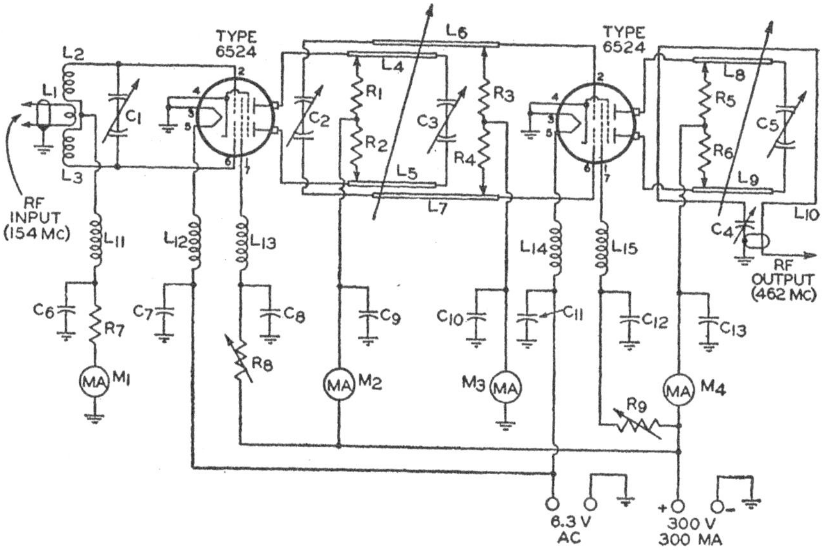

C_1 C_2=2.2-8.0 μf per section, variable, butterfly, air gap 0.017 inch, Johnson 9MB11 or equivalent

C_3 C_5=2.7-10.8 $\mu\mu$f per section, variable, butterfly, air gap 0.017 inch, Johnson 11MB11 or equivalent

C_4=1.5-5.0 $\mu\mu$f, variable, air gap 0.017 inch, Johnson 5M11 or equivalent

C_6 C_7 C_8 C_9 C_{10} C_{11} C_{12} C_{13}=1500 $\mu\mu$f, feed-through ceramic, Erie 362-152 or equivalent

L_1=1 turn of No. 10 base copper wire, wound on ½-inch diameter

L_2 L_3=1½ turns of No. 10 base copper wire close-wound on ½-inch diameter. L_2 and L_3 are spaced to accommodate L_1

L_4 L_5 L_8 L_9=Silver-plated copper rod 3/16-inch diameter approximately 3 inches long. Rods of each pair spaced 11/16 inch on centers

L_6 L_7=Silver-plated copper rod 3/16-inch diameter approximately 1½ inches long. Rod spaced 1 inch on centers

L_{10}=1 turn of No. 8 silver-plated copper wire approximately 1 inch square

L_{11} L_{12} L_{13} L_{14} L_{15}=RF choke, Ohmite Z-460 or equivalent

M_1 M_3=Milliammeter, 0-5 ma, dc

M_2 M_4=Milliammeter, 0-150 ma, dc

R_1 R_2 R_5 R_6=57 ohms, 1 watt

R_3 R_4=25000 ohms, 0.25 watt

R7=51000 ohms, 0.5 watt

R_8 R_9=Potentiometer, 20000 ohms, 2 watts

NOTE: Suitable tube sockets are Johnson 122-248 or equivalent mounted 9/16 inch below chassis. For detailed operating conditions of this circuit, refer to type 6524 in the *Tube Types* Section where typical operation values for Intermittent Commercial and Amateur Service (ICAS) are given for both the tripler and final at 462 Mc.

(4-14)

OSCILLATOR FOR DIELECTRIC HEATING

Frequency 27 Mc (Approx.)

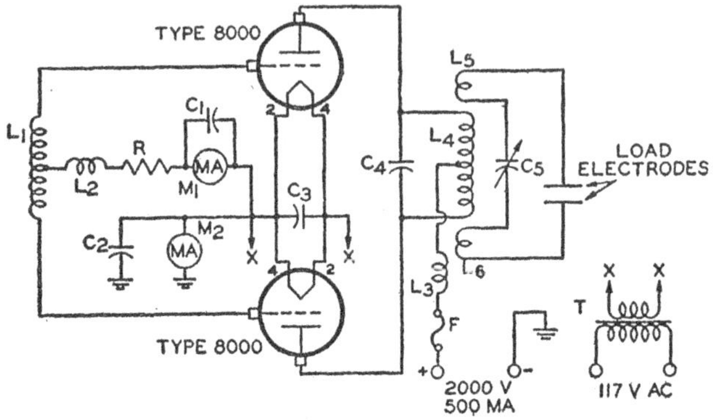

C_1 C_2 C_3=0.005 μf, mica, 600 v.
C4=2 plates 3/32-inch aluminum, 5 inches x 7 inches spaced 7/8 inch
C_5=50 μμf, max., depends on work load
F=Fuse, 0.5 amp
L_1=5 turns 3/16-inch copper tubing spaced 3/4 inch on 2 1/4-inch I.D.
L_2=RF choke, 40 ma
L_3=RF choke, 500 ma
L_4=3 turns 5/16-inch copper tubing spaced 5/8 inch on 3 3/4-inch I.D.
L_5 L_6= 2 turns 3/16-inch copper tubing with adjustable spacing between turns on 3 3/4-inch I.D.
M_1=Milliammeter, 0-100 ma, dc
M_2=Milliammeter, 0-1000 ma, dc
R=5000 ohms, 25 watts
T=Filament transformer, 10 volts rms, 9 amp

NOTE: Adequate shielding should be used to assure compliance with FCC requirements regarding spurious radiation.

(4-15)

OSCILLATOR FOR INDUCTION HEATING

Frequency 450 Kc (Approx.)

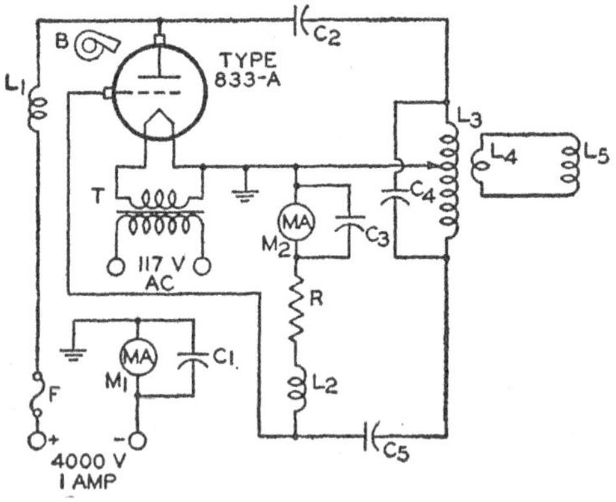

C_1 C_3=0.01 μf, mica, 600 v.
C_2 C_5=0.1 μf, paper, 5000 v., 0.6 amp rms min.
C_4=0.002 μf, mica, 8000 volts min., 15 amp rms
F=Fuse, 1 amp
L_1= 3mh, rf choke, 1 amp rms, insulated for 10000 peak volts, single-layer solenoid, 300 turns No. 18 Enam., 12 inches long on 4-inch diameter
L_2=3.5 mh, rf choke, 250 ma
L_3=63 μf, choke, 15 amp rms, insulated for 5000 peak volts, 40 turns No. 8 Enam., 8 inches on 4-inch diameter form.
L_4=Single-turn secondary, sheet copper
L_5=Work coil
M_1=Milliammeter, 0-1000 ma, dc
M_2=Milliammeter, 0-150 ma, dc
R=2500 ohms, 50 watts
T=Filament transformer, 10 volts rms, 10 amp
B=Blower, designed to supply an air flow of 40 cfm from a 2-inch-diameter nozzle directed vertically on bulb between grid and plate seals.

NOTE: Adequate shielding should be used to assure compliance with FCC requirements regarding spurious radiation.

(4-16)

VHF OSCILLATOR FOR DIELECTRIC HEATING

Frequency 160 Mc (Approx.)

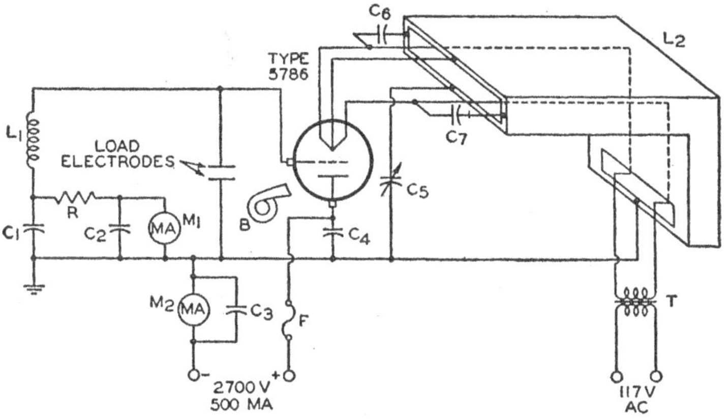

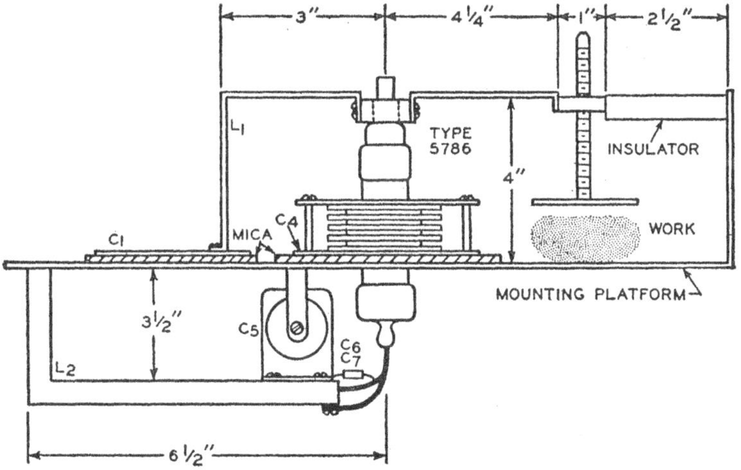

C_1=250 $\mu\mu$f, mica 0.005 inch thick, 3 inches x 3¾ inches copper plate, held to mounting platform by insulated pressure clamps
C_2 C_3=0.001 μf, mica, 600 v.
C_4=200 $\mu\mu$f, mica 0.005 inch thick, 4 inches x 5 inches copper plate, held to mounting platform by insulated pressure clamps
C_5=10-30 $\mu\mu$f, variable, consisting of copper plate 3 inches x 3½ inches mounted on L_2 and round disk 3 inches in diameter, air gap ¼ inch to 1 inch
C_6 C_7=100 $\mu\mu$f, mica ("postage stamp"), 600 v.
F=Fuse, 0.5 amp
L_1=Copper strap 1-3/16 inches wide x 1/16 inch thick
L_2=½ inch x 1 inch rectangular waveguide or equivalent
M_1=Milliammeter, 0-150 ma, dc
M_2=Milliammeter, 0-750 ma, dc
R=2000 ohms, wire-wound, 50 watts
T=Filament transformer, 11 volts rms, 12.5 amp, maximum starting surge 50 amp
B=Blower, designed to supply an air flow of at least 140 cfm through an outlet area of 6¼ square inches to the radiator and the filament and grid seals.

NOTE: Entire oscillator and load assembly is enclosed in metal box having one end open for cooling-air exit and for ease of loading work. Mounting platform divides box into two compartments. See tube data for RCA-5786 forced-air-cooling requirements. Tube and circuit must be protected from fumes or vapors that may come from work. Adequate shielding should be used to assure compliance with FCC requirements regarding spurious radiation.

Notes:

INDEX

Notes:

Notes:

RCA Tube Division Technical Publications

Copies of the publications listed below may be obtained from your RCA Tube Distributor, or direct from Commercial Engineering, Tube Division, Radio Corporation of America, Harrison. New Jersey.

Electron Tubes

• **RCA TUBE HANDBOOK**—HB-3 (7⅜"x 5"). Five deluxe 2-inch-capacity binders imprinted in gold. The bible of the industry—contains over 3100 pages of loose-leaf data and curves on RCA receiving tubes, picture tubes, cathode-ray tubes, phototubes, special tubes, and semiconductor devices. Available on subscription basis. Price $17.50* including service for first year. Write to Commercial Engineering for descriptive folder and order form.

• **RCA TRANSMITTING TUBES** — TT-4 (8⅜" x 5⅜")—256 pages. Written for the engineer, service technician, radio amateur, student, and experimenter. Contains basic information on generic tube types, on tube parts and materials, on tube installation and application, and on interpretation of tube data. Includes maximum ratings, typical operating values, and characteristics curves for power tubes having plate-input ratings up to 4 kilowatts, and maximum ratings and operating values for associated rectifier tubes. Contains sections on transmitter-design considerations and on rectifier circuits and filters. Features classification charts for quick, easy selection of tubes, and circuit diagrams for transmitting and industrial applications. Features lie-flat binding. Price $1.00.*

• **RCA RECEIVING TUBE MANUAL**—RC-17 (8⅜" x 5⅜")—336 pages. Revised, expanded, and brought up to date. Contains the latest receiving tubes, including types for black-and-white and color television applications. Features tube theory written for the layman, application data, Resistance-Coupled Amplifier Section, and several new circuits for high-fidelity audio amplifiers. Features lie-flat binding. Price 60 cents.*

• **RADIOTRON† DESIGNER'S HANDBOOK** —4th Edition (8¾" x 5½")—1500 pages. Comprehensive reference thoroughly covering the design of radio and audio circuits and equipment. Written for the design engineer, student, and experimenter. Contains 1000 illustrations, 2500 references, and cross-referenced index of 7000 entries. Edited by F. Langford-Smith of Amalgamated Wireless Valve Co., Pty., Ltd. in Australia. Price $7.00.*

• **RCA POWER AND GAS TUBES**—PG-101C (10⅞" x 8⅜")—24 pages. Completely revised and brought up to date. Technical information on 174 RCA vacuum power tubes, rectifier tubes, thyratrons, ignitrons, magnetrons, and vacuum-gauge tubes. Includes terminal connections. Price 20 cents.*

• **RECEIVING-TYPE TUBES FOR INDUSTRY AND COMMUNICATIONS** — RT-104 (10⅞" x 8⅜")—20 pages. Technical information on 130 RCA "special red" tubes, premium tubes, computer tubes, pencil tubes, glow-discharge tubes, small thyratrons, low-microphonic amplifier tubes, and other special types. Includes socket-connection diagrams. Price 20 cents.*

• **RCA RECEIVING TUBES FOR AM, FM, AND TELEVISION BROADCAST**—1275-G (10⅞" x 8⅜")—28 pages. New booklet contains classification chart, characteristics chart, and base and envelope connection diagrams on more than 600 entertainment receiving tubes and picture tubes. Price 25 cents.*

• **RCA PHOTOTUBES**—PT-20R1 (10⅞" x 8⅜")—16 pages. Phototube theory, data on 15 types, curves and circuits for light-operated relays, light measurements, and sound reproduction. Single copy free on request.

†Trade Mark Reg. U. S. Pat. Off.

*Prices shown apply in U.S.A. and are subject to change without notice.

• RCA PHOTOSENSITIVE DEVICES AND CATHODE-RAY TUBES — CRPD-105 (10⅞" x 8⅜")—24 pages. Contains technical information on 109 RCA tubes including single-unit, twin-unit, and multiplier phototubes; flying spot tubes; monitor, projection, transcriber, and view-finder kinescopes; and storage tubes. Price 20 cents.*

• RCA PICTURE TUBES—KB-106 (10⅞" x 8⅞")—16 pages. Contains characteristics and base-connection diagrams for RCA's complete line of picture tubes. Features an interchangeability directory on more than 150 types. Price 20 cents.*

• RCA TUBE PICTURE BOOK—TPB-1 (10⅞" x 8⅜")—16 pages. Collection of photographs and cutaway drawings of representative tube types. Prepared especially for use by students. A visual aid for the details of tube construction. Price 25 cents.*

• RCA POWER-TUBE FITTINGS—PTF-1012A (10⅞" x 8⅜")—24 pages. Lists 39 power-tube fittings designed for supporting and cooling power tubes, and illustrates their use with power tubes made by RCA and other manufacturers. Includes exploded-view assembly drawings as well as detail drawings of all fittings. Price 25 cents.*

• HEADLINERS FOR HAMS—HAM-103B (10⅞" x 8⅜")—4 pages. Technical information and terminal-connection diagrams for 48 RCA "HAM" PREFERENCE TYPES: modulators, class C amplifiers and oscillators, frequency multipliers, rectifier tubes, thyratrons, cold-cathode (glow-discharge) tubes, and cathode-ray tubes. Single copy free on request.

• TECHNICAL BULLETINS — Complete authorized information on RCA transmitting tubes and other tubes for communications and industry. Be sure to mention tube-type bulletin desired. Single copy on any type free on request.

• RCA PREFERRED TYPES LIST—PTL-501-B (10⅞" x 8⅜")—4 pages. Lists RCA Preferred Tube Types, both receiving and non-receiving, by function. An aid to equipment designers in the selection of tube types for new equipment design. Single copy free on request.

• RCA INTERCHANGEABILITY DIRECTORY OF INDUSTRIAL-TYPE ELECTRON TUBES—ID-1020A (10⅞" x 8⅜")—16 pages. Lists more than 2000 type designations of 26 different manufacturers arranged in alphabetical-numerical sequence; shows the RCA Direct Replacement Type or the RCA Similar Type, when available. Price 20 cents.*

Test and Measuring Equipment

INSTRUCTION BOOKLETS — Illustrated instruction booklets, containing specifications, operating and maintenance data, application information, schematic diagrams, and replacement parts lists, are available for all RCA test instruments. Booklets for the following popular instruments are available at the prices indicated. Prices for booklets on other instruments are available on request.

25 cents each*

WO-55A (3" Oscilloscope)
WR-39A (TV Calibrator)
WR-59A (TV Sweep Generator)
WR-67A (Test Oscillator
WV-65A (VoltOhmyst†)
WV-75A (VoltOhmyst†)
WV-77A (VoltOhmyst†)
WV-77B (VoltOhmyst†)
WV-84A (Microammeter)
WV-95A (VoltOhmyst†)
165 (VoltOhmyst†)
165-A (VoltOhmyst†)
195-A (VoltOhmyst†)

†Trade Mark Reg. U. S. Pat. Off.

*Prices shown apply in U.S.A. and are subject to change without notice.

50 cents each*

WA-44A (Audio Oscillator)
WO-56A (7″ Oscilloscope)
WO-57A (3″ Oscilloscope)
WO-57B (3″ Oscilloscope)
WO-60C (5″ Oscilloscope)
WO-78A (5″ Oscilloscope)
WO-79A (3″ Oscilloscope)
WO-79B (3″ Oscilloscope)
WO-88A (5″ Oscilloscope)
WO-91A (5″ Oscilloscope)
WR-36A (Dot-Bar Generator
WR-39B (TV Calibrator)
WR-39C (TV Calibrator)
WR-40A (UHF Generator)
WR-41A (UHF Generator)
WR-41B (UHF Generator)
WR-49A (RF Generator)
WR-59B (TV Sweep Generator)
WR-59C (TV Sweep Generator)
WR-61A (Color-Bar Generator)
WR-61B (Color-Bar Generator)
WR-86A (UHF Sweep Generator)
WR-89A (Marker Generator)
WV-87A (VoltOhmyst†)
WV-97A (VoltOhmyst†)

75 cents each*

WR-46A (Video Dot/Crosshatch Generator)
WV-98A (VoltOhmyst†)

$1.00 each*

WT-100A (Electron-Tube MicroMhoMeter)

Batteries

• RCA RADIO BATTERIES FOR FLASHLIGHT, RADIO, AND INDUSTRIAL APPLICATIONS —BAT-134B (10⅞″ x 8⅜″)—8 pages. Contains characteristics, terminal connections, and socket patterns of 82 RCA dry batteries for radio, flashlight, and industrial applications. Includes interchangeability directory, and a battery replacement guide for 1948 to 1954 inclusive for portable radios. Single copy free on request.

†Trade Mark Reg. U. S. Pat. Off.

*Prices shown apply in U.S.A. and are subject to change without notice.

Reading List

The publications listed represent both elementary and advanced treatments of power and rectifier tube theory, applications, and circuit design. The list, obviously, is not inclusive, but additional references are given in the publications listed.

ARRL Antenna Book. American Radio Relay League.

BENEDICT, R. R. *Industrial Electronics*. Prentice-Hall, Inc.

CHUTE, G. M. *Electronics in Industry*. McGraw-Hill Book Co., Inc.

DAVID AND WEED. *Industrial Electronic Engineering*. Prentice-Hall, Inc.

DOME, R. B. *Television Principles*. McGraw-Hill Book Co., Inc.

EVERITT, W. L. *Communication Engineering*. McGraw-Hill Book Co., Inc.

FINK, D. G. *Engineering Electronics*. McGraw-Hill Book Co., Inc.

GRAY, T. S. *Applied Electronics*. John Wiley & Sons, Inc.

KLOEFFLER, R. G. *Industrial Electronics and Control*. John Wiley & Sons, Inc.

KOLLER, L. R. *Physics of Electron Tubes*. McGraw-Hill Book Co., Inc.

MARKUS AND ZELUFF. *Electronics for Communication Engineers*. McGraw-Hill Book Co., Inc.

MARKUS AND ZELUFF. *Handbook of Industrial Electronic Circuits*. McGraw-Hill Book Co., Inc.

PENDER, DELMAR, AND MCILWAIN. *Handbook for Electrical Engineering—Communications and Electronics*. John Wiley & Sons, Inc.

PREISMAN, A. *Graphical Constructions for Vacuum Tube Circuits*. McGraw-Hill Book Co., Inc.

PRINCIPLES OF ELECTRICAL ENGINEERING SERIES. *Applied Electronics*. John Wiley & Sons, Inc.

RADIATION LABORATORY SERIES. Vol. 18—*Vacuum-Tube Amplifiers;* Vol. 19—*Wave-forms*. McGraw-Hill Book Co., Inc.

RADIO RESEARCH LABORATORY, HARVARD UNIVERSITY. *Very-High-Frequency Techniques*. McGraw-Hill Book Co., Inc.

REICH, H. J. *Theory and Applications of Electron Tubes*. McGraw-Hill Book Co., Inc.

RICHTER, WALTHER. *Fundamentals of Industrial Electronic Circuits*. McGraw-Hill Book Co., Inc.

Single Sideband for the Radio Amateur. American Radio Relay League.

SPANGENBERG, K. R. *Vacuum Tubes*. McGraw-Hill Book Co., Inc.

TERMAN, F. E. *Electronic and Radio Engineering*. McGraw-Hill Book Co., Inc.

TERMAN, F. E. *Radio Engineers Handbook*. McGraw-Hill Book Co., Inc.

TERMAN AND PETTIT. *Electronic Measurements*. McGraw-Hill Book Co., Inc.

The Radio Amateurs Handbook. American Radio Relay League.

The Radio Handbook. Editors & Engineers, Ltd.

FEDERAL COMMUNICATIONS COMMISSION

Part 12: *Rules Governing Amateur Radio Service.*

Part 18: *Rules and Regulations Relating to Industrial, Scientific, and Medical Service.*

RCA Transmitting Tubes NOT Recommended For New Equipment Design

Certain transmitting tube types should be avoided in the design of new equipment because they are approaching obsolescence or have limited or dwindling demand. Such RCA types are listed below. For a guide to the selection of tube types recommended for new equipment design, refer to the Charts Section.

2C40	800	841	1623
2C43	801-A	842	1624
3E22	803	851	1626
4-1000A	804	860	1635
10-Y	806	861	5588
211	830-B	865	8003
217-C	838	1619	8012-A

RCA Preferred Types List

A list of preferred tube types is available to assist equipment designers and manufacturers in formulating their plans for future production of electronic equipment. This list is based on periodic surveys of the needs of the engineering and manufacturing fields and keeps abreast of technological advances in tube design and application.

A copy of the current list will be gladly furnished on request. Write to Commercial Engineering, Tube Division, Radio Corporation of America, Harrison, N. J.

Legend for Base and Envelope Connection Diagrams

Diagrams show terminals viewed from base or filament end of tube

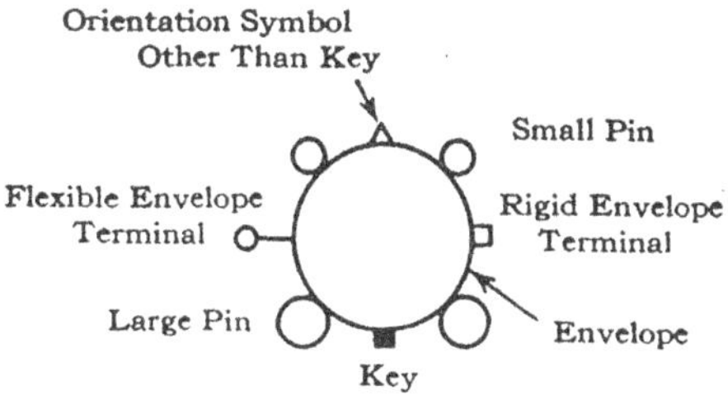

• = Gas-Type Tube	G = Grid	IS = Internal Shield
BC = Base Sleeve	H = Heater	K = Cathode
CP = Center Pin	H_M = Heater Mid-Tap	NC = No Connection
F = Filament	IC = Internal Connection—Do not use	P = Plate or Anode
F_M = Filament Mid-Tap		S = Shell

Alphabetical Subscripts B, D, P, T, and TR indicate, respectively, beam unit, diode unit, pentode unit, triode unit, and tetrode unit in multi-unit tubes.

9 781882 580200

Made in the USA
Monee, IL
04 November 2026

67382854R10144